DATE DUE

OC 28 04			
NO 18 04			

DEMCO 38-296

METHODS IN
ecological
and
agricultural
ENTOMOLOGY

Edited by

D.R. Dent

International Institute of Biological Control
Ascot
UK

and

M.P. Walton

Agriculture Victoria
Australia

CAB INTERNATIONAL

CAB INTERNATIONAL
Wallingford
Oxon OX10 8DE
UK

Tel: +44 (0)1491 832111
Fax: +44 (0)1491 833508
E-mail: cabi@cabi.org

CAB INTERNATIONAL
198 Madison Avenue
New York, NY 10016-4341
USA

Tel: +212 726 6490
Fax: + 212 686 7993
E-mail: cabi-nao@cabi.org

A catalogue record for this book is available from the British Library, London, UK
A catalogue record for this book is available from the Library of Congress,
Washington DC, USA

ISBN 0 85199 131 9 (Hbk)
ISBN 0 85199 132 7 (Pbk)

Typeset in Palatino by Columns Design Ltd, Reading
Printed and bound in the UK at the University Press, Cambridge

Contents

8. Injury, Damage and Threshold Concepts

J.D. Mumford and J.D. Knight

9. Techniques in the Study of Insect Pollination

S.A. Corbet and J.L. Osborne

10. Techniques to Evaluate Insecticide Efficacy

G.A. Matthews

Contributors

N.J. Armes, *Natural Resources Institute, Central Avenue, Chatham Maritime, Chatham, Kent ME4 4TB, UK.*

E. Bernays, *Department of Entomology, College of Agriculture, The University of Arizona, Forbes Building, Room 410, Tucson, Arizona 85721, USA.*

R.A. Cheke, *Natural Resources Institute, Central Avenue, Chatham Maritime, Chatham, Kent ME4 4TB, UK.*

J. Colvin, *Natural Resources Institute, Central Avenue, Chatham Maritime, Chatham, Kent ME4 4TB, UK.*

R.J. Cooter, *Natural Resources Institute, Central Avenue, Chatham Maritime, Chatham, Kent ME4 4TB, UK.*

S.A. Corbet, *Department of Zoology, University of Cambridge, Downing Street, Cambridge CB2 3EJ, UK.*

D.R. Dent, *International Institute of Biological Control, Silwood Park, Buckhurst Road, Ascot, Berkshire SL5 7TA, UK.*

S.D. Eigenbrode, *Department of Plant, Soil and Entomological Sciences, University of Idaho, Moscow, Idaho 83844-2339, USA.*

J. Hemingway, *School of Pure & Applied Biology, University of Wales, Cardiff, PO Box 915, Cardiff CF1 3TL, UK.*

J. Holt, *Natural Resources Institute, Central Avenue, Chatham Maritime, Chatham, Kent ME4 4TB, UK.*

J.D. Knight, *Centre for Environmental Technology, Imperial College of Science, Technology and Medicine, Silwood Park, Ascot, Berkshire SL5 7PY, UK.*

G.A. Matthews, *International Pesticide Application Research Centre, Imperial College of Science, Technology and Medicine, Silwood Park, Ascot, Berkshire SL5 7PY, UK.*

P.K. McEwen, *Insect Investigations Ltd, School of Pure & Applied Biology, University of Wales, Cardiff, PO Box 915, Cardiff CF1 3TL, UK.*

N. Mills, *Center for Biological Control, Department of Environmental Science,*

Policy and Management, University of California at Berkeley, 201 Wellman Hall, Berkeley, California 94720-3112, USA.

J.D. Mumford, *Centre for Environmental Technology, Imperial College of Science, Technology and Medicine, Silwood Park, Ascot, Berkshire SL5 7PY, UK.*

J.L. Osborne, *Department of Entomology and Nematology, Institute of Arable Crops Research, Rothamsted, Harpenden, Herts AL5 2JQ, UK.*

J.N. Perry, *Department of Entomology and Nematology, Rothamsted Experimental Station, Harpenden, Herts AL5 2JQ, UK.*

D.R. Reynolds, *Natural Resources Institute, Central Avenue, Chatham Maritime, Chatham, Kent ME4 4TB, UK.*

J.R. Riley, *Natural Resources Institute, Central Avenue, Chatham Maritime, Chatham, Kent ME4 4TB, UK.*

W.O.C. Symondson, *School of Pure & Applied Biology, University of Wales, Cardiff, PO Box 915, Cardiff CF1 3TL, UK.*

M.R. Tucker, *Natural Resources Institute, Central Avenue, Chatham Maritime, Chatham, Kent ME4 4TB, UK.*

M.P. Walton, *Program Manager – Target 10, Agriculture Victoria, RMB 2460, Hazeldene Road, Ellinbank, Victoria 3820, Australia.*

T.D. Wyatt, *Department of Zoology (and Department for Continuing Education), University of Oxford, South Parks Road, Oxford OX1 3PS, UK.*

Preface

Ecological and agricultural entomology need rigorous measurement and critical analysis in experimental research, combined with the best use of the increasingly sophisticated techniques that are now available. These techniques have revolutionized some aspects of the collection and handling of ecological data although the fundamental need for reliable data and a clear understanding of aims and analytical procedures remains as important as ever. This has created a need for an up-to-date and comprehensive textbook on experimental and analytical methods in both ecological and agricultural entomology to supplement the range of standard texts upon which insect ecologists have relied for a number of years.

Comprehensive texts on specific aspects of insect ecology are available which provide a review of the approaches and techniques that are most appropriate for the study of a particular specialist subject, e.g. insect natural enemies. However, there is a need to cover the broader aspects of studies in ecological and agricultural entomology, with special reference to those advances in technology which have practical implications for these disciplines. Advances in technology have created opportunities: (i) for studies which were previously impossible; and (ii) to carry out studies with greater power, precision and reliability. A greater emphasis is now placed on both modelling and biochemical techniques, with important new approaches to the study of insect behaviour and insecticide efficacy also making inroads into traditional approaches. This book therefore seeks to integrate new technologies and approaches to research with traditional and well proven methods; to provide a balanced view of the subject for use by final year undergraduates and postgraduates; and to be a useful source of reference for established research workers and those who teach the practical aspects of ecological and agricultural entomology.

D.R. Dent
M.P. Walton

1 Introduction

M.P. Walton[1] and D.R. Dent[2]
[1]*Agriculture Victoria, RMB 2460, Hazeldene Road, Ellinbank, Victoria 3820, Australia;* [2]*International Institute of Biological Control, Silwood Park, Buckhurst Road, Ascot, Berkshire SL5 7TA, UK*

The study of animal ecology is probably the oldest science (Begon *et al.*, 1990) and has formed the foundation of much of mankind's exploitation of the natural world. This study has ranged from simply the identification of edible plants and animals to more complex and subtle studies of the interactions between organisms and their physical, chemical and biological environments. If we are to make sustainable use of our environment and cater to the needs of society, i.e. produce food, clothing, shelter and fuel, reliably and efficiently, then we must study and understand the ecology of both the natural world and the man-made ecosystems which result from our use of this natural world. As we intensify the demands on our environment, agriculture has become ever more important for feeding and clothing us; hence any agents (e.g. insects) which cause injury to man, his crops, animals or property are deemed to be pests (Dent, 1991).

Information about insect ecology is sought for a variety of purposes. Population ecology has tended to be dominated by studies with insects which are used as models in the development of population dynamic theory. Insects have been used in this context mainly because their short generation time makes experimentation productive, but also because insects are important as pests (Putman and Wratten, 1984). Ecological studies of potential pests have become central to the understanding necessary for us to meet the challenges of obtaining sufficient protection of our crops and livestock. Hence, the continued interest in agricultural entomology, where the purpose is the management of insect populations in such a way that these populations do not become damaging to the interests of mankind.

Entomology as a branch of biological science has undergone rapid expansion and development, especially during the last 20 years, and there have been major advances in the technologies associated with pest management and the ecological studies that underpin so much of this work.

The methodologies used in ecological and agricultural entomology have contributed significantly to this advancement.

Progress in a science is often limited by the constraints of a particular methodology and the state of development in a discipline is reflected in the balance of the utilization of new and established methodologies. A young science may be characterized by the establishment of new methodologies or analytical techniques, i.e. there is a 'vertical leap' in methodologies; older more established sciences, and especially those that are stagnating, may be characterized by an overwhelming expansion in the use of established methodologies, i.e. a 'horizontal expansion', and a dearth of vertical methodologies. A good balance between vertical and horizontal type methodologies is indicative of a healthy and thriving discipline. In such a thriving discipline new techniques are continually being developed, moving the subject forward to new possibilities, but balanced by extensive evaluation and testing of existing methods in terms of their use with different species, conditions and systems. Such a balance is evident at this present time in ecology and agricultural entomology, providing scientists with a wealth of new approaches combined with an array of tried and tested techniques relevant to a whole range of pest species and situations.

The following text provides a critical analysis and evaluation of methods available in ecology and agricultural entomology through reference to general principles, but with an emphasis placed on providing descriptions of relevant methods and their applications. The text is divided into 13 chapters according to types/categories of research performed and each is written by recognized authorities in their field. Relevant background information is reviewed briefly in each topic, but some prior knowledge of the general subject area is assumed. Each chapter has its own reference list and provides a primary source for information on available literature.

The techniques of sampling, handling and rearing insects – dealt with in Chapter 2 – are fundamental to our ability to carry out research. Sampling may be carried out for a variety of reasons. Insects may be collected in order to identify which species are present at a particular locality or time of year, or to quantify their distribution and abundance. Ultimately such approaches may lead to a formal monitoring and forecasting technique. Collected specimens may be killed or maintained and used to establish or 'refresh' a laboratory culture. A great deal of research has been conducted into the techniques of insect rearing to ensure that cultured insects meet requirements for experimental work. The problem is to maintain cultured populations that do not differ significantly in any important trait from their wild counterparts. Differences often occur due to the constraints imposed on insect populations caused by confinement, the environmental conditions and host or diet. The effects may be obvious differences in size and fitness or more subtle effects in the way in which individuals behave.

Studies of insect behaviour occur in a number of areas of ecological and agricultural entomology. For instance, in the identification of host plant resistance and the interaction between natural enemies, their environment and their prey. Insect behavioural studies are vital for increasing our understanding of insect, environmental and host interactions. They provide a direct means of establishing what occurs and can provide invaluable insights into the way an event occurs. In addition, there can be no substitute for the experience gained by direct observation of the insect in its habitat (Dent, 1991). Insect behavioural studies have wide ranging value, and hence due emphasis is given to this subject in Chapter 3 and also in Chapter 6.

The potential number of interactions involved in any insect life system is enormous, involving a multitude of abiotic and biotic factors. Thus, the study of population dynamics and interactions is by necessity a process of ecological abstraction, involving the identification of the key factors that describe and predict ecological patterns (Tilman, 1989). Insect population growth may be defined by the action of five components: the physical environment, the host or food source, space, the population itself and other species (Southwood and Way, 1970). All five components act through the pathways of natality, mortality, immigration and emigration. The methods used to quantify natality and mortality are considered in Chapter 4 and followed by the techniques available for studying insect migration in Chapter 5. In Chapter 6 the methods used for evaluating the factors affecting insect host plant selection are reviewed. These methods are of value to those interested in host plant selection from an evolutionary or ecological standpoint or from the practical need to select crop cultivars expressing antixenotic resistance.

The guiding principles and methods of field experimentation are considered in Chapter 7. So much of agricultural entomology involves the use of field experiments utilizing classical agricultural experimental designs. Their correct use is integral to the development of meaningful crop protection and integrated pest management (IPM) strategies. The economic injury threshold concept introduced by Stern *et al.* (1959) is often regarded as central to IPM, the aim of many pest management strategies, but without a clear understanding of what constitutes economic injury, the goal of IPM remains distant. Chapter 8 covers the concepts of what constitutes injury to the interests of man which render the organism a pest in the eyes of a farmer or grower. Furthermore, since the treatment, or application of a control measure is often triggered when some form of injury level is exceeded, the methods by which injury to man's interests are measured and a pest–damage relationship established are also discussed.

Crop loss is often caused through the inadvertent use of insecticides which decimate beneficial insect populations particularly pollinators and natural enemies. The techniques used in the study of insect pollination are discussed in Chapter 9. This is followed by chapters dealing with the techniques which are used to evaluate insecticide efficacy (Chapter 10) and natural enemy efficacy (Chapter 11). The use of both insecticides and natural enemies is considered central to many pest control strategies. Methods for evaluating their efficacy are now well established, especially those developed for testing the effectiveness of chemical pesticides, but increasingly more sophisticated analytical techniques are now being used.

The methods described and discussed in the penultimate and final chapters are heavily reliant on recent, and continuing, advances in technology. Chapter 12 describes the biochemical techniques available to the modern invertebrate ecologist, especially the rapidly expanding fields of molecular biology and monoclonal antibody technology. The more established techniques of electrophoresis are described and discussed and the relative merits of each technique compared and contrasted for a range of studies including systematics, taxonomy, predation, parasitism, insecticide resistance and population genetics.

The final chapter deals with the various modelling options available to the ecologist, ranging from those based on solutions of complex differential equations to those decisions based on simple, qualitative differences. Southwood (1978) recognized that ecology is often concerned with quantitative interactions in a complex system and computers have proved to be very powerful tools for solving the differential equations that constitute the models often used in such studies. These models can be used for a range of purposes including forecasting pest damage, plant growth and pest damage and the interactions between pests and their natural enemies (Dent, 1991, 1995; Norton and Mumford, 1993).

References

Begon, M., Harper, J.L. and Townsend, C.R. (1990) *Ecology: Individuals, Population and Communities*. Blackwell Scientific Publications, Oxford.

Dent, D.R. (1991) *Insect Pest Management*. CAB International, Wallingford, 604pp.

Dent, D. (ed.) (1995) *Integrated Pest Management*. Chapman & Hall, London.

Norton, G.A. and Mumford, J.D. (1993) *Decision Tools for Pest Management*. CAB International, Wallingford, Oxon, 279pp.

Putman, R.J. and Wratten, S.D. (1984) *Principles of Ecology*. Croom Helm, London.

Southwood, T.R.E. (1978) *Ecological Methods with Particular Reference to the Study of Insect Populations*, 2nd edn. University Printing House, Cambridge.

Southwood, T.R.E. and Way, M.J. (1970) Ecological background to pest management. In: Rabb, R.L. and Guthrie, F.E. (eds) *Concepts in Pest Management*. North Carolina State University Press, Raleigh, pp. 6–29.

Stern, V.M., Smith, R.F., Bosch, R. van den and Hagen, K.S. (1959) The integrated control concept. *Hilgardia* **29**, 81–101.

Tilman, D. (1989) Population dynamics and species interactions. In: Roughgarden, J. May, R.M. and Levin, S.A. (eds) *Perspectives in Ecological Theory*. Princeton University Press, New York, pp. 89–100.

2 Sampling, Handling and Rearing Insects

P. McEwen
Insect Investigations Ltd, School of Pure & Applied Biology, University of Wales, PO Box 915, Cardiff CF1 3TL, UK

2.1. Introduction

The title of this chapter covers a potentially vast area of entomology and it is only possible for me to give the bare outlines in the space available. I have therefore provided the reader with a list of insect sampling methods (plus information on each method), a brief summary of handling techniques, and rather more detail on the requirements for successful insect rearing.

2.2. Insect Sampling

With rare exceptions it is not possible to count every individual insect. It is therefore necessary to take samples and then to use statistical methods to decide what the sample tells us about the population as a whole. The statistics of insect sampling has been well covered by, among others, Southwood (1978). Insect sampling methods can be divided for convenience into those designed to intercept flying insects and those designed to sample insects from the ground or from vegetation. More detail on many of the sampling techniques described below can be found in Muirhead-Thomson (1991), and Powell *et al.* (1996).

2.2.1. Flight interception traps

IMPACTION OR STICKY TRAPS

Usually used as a monitoring device (see pheromone traps below), insects are trapped in flight on a surface coated with long-lasting glue. Effectiveness of the traps can be increased by using attractive colours or odours. Coloured traps will greatly increase the rate of insect catch with the colour selected depending on the species that one wishes to attract (Kirk, 1984), but care must be taken not to disproportionally catch beneficial insects. For example, Neuenschwander (1982) found that 3–5 yellow sticky traps per olive tree were sufficient to almost eliminate the tree's beneficial fauna.

Trap catches are heavily affected by weather conditions and by the position and height of the trap in the field. Trap catches can be corrected according to the methods described by Powell *et al.* (1996). In addition, traps can soon become clogged up with insects and dust, thus reducing their effectiveness, and regular monitoring is essential. Rohitha and Stevenson (1987) describe an automatic sticky trap that segregates each day's catch and operates for seven days. Traps can be removed from the field using the method described by McEwen *et al.* (1993) and insects subsequently removed from the traps using a number of solvents (Murphy, 1985; Miller *et al.*, 1993) although white spirit usually suffices (personal observation). Sticky traps are a widely used entomological tool with recent examples from the literature including the monitoring of thrips (Cho *et al.*, 1995); stored wheat insects (Hagstrum *et al.*, 1994); sand flies (Oshagi *et al.*, 1994); and whitefly (Liu *et al.*, 1994).

LIGHT TRAPS

Light traps are a highly efficient means of catching large numbers of flying insects. The basic light trap comprises an electric bulb as an attractant, and a funnel to direct attracted insects into a container or collecting bag (Muirhead-Thomson, 1991). Many insects are attracted to light in the near ultraviolet region (320–400 nm) so that light traps often use an ultraviolet lamp (the so-called black-light traps) (Matthews, 1984). Rapid kill of the catch is essential in order to have good specimens for examination.

Temperature, windspeed, rain, moonlight, cloud conditions and open and woodland situations have a major impact on trap catches. Bowden (1982) analysed published data on light trap catches finding that all conformed to the model catch = constant $X\sqrt{W/I}$ where W = trap illumination and I = background illumination. In a comparison of *Chrysoperla carnea* catches from light and suction traps, Bowden (1981) found that where light catch numbers were adjusted to allow for variation in illumination during the trapping period, the light trap catch was very similar to suction trap catches suggesting that the effects of illumination must be compensated for. Recent studies where light traps were used include trapping of *Anopheles* spp. (Githeko *et al.*, 1994; Ritchie and Kline, 1995) and of Lepidoptera (Coombs *et al.*, 1993; Gregg *et al.*, 1994).

MALAISE TRAPS

Useful for sampling certain highly active adult Diptera and Hymenoptera and thus useful in estimating populations of certain parasites within crops (Matthews, 1984), the malaise trap takes advantage of the observation that in most cases a flying insect that meets an obstacle attempts to fly up and around it (British Museum, 1974). The trap is a netting tent with one open side into which is placed a small container at the highest point. Insects flying or crawling through the opening, move up the netting and are collected in the container. Recent studies have used malaise traps to capture Coleoptera (Dutra and Miyazaki, 1995; Jackman and Nelson, 1995; Sarospataki and Marko, 1995), Lepidoptera (Jeanneret and Charmillot, 1995), Diptera (Roller, 1995) and Planipennia (Vidlicka, 1995).

McPHAIL TRAPS

The McPhail trap is a simple invaginated glass trap which is used principally to sample Diptera (Hennessey, 1994; Celedonio Hurtado *et al.*, 1995; Robacker, 1995). These traps have recently been improved as a monitoring device for a number of species of fruit fly *Bactrocera* spp. For example, *Bactrocera oleae*, the olive fruit fly, is attracted to the colour yellow as well as to chemical attractants comprising protein hydrolysates and ammonium carbonates. These two factors have been combined in a new, robust, yellow-coloured PVC bottle (Fig. 2.1), which results in markedly increased catches of *B. oleae* for most attractants when compared with the traditional McPhail trap. As might be expected from a trap for flying insects, trap efficiency is greatly affected by air temperature which accounts for 85% of variation in catches (Kapatos and Fletcher, 1983).

PHEROMONE TRAPS

Sex pheromones of female insects can attract males over distances of several hundred metres (Schlyter, 1992) and are used to monitor pest populations either as a means of detecting early infestations or to determine when to make pesticide applications. Pheromone traps are generally species specific

Fig. 2.1. Pheromone trap. (Courtesy of Agrisense BCS Ltd.)

and are easy and cheap to operate. A typical pheromone trap is that used for monitoring the olive moth (*Prays oleae*), where a delta-shaped trap, open at both ends (Fig. 2.2), protects a horizontally placed sticky trap. A sealed polyethylene vial containing the pheromone is placed upright on the sticky trap and acts as a slow release mechanism for the chemical. Potential variables affecting the efficiency of pheromone traps include wind speed and direction, height of trap relative to crop, and adhesive used (Matthews, 1984). Pheromone traps have generally been used to monitor Lepidoptera (McVay *et al.*, 1995; Sharov *et al.*, 1995, Shirai and Nakamura, 1995), but also find an application for other groups of insects such as the Hymenoptera (Simandl and Anderbrant, 1995).

RADAR

Entomological radars are usually small, mobile, incoherent pulse systems which use a wavelength of 3.2 cm. These radars transmit short pulses from their antennae in a narrow, conical beam. Objects illuminated by the pulse reflect or scatter the pulse energy returning part of the scattered energy (an echo) to the radar. If strong enough, the echo is detected and amplified at the radar receiver and the presence of the target displayed. Insects of around 100 mg can be detected using this method at up to 2.5 km (Riley, 1989). Other sorts of radar of interest to the entomologist include the ground-based scanning radar, airborne radar, tracking radar, frequency-modulated continuous wave radar, harmonic radar, and bistatic and doppler radars, all of which are described by Riley (1989), and the vertical looking radar (VLR) (Riley and Reynolds, 1993). Entomological radar is discussed in more detail elsewhere (see this volume, Chapter 5).

Fig. 2.2. Pheromone trap for the olive moth, *Prays oleae*. (Courtesy of Agrisense BCS Ltd.)

Radar has also been used to monitor moth movements in relation to light traps, as well as assessing the efficiency of suction traps (Schaefer *et al.*, 1985).

SUCTION TRAPS

Suction traps represent a non-attractant method of sampling aerial populations of insects, and were originally designed to sample aphids and other aerial pests of agricultural crops. Suction traps consist of machines with engine-driven fans which suck insects into a fine mesh net, filtering out the insects which can then be collected in a container. Traps may be fitted with a segregating device to separate the catch according to time, thus providing information on the periodicity of flight. They can be mobile or fixed. Most of the work on the development and use of the suction trap has been carried out at the Rothamsted Experimental Station, Harpenden, Herts, UK, where suction trap data are used to forecast outbreaks of aphids. Wind speed will affect trap catches. Recent studies where suction traps have been used include the sampling of tabanids (Braverman *et al.*, 1995) and aphids (Zhou *et al.*, 1995).

WATER TRAPS

Yellow water traps, also called yellow pan traps, are used to sample a wide range of species including aphids (Webb *et al.*, 1994; Costello, 1995), thrips (Felland *et al.*, 1995), and cabbage root fly (Finch, 1995). The traps are usually most effective if raised above the ground. Vertical baffles placed in the trap can increase the number of insects caught (Matthews, 1984) and the addition of a few drops of detergent to the water decreases surface tension thus facilitating the drowning of insects (Powell *et al.*, 1996). Volatile chemicals may be added to the water in order to increase their effectiveness. Thus, in one study, the addition of ethyl nicotinate to the water increased the trap capture of the New Zealand flower thrips (*Thrips obscuratus*) by 100 times (Teulon *et al.*, 1993). As with all yellow traps there is a risk that beneficial insects will be caught alongside the target pest, however Finch (1991) discovered that by painting the inner wall of the trap black, 95% fewer syrphids were caught, with little effect on pest catches.

2.2.2. *Sampling insects from the ground or from vegetation*

BAIT SPRAYS

Matthews (1984) notes that insecticide-laced molasses can be sprayed along a row crop in order to sample moths, but the moths must be removed quickly before ants are able to remove the bodies. Fish-meal-baited traps can be used to sample female shoot flies which require the protein for egg production.

DIRECT INSPECTION

Regular inspection of the crop is an essential part of evaluating a pest problem. Referred to as scouting, an entomologist visits the field regularly, samples eggs or early larval instars, and advises on appropriate action. Foliage, fruit or branches may be sampled and counts made back in the laboratory (this is particularly useful for making egg counts of species such as the olive moth (*Prays oleae*) or for counting galleries of bark beetles such as the olive bark beetle *Phloeotribus scarabeoides*. personal observation). The main drawback with crop inspection is that it is usually too labour intensive to be economic. In addition, except in the case of large, conspicuous insects, the accuracy of insect counts depends on the observer (Shufran and Raney, 1989) and much time is required to get a sufficiently large sample for accuracy. However, Merchant and Teetes (1992) evaluated a number of sampling techniques for panicle infesting insect pests of sorghum and found a high correlation ($R^2 = 0.92$) between *in situ* visual estimates of the sorghum midge and estimates obtained by whole panicle samples. In this case scouting detected almost 89% of insects. Furthermore, scouting efficiency can be improved by a good understanding of the target pest. Mack *et al.* (1993) developed a system based on 'borer days' (a running total of hot and dry weather and cooler and wetter weather) to predict when to scout for the lesser cornstalk borer (*Elasmopalpus lignosellus*), thus cutting down on the scouting effort.

KNOCKDOWN OR SONDAGE SAMPLING

This is a useful technique for sampling insects in trees. An irritant spray such as natural pyrethrin is applied using a knapsack mistblower which projects the spray high into the tree. The tree is shrouded with a sheet or screen and the insects are collected on a groundsheet below. Enough time must be allowed to ensure that insects are knocked down and Southwood (1978) recommends that this period is at least one hour. Efficiency can be improved by knocking the tree after about half an hour. Southwood *et al.* (1982) compared the fauna of six British tree species sampled by pyrethrum knockdown with detailed faunal lists for the same species and found about 40% of the entirely phytophagous species in the faunal lists were found in the knockdown samples.

Knockdown resistance resulting from insensitivity of the nervous system to pyrethroids, DDT and some sodium channel neurotoxins has been reported in a number of species, such as the house fly (*Musca domestica*) and the German cockroach (*Blattella germanica*) and has obvious implications for this sampling method (Scott and Dong, 1994). In moist conditions the effectiveness of knockdown is greatly reduced.

PITFALL TRAPS

The pitfall trap consists of a small metal, plastic or glass container which is sunk in the ground such that the rim is level with the soil surface. Mobile insects fall into the trap and cannot escape. A suction apparatus may be used to remove insects without disturbing the trap. Baits or preservatives may be used to increase the attractivity of the trap although Zumr and Stary (1991) reported significant adverse effects of such traps on non-target insects, most notably carabid beetles and some scolytids. Pitfall traps require protection from rainfall and from larger animals and should be covered with brushwood or a large stone supported on smaller stones to allow insects access to the trap (British Museum, 1974).

Pitfall traps are cheap and easy to use, however catch size can be influenced by a range of factors other than population size. These include weather conditions, size of the trap, age, sex and condition of the insect. Pitfall traps monitor both insect abundance and insect activity and this introduces a potential bias into the sample, for example males are often more active than females and thus more likely to be caught. Problems associated with the use of pitfall traps are highlighted by the study of Topping and Sunderland (1992) where results from pitfall traps and absolute density sampling were found to be inconsistent. Pitfall traps are discussed in some detail elsewhere (e.g. Sunderland *et al.*, 1995; Powell *et al.*, 1996).

QUADRATS

In the case of highly mobile insects, quadrats should be placed in the field and left in place for some time before counts of insects are made. The quadrat is approached carefully and the number of insects counted as they fly or hop out. In the case of less mobile insects the count can be made as soon as the quadrat is in position. Wooden or metal sides may be needed to stop insects from running in to or out of the collecting area (Southwood, 1978).

Sunderland *et al.* (1995) describe a system referred to as soil flooding where 0.1 m^2 quadrats are sunk into the soil, vegetation is examined, excised and removed, then 2 litres of water poured into each quadrat. Predators on the surface are removed and 2 more litres of water subsequently added, any additional predators that appear are removed. Total collection time is 10 min per quadrat.

SHAKING AND BEATING

A sheet of cloth or plastic, or a tray, is laid out under the plant which is then shaken or beaten with a stick and insects collected quickly before they can escape. While this can be a good system for seeing what is there, the amounts of foliage and the intensity/frequency of beating/shaking affect the proportion of insects dislodged resulting in poor correlations with insect population estimates from other sampling techniques (Matthews, 1984), and some small animals may be overlooked entirely. Nevertheless, Southwood (1978) considers that for some species, such as leaf beetles (Chrysomelidae), many weevils, and Lepidoptera larvae that fall from the host plant if disturbed, a sufficiently high proportion may be collected to consider this an absolute method. More accurate counts may be obtained by fastening a screen over a large funnel – falling insects are funnelled into a container for examination. A further disadvantage of this method is that the host plant may be damaged by the beating.

SOIL AND LEAF LITTER SAMPLES

A number of insects inhabit the soil and leaf litter, or overwinter there, and soil coring or sieving can be used to sample such populations (British Museum, 1974; Southwood, 1978; Matthews, 1984). Insects can be extracted either by sieving or flotation, or by the use of chemical irritants (British Museum, 1974). A Berlese collector may be used to collect insects from litter. The litter is placed in a screen tray on top of a large metal funnel surrounded by a copper water jacket. The water in the jacket is heated until it warms the sample of litter. This heat produces activity in the insects in the litter which respond by going down into the litter and through the screen into the funnel where they can be caught in a glass bottle placed at the end of the funnel (Peterson, 1964). Alternatively a light can be shone upon the litter sample to provide heat.

SWEEP NETS

An inexpensive and commonly used means of sampling insects from vegetation (although also used for sampling flying insects), the net is swung a set number of times through the crop with the net twisted around at the end of each sweep in order to keep the mouth of the net in contact with the vegetation (British Museum, 1974). To prevent insects escaping the mouth of the net must be closed as soon as sweeping is completed (Matthews, 1984). The efficiency of the net is affected by the individual user, size and specifications of the net, and other conditions such as height of crop, diurnal rhythm of the insect, and weather. Sweep netting should not be attempted when the weather is wet as the insects adhere to the net and to each other (British Museum, 1974). Fast flying insects escape easily, and this method is not suitable for species living near the base of the plant.

2.3. Handling

Very little has been written specifically on insect handling although most general entomology texts do cover the subject briefly. Insects need to be handled for a variety of reasons and methodology will depend to a large extent on whether one is handling live or dead specimens.

2.3.1. Dead insects

Insects can be killed by placing them in an airtight jar, the bottom of which is lined with a layer of plaster of Paris, into which the killing agent is introduced (Youdeowei, 1977). There are a number of killing agents available to the entomologist and these are listed (British Museum, 1974) as cyanide, crushed laurel leaves, chloroform, carbon tetrachloride, ethylene tetrachloride, ethyl acetate, ether, ammonia, and hot water, although some of these obviously need to be handled with care. Beetles can be kept in sealed vials containing cotton wool moistened with ethyl acetate and will remain relaxed and retain their colour for months or years. Insects may also be killed by placing them in the deep freeze. Using this method moths will not damage themselves by fluttering as they do when being gassed. Carbon dioxide and cigarette smoke will also both kill insects. Preservation of killed specimens (Peterson, 1964; Little, 1972; British Museum, 1974) basically falls into the categories of pinning, preserving in alcohol (with the addition of glycerine preventing stiffening), papering (transport in paper envelopes) and layering (storage between layers of soft material such as cellulose wadding, in boxes). In all cases adequate labelling is required to record details of the insect (species), collection site (crop, locality), date and collector. There are some excellent software packages available to allow the production of very high quality insect labels (e.g. EntoPrint, available by e-mailing the producers at entomation@aol.com), though the reader should be aware that not all computer-generated labels are stable in preserving fluids.

2.3.2. Live insects

In the case of live insects handling can be required for experimental purposes, for starting new colonies, for biological control programmes or for other purposes such as photography. Where the requirement is to handle small numbers of insects the aspirator or pooter can be used to remove insects from sweep nets, to collect directly from vegetation or to move insects around the laboratory, for example to introduce insects to experimental arena (although some fragile insects can be damaged by using an aspirator). There are two fundamental designs of aspirator: those relying on direct suction of air through the aspirator which draw insects into a collecting chamber; and those where the operator blows air into the aspirator and draws insects into a container on a current of air. Important points to consider are firstly that the direct suction type may result in insect frass or other material being drawn directly into the operator's lungs, and the use of a suction bulb is recommended to avoid this possibility, and secondly that in dry or dusty conditions a dust trap should be placed between the mouthpiece and the aspirator (British Museum, 1974).

Different stages of the same species will vary in their ease of handling, hence Nielson and Lehman (1980) used adult weevils in their work in preference to larvae because the former were easier to collect, store and handle. In some cases insects can be so sensitive to handling that it is very difficult to work with them (Pathak and Saxena, 1980).

In order to minimize damage to the insect fine-haired paint-brushes (camel hair is often recommended) should be used to handle small soft-bodied insects such as Lepidoptera eggs and larvae (Ortega *et al.*, 1980). Entomological forceps are appropriate for handling larger insects although care must be taken not to cause them any damage. Soft forceps are available to handle pupae.

Van den Assem (1996) notes that handling of insects before experiments should be kept to a minimum and indicates that parasitic wasps can easily be coaxed into moving from one glass vial to another by holding the destination vial in the direction of a bright light source

Often, it is necessary to immobilize individual living insects for closer examination. Carbon dioxide can be used to immobilize a wide range of insects to allow easier handling although it should be noted that carbon dioxide is also used in insect pest control (Hashem and Reichmuth, 1994) and too long exposure leads to death. The use of carbon dioxide is nevertheless of great value for handling fragile insects such as adult green lacewings which are otherwise inevitably damaged by handling, although other species, such as whitefly, are not immobilized for long enough for this method to be recommended (L. Senior, Cardiff, 1995, personal communication). In some cases exposure to carbon dioxide can have adverse and subtle effects on the insect. For example, anaesthetizing the cat flea (*Ctenocephalides felis*) with carbon dioxide for periods of 5–90 min does not cause significant mortality within 24 h of exposure. However, when such fleas are exposed to chlorpyrifos there is a direct correlation between susceptibility to this chemical and exposure to the carbon dioxide (El-Gazzar *et al.*, 1988).

Some insects can also be immobilized by placing them for a few minutes in a refrigerator at 4°C. As with carbon dioxide, cold anaesthetization causes significantly higher mortality in cat fleas after treatment with chlorpyrifos (El-Gazzar *et al.*, 1988).

Electro-immobilization (EI) is a new technique described by Chesmore and Monkman (1994). Developed to allow the handling of live Lepidoptera specimens in the field, the insect is first anaesthetized then secured on electrostatically charged pads once the wings have been arranged as required. The insect remains immobile until the pads are switched off, allowing detailed image analyses to be carried out on the live specimen. Possible adverse effects of this system on the insect have still to be investigated.

In many cases, particularly in biological control programmes, the requirement is to handle large numbers of insects. Biological control must be economically competitive with other control methods in order to succeed. In many cases economic feasibility will only come with automated techniques to allow the rapid and efficient release of large numbers of insect natural enemies (Ables *et al.*, 1979). There has been considerable movement towards the mechanization of handling procedures for the distribution of insects in containers, liquid sprays (see Jones and Ridgway, 1976 and McEwen, 1996 for further discussion), and granular mixes, as well as aerial releases (see Ables *et al.*, 1979, for a review). Increasingly sophisticated technology is employed in insect handling techniques. For example, Shuman (personal communication, 1995) describes a beneficial insect counting and packaging system that counts and packages the leafminer parasitoid *Opius dissitus*. An infrared beam counts parasitized laboratory reared leafminer larvae as they drop from a leaf. The larvae pass through the beam and slide down a funnel into cups on a turntable. A computer rotates the turntable when a cup has received the allotted number of larvae. Larvae are then placed either in the glasshouse or outdoors where emerging wasps contribute to pest control.

In some cases handling may be required to artificially infest plants for experiments. Technique will vary according to stage of insect, numbers to be handled, site of release and growth stage of plant. Dent (1986) describes a system whereby aphid nymphs are transferred from clip cages to plants simply by allowing them to drop from cage to hand and then walk out on to the plant surface, and Panda and Khush (1995) suggest that infestation can be achieved simply by shaking test insects from host plants. Panda and Khush (1995) also describe different techniques for infesting plants with eggs or larvae. In the case of the maize stem borer egg masses are placed in the whorl of the plant or, alternatively, discs with egg masses are pinned to the leaves. A number of examples are given where eggs are sprayed suspended in agar. In the case of larvae, Panda and Khush (1995) describe a dispenser whereby newly hatched larvae are mixed with corncob grits and are dispensed automatically by a device known as a bazooka.

2.4. Rearing

Once insects have been collected from the field it is often necessary to establish laboratory colonies for further study. Insect rearing is carried out for a variety of reasons including research, pest control, commercial gain, and to meet quarantine requirements. There are several excellent works on insect rearing to which the reader is referred for an in-depth treatment of the subject (e.g. Peterson, 1964; Singh and Moore, 1985; Anderson and Leppla, 1992). Rearing insects can be one of the most difficult tasks faced by the entomologist. Establishing and maintaining insect colonies requires a great deal of time, patience, and expertise and good technical assistance is essential. Important elements in successful insect rearing are described below.

2.4.1. Artificial diets

While many plant-feeding insects (Diptera, Hemiptera, Homoptera, Lepidoptera) can be mass-reared on their natural hosts (Panda and Khush, 1995) others require the provision of an artificial substitute. A great deal of effort has gone into developing such diets and many standard artificial diets are now available commercially.

In developing an artificial diet it is necessary to understand the nutritional requirements of the insect involved. These requirements will vary from species to species and the reader is directed to some of the excellent works on this subject (e.g. Vanderzant, 1973; House, 1977; Hagen *et al.*, 1984; Hagen, 1987; Niijima, 1989, 1993a, b; Cohen, 1992). Moore (1985) points out that not only must the diet have adequate nutrients but the insect must also be willing to ingest it. This means that both the physical (hardness, texture, water content, homogeneity) and chemical (nutrients, compounds associated with the host plant) factors must be considered.

The development of artificial diets is economically attractive as it allows large numbers of insects to be produced without the need to establish host populations, and reduces the risks of diseases being introduced to the colony (Waage et al., 1985). Such diets can be species specific or suitable for a number of species. In discussing multi-species diets Singh (1985) states that they should be inexpensive, easily produced from locally available ingredients, supply all the nutritional requirements to complete all stages of the life cycle, be innocuous to use, acceptable to a wide range of species, have a long shelf life and produce at least 75% average yield of adults from initial viable eggs (i.e. produce insects of high quality). Diets must also allow a large number of generations to develop without the loss of vitality. Panda and Khush (1995) describe production systems for a number of species that utilize artificial foods.

In the case of entomophagous insects the lack of suitable artificial diets, and the subsequent necessity to maintain colonies of food insects (requiring extra time and space), is a major constraint on successful insect rearing, especially where the host or host plant is difficult to rear. Most species still have to be reared on natural hosts. Waage et al. (1985) note that most success with artificial diets for entomophages has been with parasitoids (e.g. Guerra et al., 1993), with much less success with predators. No predator has been successfully reared on a totally synthetic diet although there has been some success with rearing predators on semi-artificial diets, for example the green lacewing Chrysoperla carnea (Niijima and Matsuka, 1990) and other species (Matsuka and Niijima, 1985) on drone powder derived from bee brood.

2.4.2. Design of insectary: environmental conditions

An important factor in the successful rearing of insects is to provide the right conditions for the insects to thrive (Fisher and Leppla, 1985; Goodenough and Parnell, 1985). The insectary must be able to provide control and monitoring of temperature, humidity and lighting, as well as allowing quarantine of the insects if necessary, and must also provide adequate working conditions for the personnel. As such, a combination of biological, engineering and management needs influence insectary design. The insectary requires an appropriate ventilation system the design of which will depend on such factors as whether insect scales and other parts need to be removed, whether odours need to be removed, and how many air changes are required in a given time – a low number avoids desiccation, a high number most efficiently removes airborne contaminants (Goodenough and Parnell, 1985). External contaminants such as microorganisms can be kept out by maintaining a slight positive air pressure inside the insectary. Each insect production programme will require slightly different conditions.

Diapause occurs in univoltine and multivoltine species and may occur in any life stage according to species (Waage *et al.*, 1985). In multivoltine species diapause may be under environmental control and careful monitoring of the conditions inducing diapause (daylength, temperature, food availability) can prevent diapause occurring. However, there is the possibility that diapause could actually be advantageous if it could be harnessed to allow the long-term storage of insects (see below).

2.4.4. Health hazards

Apart from the obvious health hazards that cause cuts, burns and abrasions, insectaries also have hazards caused by chemical vapours and airborne particles (Wolf, 1985). Such hazards can result in respiratory illness and allergies in insectary workers. Mechanical filters, ducts, containers, blowers and hoods can all help to solve the problem and such controls are best established close to the source of contamination (Wolf, 1985). Oral aspirators (pooters) used to suck insects into collecting jars can result in insect particles getting into the lungs and should not be used without adequate filters.

2.4.5. Hygiene

Good hygiene is essential to minimize the risk of disease which can wipe out insect colonies and be expensive to eradicate. A wide range of microorganisms can infect insects including viruses, bacteria, fungi, protozoa, rickettsia and microsporidians, and some of these are described by Soares (1992). Fisher and Leppla (1985) state that a successful insectary should either exclude, or have the facilities for the elimination of, microorganisms that might harm the colony and point out that even when they are not lethal, such microorganisms together with dietary contaminants, may affect insect quality. There are a number of potential sources of contamination for insectaries including air conditioning, outside air, recycled air, auxiliary air from fume cupboards, personnel, dust and dirt on the floor, leakage of contaminants, dietary ingredients and other miscellaneous materials, and insect containers (Fisher and Leppla, 1985). Sikorowski and Goodwin (1985) recommend that contamination be minimized by sterilization of insect eggs using formaldehyde or sodium hypochlorite (noting that formaldehyde is a carcinogen and must be handled under a fume hood); sterilization of pupae (using NaOCl; Soares, 1992); sterilization procedures for diets (which are a rich food source for microorganisms) to destroy bacteria and fungi (using an autoclave or flash sterilization for example); sterilization of equipment; and the use of antibiotics mixed with diets for their antimicrobial activity. Personnel should take sanitation methods seriously, including the wearing of gloves, gowns and masks, and a generally clean environment should be maintained (surfaces should be disinfected for example). Soares (1992) adds that colonies should be established using disease-free stock, field collected stock should be quarantined and insectaries should be designed with hygiene in mind.

2.4.6. Management

Singh and Ashby (1985) define insect rearing management (IRM) as 'the efficient utilization of resources for the production of insects of standardized quality to meet programme goals' and describe in some detail the elements of IRM. Phases of IRM associated with insect production programmes are summarized by Singh and Clare (1992) as:

1. Clearly defining objectives.
2. Laboratory design taking into account types and numbers of insects to be reared, their behaviour and food requirements, space and equipment needed, environmental conditions, and standards of hygiene.
3. Colony establishment and maintenance, requiring knowledge of each species behaviour and selecting the best food and rearing method. Initial rearing to be done in quarantine.
4. Research and development techniques. Thorough research of the life cycle in the laboratory to provide developmental data for use in colony maintenance and production.
5. Resources. Finance, personnel, equipment, materials, facilities and insect colonies.
6. Quality control and biological performance.
7. Production.

Ochiengodero and Singh (1993) note that IRM is a valuable system for regulating production of different life stages as it produces insects of specified quality as required. IRM is discussed further elsewhere (Ochiengodero, 1990; Ochiengodero *et al.*, 1991).

2.4.7. Quality of insects

It is important to sustain the quality of insects maintained in insectaries or in the laboratory, especially if they are to be used in behavioural studies, or released back to the field in pest management programmes. Quality can be determined by a number of factors including host quality, crowding (which can affect development rates and sex ratio), other environmental conditions, and inbreeding.

Early on in the establishment of a laboratory colony performance traits of the wild population should be identified, quantified and subsequently used to assess the condition of the laboratory population. In the interests of more reliable biological control, attempts have been made to set quality standards for some insect species, specifying for example in the case of the green lacewing, *C. carnea*, expected longevity, hatching rate, numbers reaching the second instar, etc. (Van Lenteren, 1994). Other parameters include development rate, pupal weight, fecundity, LD_{50s} for known pesticides, flight propensity, irritability, locomotor response to light or gravity, mating competitiveness, orientation to host, and survival due to heat stress and gravity. One might add to this list response to chemical cues as measured by an electroantennograph.

Some observed changes in insect populations kept in the laboratory are under genetic control associated with the effects of selection on existing genotypes and the random loss of genotypes due to genetic drift (Bartlett, 1985; Mangan, 1992). Bartlett (1985) refers to the types of genetic changes that can be expected in insect colonies as a process of domestication. During colony establishment the principal factors influencing the genetic makeup of the population will be size of the sample taken from the field population, natural selection within the new environment making insects better adapted to the laboratory environment but less well adapted to the field environment, inbreeding, and mutation.

To minimize these effects new colonies should be established with the largest founder population that can possibly be handled (preferably several hundred individuals), insects should be reared for the minimum number of generations possible, individuals should be drawn from as wide an area as possible (this is particularly important in the case of sedentary species with a patchy distribution, although this might also have dangers where individuals are taken from well adapted local populations), environmental conditions should be varied to best match those experienced by the field population, genetic change should be monitored using electrophoresis (Joslyn, 1984), new individuals collected from the field should periodically be added to the culture, or better still, colonies should be completely replaced with new insects from the field (Bartlett, 1985; Waage *et al.*, 1985).

2.4.8. Storage

The storage of entomophages without the loss of viability and effectiveness has been identified as a priority by King *et al.* (1985) and Morrison and King (1977) observed that the ability to store entomophages is a key factor in the development of the augmentation method of biological control. Storage of insects allows insectaries to close down production for parts of the year, when predators and parasitoids are not required for example, with savings in production costs which will make biological control of pests more economically attractive. Recent work by Osman and Selman (1993) and Tauber *et al.* (1993) have shown the potential of storage techniques for the green lacewing *Chrysoperla carnea* by taking advantage of diapause and low temperature, and various parasitoids can be stored as mummies or as pupae (Stinner, 1977).

2.4.9. Water

Duncan and Pickwell (1939) emphasize that water supply is as important as food supply in the rearing of insects and suggest that rearing often fails through overlooking this important point. Water can either be provided as droplets or on a piece of damp material such as cotton wool. This will also serve to raise humidity.

2.4.10. Other

For regular information on insect rearing the reader is referred to the insect rearing group newsletter, FRASS, which can be obtained electronically by contacting Richard_mcdonald@ ncdamail.agr.state.nc.us with your user ID, and requesting either a Wordperfect 5.1 version *or* Word for Windows version.

2.5. Acknowledgements

Thanks to Enzo Casagrande of Agrisense BCS for providing the figures, and to Mike Wilson, Brian Levey, John Deeming at the National Museum of Wales, and Lara Senior and David Dent at the University of Wales, Cardiff, for reading through the text.

2.6. References

Ables, J.R., Reeves, B.G., Morrison, R.K., Kinzer, R.E., Jones, S.L., Ridgway, R.L. and Bull, D.L. (1979) Methods for the field release of insect parasites and predators. *Transactions American Society Agricultural Engineers* 18(6), 59–62.

Anderson, T.E. and Leppla, N.C. (1992) *Advances in Insect Rearing for Research and Pest Management.* Westview Press, Oxford.

Bartlett, A.C. (1985) Guidelines for genetic diversity in laboratory colony establishment and maintenance. In: Singh, P. and Moore, R.F. (eds) *Handbook of Insect Rearing, vol 1.* Elsevier, Oxford, pp 7–17.

Bowden, J. (1981) The relationship between light- and suction-trap catches of *Chrysoperla carnea* (Stephens) (Neuroptera: Chrysopidae), and the adjustment of light-trap catches to allow for variation in moonlight. *Bulletin of Entomological Research* 71, 621–629.

Bowden, J. (1982) An analysis of factors affecting catches of insects in light-traps. *Bulletin of Entomological Research* 72, 535–556.

Braverman, Y., Chizovginzburg, A. and Galker, F. (1995) *Tabanus arenivagus* (Diptera: Tabanidae) attracted to ultra-violet light suction traps in Israel. *Journal of the American Mosquito Control Association* 11 (4), 489–490.

British Museum (1974) *Instructions for Collectors No. 4a.* British Museum Publication Number 705.

Celedonio Hurtado, H., Aluja, M. and Liedo, P. (1995) Adult population fluctuations of *Anastrepha* species (Diptera: Tephritidae) in tropical orchard habitats of Chiapas, Mexico. *Environmental Entomology* 24 (4), 861–869.

Chesmore, D. and Monkman, G. (1994) Automated analysis of variation in Lepidoptera. *The Entomologist* 111 (3 & 4), 171–182.

Cho, K.J., Eckel, C.S., Walgenbach, J.F. and Kennedy, G.G. (1995) Comparison of colored sticky traps for monitoring thrips populations (Thysanoptera: Thripidae) in staked tomato fields. *Journal of Entomological Science* 30 (2), 176–190.

Cohen, A.C. (1992) Using a systematic approach to develop artificial diets for predators. In: Anderson, T.E. and Leppla, N.C. (eds) *Advances in Insect Rearing for Research & Pest Management*. Westview Press, Oxford, Ch. 6.

Coombs, M., Delsocorro, A.P., Fitt, G.P. and Gregg, P.C. (1993) The reproductive maturity and mating status of *Helicoverpa armigera*, *Helicoverpa punctigera* and *Mythimna convecta* (Lepidoptera: Noctuidae) collected in tower mounted light traps in Northern New South Wales, Australia. *Bulletin of Entomological Research* 83 (4), 529–534.

Costello, M. J. (1995) Spectral reflectance from a broccoli crop with vegetation or soil as background – influence on immigration by *Brevicoryne brassicae* and *Myzus persicae*. *Entomologia Experimentalis et Applicata* 75 (2), 109–118.

Dent, D.R. (1986) Resistance to the aphid *Metopolophium festucae cerealium*: Effects of the host plant on flight and reproduction. *Annals of Applied Biology* 108, 577–583.

Duncan, C.D. and Pickwell, G. (1939) *The World of Insects*. McGraw-Hill Book Company, London.

Dutra, R.R.C. and Miyazaki, R.D. (1995) Families of Coleoptera captured with Malaise trap in 2 sites of Ilha-do-mel, Paranagua Bay, Parana, Brazil. *Arquivos de Biologia e Tecnologia* 38 (1), 175–190.

El-Gazzar, L.M., Koehler, P.G. and Patterson, R.S. (1988) Factors affecting the susceptibility of the cat flea, *Ctenocephalides felis* Bouche, to chlorpyrifos. *Journal of Agricultural Entomology* 5 (2), 127–130.

Felland, C.M., Teulon, D.A. J., Hull, L.A. and Polk, D.F. (1995) Distribution and management of thrips (Thysanoptera: Thripidae) on nectarine in the mid-Atlantic region. *Journal of Economic Entomology* 88 (4), 1004–1011.

Finch, S. (1991) Influence of trap surface on the numbers of insects caught in water traps in brassica crops. *Entomologia Experimentalis et Applicata* 59 (2), 169–173.

Finch, S. (1995) Effect of trap background on cabbage root fly landing and capture. *Entomologia Experimentalis et Applicata* 74 (3), 201–208.

Fisher, W.R. and Leppla, N.C. (1985) Insectary design and operation. In: Singh, P. and Moore, R.F. (eds) *Handbook of Insect Rearing, vol 1*. Elsevier, Oxford, pp. 167–183.

Githeko, A.K., Service, M.W., Mbogo, C.M., Atieli, F.A. and Juma, F.O. (1994) Sampling *Anopheles arabiensis*, *Anopheles gambiae sensu lato* and *Anopheles funestus* (Diptera: Culicidae) with CDC light traps near a rice irrigation area and a sugarcane belt in Western Kenya. *Bulletin of Entomological Research* 84 (3), 319–324.

Goodenough, J.L. and Parnell, C.B. (1985) Basic engineering design requirements for ventilation, heating, cooling, and humidification of insect rearing facilities. In: Singh, P. and Moore, R.F. (eds) *Handbook of Insect Rearing, vol 1*. Elsevier, Oxford, pp. 137–155.

Gregg, P.C., Fitt, G.P., Coombs, M. and Henderson, G.S. (1994) Migrating moths collected in tower mounted light traps in Northern New South Wales, Australia – influence of local and synoptic weather. *Bulletin of Entomological Research* 84 (1), 17–30.

Guerra, A.A., Robacker, K.M. and Martinez, S. (1993) In-vitro rearing of *Bracon mellitor* and *Catolaccus grandis* with artificial diets devoid of insect components. *Entomologia Experimentalis et Applicata* 68 (3), 303–307.

Hagen, K.S. (1987) Nutritional ecology of terrestrial insect predators. In: Slansky F. Jr and Rodriguez J.G. (eds) *Nutritional Ecology of Insects, Mites, Spiders, and Related Invertebrates*. John Wiley & Sons, Chichester, pp. 533–575.

Hagen, K.S., Dadd, R.H. and Reese, J. (1984) The food of insects. In: Huffaker, C.B. and Rabb, R.L. (eds) *Ecological Entomology*. John Wiley & Sons, New York, Ch. 4.

Hagstrum, D.W., Dowdy, A.K. and Lippert, G.E. (1994) Early detection of insects in stored wheat using sticky traps in bin headspace and prediction of infestation level. *Environmental Entomology* 23 (5), 1241–1244.

Hashem, M.Y. and Reichmuth, C. (1994) Interactive effects of high carbon or low oxygen atmospheres and temperatures on hatchability of eggs of three stored product moths. *Journal of Plant Diseases and Protection* 101 (2), 178–182.

Hennessey, M.K. (1994) Analysis of Caribbean fruit fly (Diptera: Tephritidae) trapping data, Dade County, Florida, 1987–1991. *Florida Entomologist* 77 (1), 126–135.

House, H.L. (1977) Nutrition of natural enemies. In: Ridgeway, R.L. and Vinson, S.B. (eds) *Biological Control by Augmentation of Natural Enemies*. Plenum, New York, Ch. 5.

Jackman, J.A. and Nelson, C.R. (1995) Diversity and phenology of tumbling flower beetles (Coleoptera: Mordellidae) captured in a malaise trap. *Entomological News* 106 (3), 97–107.

Jeanneret, P. and Charmillot, P. J. (1995) Movements of tortricid moths (Lepidoptera: Tortricidae) between apple orchards and adjacent ecosystems. *Agriculture Ecosystems and Environment* 55 (1), 37–49.

Jones, S.L. and Ridgway, R.L. (1976) Development of methods for field distribution of eggs of the insect predator *Chrysopa carnea* Stephens. *US Department of Agriculture, Agricultural Research Service*, ARS-S-124.

Joslyn, D.J. (1984) Maintenance of genetic variability in reared insects. In: *Advances and Challenges in Insect Rearing*. US Department of Agriculture Handbook, pp. 20–29.

Kapatos, E. and Fletcher, B.S. (1983) Seasonal changes in the efficiency of McPhail traps and a model for estimating olive fly densities from trap catches using temperature data. *Entomologia Experimentalis et Applicata* 33, 20–26.

King, E.G., Hopper, K.R. and Powell, J.E. (1985) Analysis of systems for biological control of crop arthropod pests in the US by augmentation of predators and parasites. In: Herzog, D.C. and Hoy, M.A. (eds) *Biological Control in Agricultural IPM Systems*. Academic Press Inc., London, pp. 214–215.

Kirk, W.D.J. (1984) Ecologically selective coloured traps. *Ecological Entomology* 9, 35–41.

Little, V.A. (1972) *General and Applied Entomology*, 3rd edn. Harper & Row, London.

Liu, T.X., Oetting, R.D. and Buntin, G.D. (1994) Temperature and diel catches of *Trialeurodes vaporarium* and *Bemisia tabaci* (Homoptera: Aleyrodidae) adults on sticky traps in the greenhouse. *Journal of Entomological Science* 29 (2), 222–230.

Mack, T.P., Davis, D.P. and Lynch, R.E. (1993) Development of a system to time scouting for the lesser cornstalk borer (Lepidoptera: Pyralidae) attacking peanuts in the South Eastern United States. *Journal of Economic Entomology* 86 (1), 164–173.

Mangan, R. L. (1992) Evaluating the role of genetic change in insect colonies maintained for pest management. In: Anderson, T.E. and Leppla, N.C. (eds) *Advances in Insect Rearing for Research & Pest Management*. Westview Press, Oxford, Ch. 17.

Matsuka, M. and Niijima, K. (1985) *Harmonia axyridis*. In: Singh, P. and Moore, R.F. (eds) *Handbook of Insect Rearing, vol. 1*. Elsevier, Oxford, pp. 265–268.

Matthews, G.A. (1984) *Pest Management*. Longman, London.

McEwen, P.K. (1996) Viability of green lacewing (*Chrysoperla carnea*) eggs stored in potential spray media, and subsequent effects on survival of first instar larvae. *Journal of Applied Entomology*, 120, 171–173.

McEwen, P.K., Jervis, M.A. and Kidd, N.A.C. (1993) A convenient method of handling and shipping insect-bearing sticky traps. *Entomologists Monthly Magazine* 129, 237–238.

McVay, J.R., Eikenbary, R.D., Morrison, R.D., Kauskolekas, C.A. and Dennison, M. (1995) Effects of pheromone trap design and placement on capture of male *Cydia caryana* (Lepidoptera: Tortricidae, Olethreutinae) in Alabama pecan orchards and the relationship of trap captures to fruit infestation. *Journal of Entomological Science* 30 (2), 165–175.

Merchant, M.E. and Teetes, G.L. (1992) Evaluation of selected sampling methods for panicle-infesting insect pests of sorghum. *Journal of Economic Entomology* 85 (6), 2418–2424.

Miller, R.S., Passoa, S., Waltz, R.D. and Mastro, V. (1993) Insect removal from sticky traps using a citrus oil solvent. *Entomological News* 104 (4), 209–213.

Moore, R.F. (1985) Artificial diets: Development and improvement. In: Singh, P. and Moore, R.F. (eds) *Handbook of Insect Rearing, vol 1*. Elsevier, Oxford, pp 67–83.

Morrison, E.G. and King, E.G. (1977) Mass production of natural enemies. In: Ridgway, R.L. and Vinson, S.B. (eds) *Biological Control by Augmentation of Natural Enemies*. Plenum Press, London, Ch. 6.

Muirhead-Thomson, R.C. (1991) *Trap Responses of Flying Insects*. Academic Press, London.

Murphy, W.L. (1985) Procedure for the removal of insect specimens from sticky trap material. *Annals of the Entomological Society of America* 78 (6), 881.

Neuenschwander, P. (1982) Beneficial insects caught by yellow traps used in mass-trapping of the olive fly *Dacus oleae*. *Entomologia Experimentalis et Applicata* 32, 286–296.

Nielson, M. W. and Lehman, W. F. (1980) Breeding approaches in alfalfa. In: *Breeding Plants Resistant to Insects*. John Wiley & Sons, Chichester, Ch. 13.

Niijima, K. (1989) Nutritional studies on an aphidophagous chrysopid, *Chrysopa septempunctata* Wesmael I. Chemically-defined diets and general nutritional requirements. *Bulletin of the Faculty of Agriculture*, Tamagawa University, No. 29, 22–30.

Niijima, K. (1993a) Nutritional studies on an aphidophagous chrysopid, *Chrysopa septempunctata* Wesmael (Neuroptera: Chrysopidae) II. Amino acid requirement for larval development. *Applied Entomology and Zoology* 28 (1), 81–87.

Niijima, K. (1993b) Nutritional studies on an aphidophagous chrysopid, *Chrysopa septempunctata* Wesmael (Neuroptera: Chrysopidae) III. Vitamin requirement for larval development. *Applied Entomology and Zoology* 28 (1), 89–95.

Niijima, K. and Matsuka, M. (1990) Artificial diets for the mass production of chrysopids (Neuroptera). In: *The Use of Natural Enemies to Control Agricultural Pests*. FFTC Book Series No. 40. pp. 190–198.

Ochiengodero, J.P.R. (1990) New strategies for quality assessment and control of insects produced in artificial rearing systems. *Insect Science and its Application* 11 (2), 133–141.

Ochiengodero, J.P.R. and Singh, P. (1993) The rearing of *Cnephasia jactatana* (Walker) (Lepidoptera: Tortricidae) using the insect rearing management system. *Insect Science and its Application* 14 (3), 289–303.

Ochiengodero, J.P.R., Onyango, F.O., Kilori, J.T., Bungu, M.D.O. and Amboga, E.O. (1991) Insect rearing management as a prerequisite in the development of IPM for sustainable food production. *Insect Science and its Application* 12 (5–6), 645–651.

Ortega, A., Vasal, S.K., Mihn, J. and Hershey, C. (1980) Breeding for insect resistance in maize. In: *Breeding Plants Resistant to Insects*, John Wiley & Sons, Chichester, Ch. 16.

Oshagi, M.A., McCall, P.J. and Ward, R.D. (1994) Response of adult sandflies, *Lutzomyia longipalpis* (Diptera: Psychodidae), to sticky traps baited with host odor and tested in the laboratory. *Annals of Tropical Medicine and Parasitology* 88 (4), 439–444.

Osman, M.Z. and Selman, B.J. (1993) Storage of *Chrysoperla carnea* Steph. (Neuroptera: Chrysopidae) eggs and pupae. *Journal of Applied Entomology* 115, 420–424.

Panda, N. and Khush, G.S. (1995) *Host Plant Resistance to Insects*. CAB International, Wallingford, Ch. 8.

Pathak, M.D. and Saxena, R.C. (1980) Breeding approaches in rice. In: *Breeding Plants Resistant to Insects*, John Wiley & Sons, Chichester, Ch. 17.

Peterson, A. (1964) *Entomological Techniques: How to Work with Insects*, 10th edn. Edwards Brothers Inc., Ann Arbor, USA.

Powell, W., Walton, M.P. and Jervis, M.A. (1996) Populations and communities. In: Jervis, M.A. and Kidd N.A.C. (eds) *Insect Natural Enemies*. Chapman & Hall, London. Ch. 4.

Riley, J.R. (1989) Remote sensing in entomology. *Annual Review of Entomology* 34, 247–271.

Riley, J. and Reynolds, D. (1993) Radar monitoring of locusts and other migratory insects. In: Cartwright, A. (ed.) *World Agriculture*. Sterling Publications, London, pp. 51–53.

Ritchie, S.A. and Kline, D.L. (1995) Comparison of CDC and EVS light traps baited with carbon dioxide and octenol for trapping mosquitoes in Brisbane, Queensland (Diptera: Culicidae). *Journal of the Australian Entomological Society* 34 (3), 215–218.

Robacker, D.C. (1995) Attractiveness of a mixture of ammonia, methylamine and putrescine to Mexican fruit flies (Diptera: Tephritidae) in a citrus orchard. *Florida Entomologist* 78 (4), 571–578.

Rohitha, B.H. and Stevenson, B.E. (1987) An automatic sticky trap for aphids (Hemiptera: Aphididae) that segregates the catch daily. *Bulletin of Entomological Research* 77, 67–71.

Roller, L. (1995) Seasonal dynamics of sciomyzids (Sciomyzidae: Diptera). *Biologia* 50 (2), 171–176.

Sarospataki, M. and Marko, V. (1995) Flight activity of *Coccinella septempunctata* (Coleoptera: Coccinellidae) at different strata of a forest in relation to migration to hibernation sites. *European Journal of Entomology* 92 (2), 415–419.

Schefer, G.W., Bent, G.A. and Allsopp, K. (1985) Radar and opto-electronic measurements of the effectiveness of Rothamsted Insect Survey suction traps. *Bulletin of Entomological Research* 75, 701–715.

Schlyter, F. (1992) Sampling range, attraction range, and effective attraction radius – estimates of trap efficiency and communication distance in Coleoptera pheromone and host attractant systems. *Journal of Applied Entomology* 114 (5), 439–454.

Scott, J.G. and Dong, K. (1994) KDR-type resistance in insects with special reference to the German cockroach, *Blattella germanica*. *Comparative Biochemistry and Physiology B – Biochemistry and Molecular Biology* 109 (2–3), 191–198.

Sharov, A.A., Liebhold, A.M. and Ravlin, F.W. (1995) Prediction of gypsy moth (Lepidoptera: Lymantriidae) mating success from pheromone trap counts. *Environmental Entomology* 24 (5), 1239–1244.

Shirai, Y. and Nakamura, A. (1995) Relationship between the number of wild males captured by sex pheromone trap and the population density estimated from a mark–recapture study in the diamondback moth, *Plutella xylostella* (L.) (Lepidoptera: Yponomeutidae). *Applied Entomology and Zoology* 30 (4), 543–549.

Shufran, K.A. and Raney, H.G. (1989) Influence of inter-observer variation on insect scouting observations and management decisions. *Journal of Economic Entomology* 82 (1), 180–185.

Sikorowski, P.P. and Goodwin, R.H. (1985) Contaminant control and disease recognition in laboratory colonies. In: Singh, P. and Moore, R.F. (eds) *Handbook of Insect Rearing, vol 1*. Elsevier, Oxford, pp. 85–105.

Simandl, J. and Anderbrant, O. (1995) Spatial distribution of flying *Neodiprion sertifer* (Hymenoptera: Diprionidae) males in a mature *Pinus sylvestris* stand as determined by pheromone trap catch. *Scandinavian Journal of Forest Research* 10 (1), 51–55.

Singh, P. (1985) Multiple-species rearing diets. In: Singh, P. and Moore, R.F. (eds) *Handbook of Insect Rearing, vol 1*. Elsevier, Oxford, pp. 19–44.

Singh, P. and Ashby, M.D. (1985) Insect rearing management. In: Singh, P. and Moore, R.F. (eds) *Handbook of Insect Rearing, vol 1*. Elsevier, Oxford, pp. 185–215.

Singh P. and Clare G.K. (1992) Insect rearing management (IRM): An operating system for multiple-species rearing laboratories. In: Anderson, T.E. and Leppla, N.C. (eds) *Advances in Insect Rearing for Research & Pest Management*. Westview Press, Oxford, Ch. 9.

Singh, P. and Moore, R.F. (1985) *Handbook of Insect Rearing, vol 1*. Elsevier, Oxford.

Soares, G.G. Jr (1992) Problems with entomopathogens in insect rearing. In: Anderson, T.E. and Leppla, N.C. (eds) *Advances in Insect Rearing for Research & Pest Management*. Westview Press, Oxford, Ch. 18.

Southwood, T.R.E. (1978) *Ecological Methods*. Chapman & Hall, London.

Southwood, T.R.E., Moran, V.C. and Kennedy, C.E.J. (1982) The assessment of arboreal insect fauna: comparisons of knockdown sampling and faunal lists. *Ecological Entomology* 7, 331–340.

Stinner, R.E. (1977) Efficacy of inundative releases. *Annual Reviews in Entomology* 22, 515–531.

Sunderland, K.D., De Snoo, G.R., Dinter, A., Hance, T., Helenius, J., Jepson, P., Kromp, B., Samu, F., Sotherton, N.W., Toft, S. and Ulber, B. (1995) Density estimation for invertebrate predators in agroecosystems. *Acta Jutlandica* 70, Aarhus University Press, 133–164.

Tauber, M.J., Tauber, C.A. and Gardescu, S. (1993) Prolonged storage of *Chrysoperla carnea* (Neuroptera: Chrysopidae) *Environmental Entomology* 4, 843–848.

Teulon, D.A.J., Penman, D.R. and Ramakers, P.M.J. (1993) Volatile chemicals for thrips (Thysanoptera: Thripidae) host finding and applications for thrips pest management. *Journal of Economic Entomology* 86 (5), 1405–1415.

Topping, C.J. and Sunderland, K.D. (1992) Limitations to the use of pitfall traps in ecological studies exemplified by a study of spiders in a field of winter wheat. *Journal of Applied Ecology* 29 (2), 485–491.

Van den Assem, J. (1996) Mating behaviour, In: Jervis, M. and Kidd, N.A.C. (eds) *Insect Natural Enemies*. Chapman & Hall, London, Ch. 3.

Vanderzant, E.S. (1973) Improvements in the rearing diet for *Chrysopa carnea* and the amino acid requirements for growth. *Journal of Economic Entomology*, 66, (2), 336–338.

Van Lenteren, J.C. (1994) Sting newsletter on biological control in greenhouses. No. 14, Wageningen, The Netherlands.

Vidlicka, L. (1995) Seasonal flight activity of *Planipennia* species at the Devinska-Kobyla hill (West Carpathians). *Biologia* 50 (2), 151–156.

Waage, J.K., Carl, K.P., Mills, N.J. and Greathead, D.J. (1985) Rearing entomophagous insects. In: Singh, P. and Moore, R.F. (eds) *Handbook of Insect Rearing, vol 1*. Elsevier, Oxford, pp. 45–67.

Webb, S.E., Kokyokomi, M.L. and Voegtlin, D.J. (1994) Effect of trap color on species composition of alate aphids (Homoptera: Aphididae) caught over watermelon plants. *Florida Entomologist* 77 (1), 146–154.

Wolf, W.W. (1985) Recognition and prevention of health hazards associated with insect rearing. In: Singh, P. and Moore, R.F. (eds) *Handbook of Insect Rearing, vol. 1*. Elsevier, Oxford, pp. 157–165.

Youdeowei, A. (1977) *A Laboratory Manual of Entomology*. Oxford University Press, Ibadan, Nigeria.

Zhou, X.L., Harrington, R., Woiwod, I.P., Perry, J.N., Bale, J.S. and Clark, S.J. (1995) Effects of temperature on aphid phenology. *Global Change Biology* 1 (4), 303–313.

Zumr, V. and Stary, P. (1991) Effects of baited pitfall traps (*Hylobius abietis* L.) on non-target forest insects. *Journal of Applied Entomology* 112 (5), 525–530.

3 Methods in Studying Insect Behaviour

T.D. Wyatt

Department of Zoology (and Department for Continuing Education), University of Oxford, South Park Road, Oxford OX1 3PS, UK

3.1. Introduction

At its simplest, behaviour is what animals do – in particular their actions and reactions. Behaviour is central to the ecology of insects and thus underlies most questions in entomology. It may be an explicit part of a project but even with the most applied question, ultimately, the behaviour of insects determines if and how they become pests, whether it is how they get to the crop or what they feed on.

Where the focus of projects can differ is in the type of question being asked or addressed. The same behaviour can be looked at in complementary ways – the 'four whys' (Tinbergen, 1963): its immediate cause (or control); its development during the life of the individual (ontogeny); its function, and how it evolved. While your current question might be the mechanism or immediate stimulus for the behaviour, it may be that looking at another 'why' might provide new insight. For example, the proximate question of host plant selection by an insect might be partly explained by its evolutionary history as shown by patterns among related species. Much of the most successful behavioural work combines these approaches.

This chapter is designed for two types of researcher – those interested in behaviour for its own sake; and those whose initial question is not behavioural but who find they need to understand more about how their insects behave in a particular system. Behavioural approaches have proved useful for a wide variety of applied problems in entomology, including assessing potential biocontrol agents (see this volume, Chapter 11), studies of resistant plant varieties (see this volume, Chapter 6) and of pesticide resistance (see this volume, Chapter 10). As an example, consider the mode of action for the increased pest resistance of glossy leafed cabbage, *Brassica oleracea*, varieties over normal-wax varieties (see Eigenbrode *et al.*, 1995; this volume, Chapter 6). The resistance is not based on toxicity – indeed there was no resistance to diamondback moth, *Plutella xylostella*, in greenhouse experiments without predators. The key to resistance in the field might then be predation but was the effect due to the behaviour of the caterpillar, predators, or both? Behavioural experiments on neonate *P. xylostella* larvae showed that they spent more time searching and walking on glossy leaves, and their wax extracts, and mined into the leaves less, behaviours that left them more exposed. Predator behaviour was also important and the effects may be partly due to greater mobility of the arthropod predators on glossy leaves. By establishing the effects of the glossy waxes these behavioural studies can inform conventional plant breeding and potential genetic manipulation of surface lipids for greater pest resistance.

The study of insect behaviour should not be viewed in isolation of the behaviour of other organisms. Some of the most interesting work comes from cross-fertilization of ideas from vertebrate studies, say of food choice, to insects (and back!). The present times are characterized by an increasing specialization of both journals and scientists and there is a tendency for entomologists to look only at entomology journals, missing ideas from, for example, more general behavioural ecology (see, e.g. Alcock, 1993; Krebs and Davies, 1993). Within entomology, the division into pure and applied can be short-sighted. Applied projects can yield valuable data for wider behavioural questions. For example, the intensive study of moth pheromones, largely prompted by their importance as pests, has provided an excellent resource for studies of speciation and the evolution of mate choice (McNeil, 1991, 1992; Phelan, 1992). Equally, understanding moth mating systems better can improve the success of pest control by mating disruption with synthetic pheromone.

Insects as subjects for studies of behaviour have a number of advantages. Their small size and rapid rates of reproduction mean that one can work with statistically satisfactory numbers of animals, but at the same time most are large enough to be marked and treated as individuals. As a result, insects offer good model systems for studying many fundamental questions in animal behaviour, from foraging theory to sexual selection.

3.2. Observing and Recording Insect Behaviour

Martin and Bateson (1993) give an excellent introduction to measuring behaviour and this book would be a good starting point for any researcher. As they emphasize, preliminary observations offer the chance to become familiar with your animals and their behaviour. The best way initially is to take simple notes while watching your animals. As you watch, think about your questions and develop hypotheses. The more competing hypotheses the better. For example, is it host egg size, or just host chemicals, or both, that are important to an ovipositing parasitoid? Make specific predictions based on these hypotheses and design your experiments to distinguish between them.

3.2.1. Choosing the focus

Ideally, keep an open mind about the apparent functions of the behaviours or 'purpose' of the animal you are observing. A teleological assumption about the 'purpose' of the behaviour can blind you to underlying mechanisms (Kennedy, 1992); how the animal actually accomplishes the task, for example, of navigation from flower to nest. Yet at the same time teleological shorthand can help in thinking about a problem initially. The skill is to remain vigorously critical of hidden assumptions.

To make sense of the behaviour you will need to break the continuous stream of activity into units in some way. For example if watching a searching parasitoid wasp you might divide the sequence into: fly, land, walk, antennate the larva, probe with ovipositor, withdraw the ovipositor, drag it on the larva, walk, take off. When attempting to divide the behaviour into units to record, choose ones large enough to make recording possible (rather than, for example, the movement of each leg, when your focus is really on the searching path of the parasitoid). Choosing the appropriate units or categories of behaviour is a matter of experience. It is often possible to collapse data categories upwards but you cannot create detail afterwards if you have not recorded it. However, if your categories are too complex or fine, the patterns will be lost in the detail. Among the kinds of things your units could include are: (i) movements; (ii) the effects on other individuals; or (iii) spatial measures such as changes in the distance from a host plant or potential mate. Make the definitions of your behaviours as explicit as you can and write them down, giving them neutral terms. Even well defined units may be graded rather than all or nothing (Harris and Foster, 1995).

From the beginning, make your observations quantitative: put numbers to your observations (but be prepared not to use your preliminary data). Without quantitative data you will not be able to make useful comparisons.

3.2.2. Structuring your observations

For many experiments there may be a relatively simple end point – for example the numbers of insects landing on targets of different colours. Similarly, in experiments where only one animal is tested at a time, no decision is needed about which animal to watch. However, in other experiments sequences of behaviour may be of interest, or their frequency of occurrence, or the behaviour of individuals in groups. Martin and Bateson (1993) review some of the principal methods for sampling rules (who to watch and when) and recording rules (how the behaviours are recorded). Much work in insect behaviour involves recording all instances of a certain behaviour, such as biting a leaf, during a time period. However, instantaneous recording of the behaviour happening at regular intervals (prompted by a bleep from your timer or event recorder) can allow more individuals to be observed simultaneously. Recording the presence or absence of a behaviour within a time interval (one-zero recording) may have a role in some situations.

3.2.3. Recording the data

Having made your preliminary observations it will be time to begin collecting your data. In some cases dictating an account of the behaviours on to audiotape as you watch may be useful, in particular at the beginning of a study or if you are following elusive insects in the field. However, given the time to transcribe the tapes you may want to move to, or , most likely, start with, other ways of recording data.

CHECK SHEETS

When you have established the behavioural units to record, a simple check sheet system of columns down a page, with each line being a time unit (e.g. 1 min) for example, can help standardize your record taking. It can save time as you put ticks, frequencies or code letters in a box rather than writing out the behaviours in full, and so it may allow you to record more behaviours. Analysis is also quicker, especially if each behaviour is in a separate column. Labelled boxes at the top of each sheet can ensure that you remember to include all the important background details for each experiment including the observer, time, temperature, stimulus and other information. Such prompts are especially important if many people are making the observations. When designing your form ensure you leave some space so you can note rare events or other behaviours for which you may not have allocated codes. Of course check sheets can also be used if transcribing video/audio tape.

COMPUTER EVENT RECORDERS

When the behaviours of interest have been defined, a computer with event recorder software which allows you to tap observations directly into the machine may be useful. Small portable computers, and larger personal organizers, can be used for data collection in the field. Among the advantages of computers over paper check sheets are the greater accuracy of recording durations of behaviour, the ability to record faster and more complex sequences of behaviour, the elimination of the transcription stage from paper to computer for analysis, and the ability to store large quantities of data (Martin and Bateson, 1993). The disadvantages include the temptation to collect too much information or to proceed to analysis without inspecting the data. Much depends on the reliability of the hardware and software (losing the data can be a real danger). The ability to export the data on disc for analysis and manipulation by other software is an important feature. To be worth while, the computer event recorder needs to save you time, make difficult observations possible, or make analysis easier. Depending on your experiment a pen and check sheet may be more practical.

One of the best known commercial packages for recording animal behaviour is The Observer ™ Noldus IT (info@noldus.nl; Costerweg 5, PO Box 628, 6700 AG Wageningen, Netherlands) which will cope with different experimental situations/designs, some data analysis, and in addition other software modules can integrate with video playback equipment. However, the cost of commercial software may put it out of the reach of many researchers who instead have written their own.

PHOTOGRAPHY AND VIDEO

Video can be very useful but a good guiding principle may be: watch first, do not video. It can be very tempting to move rapidly to videoing or film-ing. The advantages of using video include the ability to record behaviours that might happen rarely, too quickly, or too slowly, or under difficult obser-vation conditions, where an observer might change the behaviour of the subject, for complex social interactions, or in the dark (where infrared sensi-tive video cameras can offer night vision). A video recorder can be more patient than a human observer – useful when there are long periods of inac-tivity between behaviours of interest. However, among the disadvantages of video or cine-film are that crucial actions can happen just off-camera, focus, depth of field and (for video especially) resolution may be limited, and analysis can be very time-consuming indeed. Ultimately the behaviour will need to be scored in much the same way as above.

Contributors in Wratten (1994), on the use of video in ecological and behavioural research, review many practical aspects including use for studying flying insects in the field (Riley, 1994) and laboratory (Young *et al.*, 1994), as well as walking insects (Varley *et al.*, 1994). One tip is to try out particular video machines yourself for the clarity of image when using sin-gle frame advance, a common use in playback.

Keeping track of time within video sequences need not be difficult. Many video cameras can create a digital clock in one corner of the image allowing you to easily return to particular sequences or help in the analysis. Young *et al.* (1994) and Varley *et al.* (1994) give other methods for adding timing signals to help synchronize cameras or enable computer handling of images. The Noldus package for video analysis exploits the hidden timing code, laid down on the tape by the video recorder itself, to keep track of time if for example you fast forward between behaviours. Other systems may be available.

If you label the sequences on the video tape by videoing the equivalent of a clapper board title sequence at the start of each experiment (with its number, stimuli details, and date) sorting out tapes will be easier and you will be able to spot the start of experiments when searching on fast forward (index marks also help). You may be able to use the sound track to record your comments during the experiment.

Analyse the tapes as you go along. Do not leave the analysis until the end of the experiments on the assumption that the tape has recorded all the details you need. You may find, for example, that the contrast is not suffi-cient to allow you to distinguish the detail you need.

ANALYSING VIDEOS TO TRACK MOVEMENT

Video is often used to record the paths of insects, for example towards or away from an odour source. However, it is worth asking if knowing the detailed path is needed to answer the question being studied. Are there simpler measures that could be observed at the time, for example distance of the insect from the target, or time spent stationary? Where the focus *is* on the underlying search mechanisms used by the animal or an explanation of how observed differences in response occur, track measurements are needed. For many other studies video tracking is a distraction.

Going from the video to tracking insect movement is time-consuming. The simplest, and most laborious, way is to go through the tape a few frames at a time, marking the position of the insects on an acetate sheet laid over the video monitor screen. These positions can later be digitized manually. Young *et al.* (1994) provide formulae for calculation of track parameters (see also Bell, 1991). Different stages can be computerized culminating in automatic tracking systems in which the computer tracks the path itself.

Automated analysis offers significant savings in time, especially when the alternative is going through a film or video manually, frame by frame. However, people are much better than a computer at following an insect on screen, especially in real time. A good compromise is to link the video player to a computer which can combine the video and computer image on the same monitor and allow hand digitizing with a mouse on screen. Varley *et al.* (1994) give details of one such system which also doubles as an event recorder.

Tele-tracking, an early system for automatic tracking, compares the incoming signal from a video, as it scans across the screen a line at a time, with a brightness threshold and gives the x–y coordinates for the first bright dot encountered in each frame. It gets confused by groups of insects or fussy backgrounds (Young *et al.*, 1994). A more sophisticated technique is image analysis, which turns the image into pixels on the computer screen, and uses software to find the insect by, for example, subtracting the previous image to detect movement (Young *et al.*, 1994). Image analysis can also be used to give the spatial distribution of activity, rather than tracking individual insects, for example, within an ant nest (Strickland and Franks, 1994).

ACTOGRAPHS AND OTHER AUTOMATIC RECORDING METHODS

For some kinds of questions, such as those on activity rhythms, automatic data collection offers major advantages, not least that it makes the investigation of circadian rhythms possible without sleep deprivation for the experimenter. The behaviour usually needs to be a fairly simple one, such as an increase or decrease in movement in a cage, which can be detected easily. By careful design of the sensors it may be possible to allow the system to distinguish different behaviours. One disadvantage of using an actograph is that one cannot see what is happening and this may limit one's scope for thinking of the next experiment.

An important feature of whatever method is used is that it should not be detectable by, or change the behaviour of, the insects themselves. The technique needs to be validated by observation to confirm the effectiveness of the recording mechanism and that it accurately reflects the insects' behaviour.

There are many different mechanisms but basically all actographs detect movement by breaking a light beam or a change of some other sort detected electronically (early actographs used levers to exaggerate the scale of movement and traced a track on a smoked drum) and are limited only by your ingenuity. New techniques are likely to be developed as new transducers are invented and electronic devices get cheaper. Among the methods used to detect movement so far are rocker-cages, infrared, ultrasound, wing beat sound, vibration, changes in capacitance, temperature, Doppler shift radar (e.g. Kyorku and Brady, 1994; Snowball and Holmqvist, 1994). As the output is often an electrical signal, most systems can now be linked to a data logger or a computer via an analog/digital converter card which allows for easier analysis (e.g. Beerwinkle *et al.*, 1995, for a flight mill, but the same principles apply).

RECORDING ENVIRONMENTAL VARIABLES

The small size of insects means that they are usually more affected by the environment than larger animals. Temperature is a major factor influencing their behaviour. It can thus be very important to be able to measure this, together with relative humidity. Meteorological station data may be inappropriate since it is not what the insects are experiencing. Rather, one needs microclimate data taken on a scale relevant to the insect. Unwin and Corbet (1991) and Unwin (1980) consider a range of methods. Data loggers may be useful for recording the data over a period of time.

Some insects are able to maintain internal temperatures well above ambient and thermoregulatory behaviour has become a new field of study (Heinrich, 1993). The standard way of recording internal temperatures, stabbing the thorax with a thermocouple for example, has been criticized by Stone and Willmer (1989). Jones (1982) used a blue pigment, which faded in sunlight, mixed in yellow paint and painted on to snail shells. Suitably calibrated, the fading from green allowed estimation of how much time the animals had spent in the sun (it works equally well with beetles).

3.2.4. Statistics and experimental design

Determining preferences, for example for one plant species over another, is a common aim of experiments in insect behaviour. In a thoughtful essay, Singer (1986) discusses the definition of preference (which is not straightforward) and the merits of a range of experimental designs: no choice, sequential choice, and simultaneous choice (where all the alternatives are offered at the same time). Different results can be obtained depending on the preference testing technique. Although Singer discusses oviposition behaviour, his conclusions have wide relevance to anyone using preference tests.

In no choice tests, ceiling effects (in which all stimuli get high responses) or, conversely, floor effects (where almost none respond) may be resolved by trying another behavioural measure, for example latency (Martin and Bateson, 1993).

Vaillant and Derridj (1992) offer an alternative, based on the Monte Carlo procedure, to traditional non-parametric analysis of two-choice experiments. In the special case of choice experiments where living food grows (e.g. plants), or dries out (e.g. evaporation from insect prey being eaten by liquid-feeding predators), during the course of the experiment you may need particular methods for the analysis (Peterson and Renaud, 1989). They also make the point that there are rarely enough replicates of controls to allow clear differentiation between treatments.

The importance of good controls cannot be overstressed. Solvent controls are needed whenever chemicals are presented in solution. Control insects need to be 'sham operated', that is handled as much and in the same way as much as possible as the experimentals. In some experiments it may be possible to use an animal as its own control (analysed as a matched pair) but there can be problems of order effects, especially in the context of the wider occurrence of experience effects in insects (see learning, Section 3.3.4).

Running experiments blind, so the observer is not aware of which treatment is being tested, is not often done in entomology but can be strongly recommended (Martin and Bateson, 1993). Assigning subject animals to treatments should be rigorously randomized, especially if it is difficult (because the treatments are obvious to the observer, for example if testing leaf shapes) to run the experiments blind.

Multiple recording of the same insects, leading to pseudo-replication, can be a problem in both field and laboratory experiments. It may be possible to design the experiment so insects can only be counted once, for example falling into a pitfall (White and Birch, 1987).

Where differences in behaviour between two types of animal may be subtle, any one variable by itself may not show a difference. While the behaviour of solitary and gregarious locusts is very different there is a continuous gradation between the two states. In 'gregarization', changes in behaviour occur more rapidly than in morphology or colour so behavioural analysis allows finer resolution of the changes. By observing the behavioural responses of individual solitary or crowd-reared nymphs when presented with a group of conspecifics, Roessingh *et al.* (1993) were able to use logistic regression to derive a predictive model based on 11 simple behavioural measures. The model distinguishes the two extreme classes of insect but, more important, gives a single index of the 'behavioural phase status' of an insect anywhere between. Analogous methods might be useful for studying other primer effects with a long time scale, for example caste determination in termites.

In some cases discovering the sequential relationship between behaviours may be of interest, for example, whether behaviour B follows behaviour A or whether the sequence of behaviours occurs at random. A well established method to test this is Markov analysis (Martin and Bateson, 1993, and references therein). If the probability of a behaviour occurring is determined by the behaviour immediately before, then the process is first order; if determined by the two behavioural events happening before, second order, and so on. Assumptions made in this analysis, and often violated, are that the transition probabilities do not differ between individuals or change over time. New 'hidden Markov models' allow such changes in transition probabilities over time to be estimated, for example in the likelihood of foraging behaviour with increasing time since the last meal (MacDonald and Raubenheimer, 1995). Simpson and Ludlow (1986) suggest a different approach using 'hazard analysis' which also provides an example of the use of GLIM in behavioural analysis.

3.3. Factors Influencing Behaviour

Any investigation of insect behaviour needs to be informed by an understanding of the factors (perhaps beyond the immediate concerns of the experiment) that influence their behaviour.

3.3.1. Internal state

The response of an insect is determined not only by the stimulus but also by changes in the internal state of the insect such as those driven by circadian rhythms. For example, in many species there are particular periods each day when activity or responsiveness to stimuli such as pheromones is greatest (for review see McNeil, 1991). Underlying such daily rhythms are other physiological changes centred on factors such as the length of time since the last meal. On a longer time scale, mating history, age, nutrition, and effects of experience (q.v.) can affect behaviour in profound ways (Barton Brown, 1993). Another potential internal factor influencing behaviour is the presence of parasites (see Bell *et al.*, 1995).

Internal stimuli may be difficult to manipulate except at a gross level, for example mated or not mated, or by testing the insect at a particular age, or by controlling access to some resource such as food or oviposition sites (Harris and Foster, 1995). It may also be useful to pre-test your insects to see if they will respond appropriately. For example, if you are investigating orientation in flight, it may make sense to include only insects which take off in a pre-test, if preliminary experiments show this is both a good predictor of later takeoff behaviour and that this selection does not bias the results (similar tests have been used to test viability of biocontrol agents before release (see Mills, this volume, Chapter 11)).

Circadian rhythms in responsiveness and day-to-day variation make it essential to carry out experiments during the active period and to use fully randomized complete blocks to ensure that all treatments are treated equally within the period. Frequently, factors such as size, age or date can be controlled for statistically in the analysis.

Experimental difficulties aside, the recognition that there might be individual variation in behaviour *within* a species, and that this variation can be interesting rather than simply a nuisance, may be one of the most important developments in modern animal behaviour. Differences between individuals could be based on differences in size, for example major and minor males of the solitary bee *Centris pallida* (Alcock *et al.*, 1977), or conditionally on the behaviour of other conspecifics at the time. Individuals could also be influenced by prior experience as well as a host of other variables affecting internal state (see, e.g. Borden *et al.*, 1986).

3.3.2. Genetics and insect behaviour

Many of the differences in behaviour between individuals may have some genetic basis as well as a developmental component (Hoffmann, 1994). Host plant selection behaviour again provides many examples, of both variation within a population, among host races in the apple maggot fly, *Rhagolitis pomonella*, and between populations of the same species in oviposition behaviour in British and Australian populations of the cabbage butterfly, *Pieris rapae* (see Bernays and Chapman, 1994; also this volume, Chapter 6).

All aspects of insect behaviour can have a genetic component, from circadian rhythms to courtship. Butlin and Ritchie (1994) give examples of the genetic basis of courtship signals in the context of speciation, and Bell *et al.* (1995) and McNeil (1991) review the genetics of pheromone production and reception in moths. Earlier work on the behaviour genetics, typically on laboratory animals such as *Drosophila,* tended to use behaviours that were easily scored in the laboratory but an increasing number of studies concern ecologically relevant behaviours (Hoffman, 1994).

Whether or not it is the focus of your project, the genetic basis of the behaviour(s) may nonetheless be important. For example, for laboratory studies you will need to be sure of the strain or biotype you are using as these can differ (and may even be defined on the basis of behaviour). Field workers will need to be aware of potential inter-population differences.

In applied entomology, selection for behavioural resistance can be important for vector control and IPM programmes. Genetic bottlenecks and inadvertent selection can have a significant influence on programmes to laboratory rear biocontrol agents and sterile males (and in routine laboratory cultures) (Prokopy and Lewis, 1993).

3.3.3. Developmental and transgenerational effects

Many insects show lasting changes in adult behaviour according to the conditions during development as a nymph or larva. The effect of crowding leading to phase change to the gregarious form in locusts is one of the most impressive examples of this type of phenotypic plasticity (Roessingh *et al.*, 1993, and references therein). In the locust, *Schistocerca gregaria*, the conditions the mother was reared under, and indeed population density at the time of mating or oviposition, can also affect the behavioural phase state of the offspring (Islam *et al.*, 1994). Such effects are an example of maternal inheritance. Maternal inheritance, where the mother influences traits of her offspring by means other than by transmission of nuclear alleles, is known in a wide range of insect taxa (reviewed by Mousseau and Dingle, 1991). In most cases the mechanisms are not known but effects mediated via egg size or egg concentrations of maternal hormones are among those demonstrated in various species so far. Maternal effects may provide adaptive phenotypic plasticity across the generations so that offspring develop appropriate phenotypes for the conditions likely to follow the conditions the mother experiences (Mousseau and Dingle, 1991). Maternal effects can be another cause of individual variation in behaviour.

3.3.4. *Prior experience and learning*

While von Frisch's early experiments on bee vision exploited associative learning to give clues to the colours bees could distinguish, learning has until recently not been thought important in other insects. Recent progress across a wide range of taxa and ecological situations has been reviewed by Papaj and Lewis (1993). There are many definitions of learning but perhaps the simplest is a change in behaviour with experience, excluding factors like changes in egg load over time. Some of the best studied examples are learning by parasitoid wasps (Turlings *et al.*, 1993; Vet *et al.*, 1995), bees (Gould, 1993) and, increasingly, phytophagous insects (Bernays and Chapman, 1994; Bernays, 1995).

Parasitoid wasps have evolved searching strategies to solve the difficult task of finding cryptic prey and many species seem able to modify their responses to foraging cues based on experience. This ability to learn profitable cues has now been demonstrated for almost 20 different species (Turlings *et al.*, 1993). Most of the learning is associative: the wasps innately recognize host-derived stimuli (unconditioned stimuli, US) on contact and associate these stimuli with surrounding stimuli (conditioned stimuli, CS), previously of little interest (Turlings *et al.*, 1993). The conditioned stimuli could be olfactory such as host frass, but may be visual, such as visual characteristics of plants, mechanosensory, or perhaps most effectively, a combination. Such learning allows the wasps to use, for example, characteristic volatiles released in relatively large quantities by host-damaged leaves as a cue to find hidden host larvae. The behaviour of a female wasp can be changed by a plant-host contact of as little as 20 s (Turlings *et al.*, 1989) but unless reinforced by successful ovipositions the preference behaviour wanes, giving a flexible response to foraging outcomes.

Changes in behaviour with experience have only been studied in a small range of situations, among them the topics above. There may be similar effects occurring in other aspects of insect behaviour, for example mate choice, which currently are not recognized. There are two points of practical relevance to all researchers. First, experience, only a small part of which may be known to the experimenter, before the observation or experiment starts will have an influence on the behaviour of the subjects. Second, insects should not be used more than once in an experiment if possible as later behaviour may be influenced by a first treatment.

The implications of insect learning to pest management are examined by Prokopy and Lewis (1993) who make the point that both beneficial insects and pests show learning of various kinds. A well established phenomenon is the effect of artificial rearing conditions on the behaviour of parasitoids for release, which may reduce their effectiveness in the field. Learning could also affect other phenomena such as pest population estimates based on mark–release–recapture, if handled insects behave differently. Pest habituation, or conversely sensitization, could affect the effectiveness of plant resistance.

3.3.5. *External stimuli*

External stimuli such as visual, auditory or olfactory stimuli are some of the most obvious influences on behaviour and are the ones that can most easily be manipulated by the experimenter. Most behavioural experiments involve changing these external stimuli (see below). While external stimuli presented to the animal might be kept constant, they will not always produce the same inputs as peripheral receptor responses are influenced by physiological variables – for example, locust taste chemoreceptors are influenced by haemolymph levels of nutrients such as amino acids, being less responsive when levels of that nutrient are high (see Simpson *et al.*, 1995).

Taking account of the way animal perceptual worlds differ from ours is always important in behaviour but perhaps especially for work on insects. Olfaction and taste are much more important in insects and even apparently familiar senses such as vision are different from our own both in spectral sensitivity and acuity.

3.3.6. *Integration*

The theme developed by Harris and Foster (1995), in an excellent review, is the integration of internal and external sensory inputs in behaviour. Internal factors include, for example, those that change with age, prior experience, and mating status. Moreover, despite the desire of the experimenter to concentrate on one external stimulus, for example host odours, the majority of insect behaviours may be driven by *combined* inputs from many senses – from vision as well as chemoreception (see also Prokopy, 1986).

While single sensory inputs, such as a chemical stimulus, can be investigated by themselves, multifactorial (and possibly multi-level) tests can tease out the stimuli involved in behaviours, and may sometimes show interactive effects of stimuli that would be missed had they each been tested alone (Harris and Foster, 1995). A good example is that of the oviposition behaviour of Hessian flies, *Mayetiola destructor*, a pest of wheat which lays its eggs on the leaves. In a three-way factorial experiment testing colour, chemical, and tactile stimuli, Harris and Rose (1990) found all three stimuli had a significant effect on numbers of eggs laid (and all possible first and second order interactions were also significant; see Fig. 3.1). From this one might conclude that the three stimuli were all integrated in the response. However, as Harris and Foster (1995) point out, a second experiment (Harris *et al.*, 1993), which investigated the behaviours leading to egg laying, as well as number of eggs laid, showed that although colour influenced approach and landing behaviour it did not affect the number of eggs laid once the female had landed, whereas chemical stimuli were important at each stage. There are some disadvantages of more complex factorial experiments: logistic problems of carrying out enough replicates and interpretation of interactions are among these.

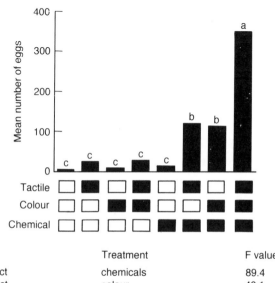

Tests	Treatment	F value	Significance
Main effect	chemicals	89.4	<0.0001
Main effect	colour	40.1	<0.0001
Main effect	tactile	47.6	<0.0001
First order interaction	chemicals × colour	37.7	<0.0001
First order interaction	chemicals × tactile	34.3	<0.0001
First order interaction	colour × tactile	8.6	<0.006
Second order interaction	chemicals × colour × tactile	8.6	<0.006

Fig. 3.1. Integration of stimuli: the role of colour, chemical, and tactile cues in oviposition by Hessian flies, *Mayetiola destructor*, on plant models. A 2 × 2 × 2 factorial gave eight combinations of stimuli. Wax coated filter-paper leaves were coloured green (compared with undyed white), dosed with wheat cuticular wax extract (compared with control solvent alone), and given vertical grooves in the wax to simulate leaf veins (or left without grooves). One of each stimulus combination was present in the simultaneous arena test with a single female. Black boxes indicate the first cue of each pair is present. The main effects of each stimulus and the higher order interactions were all significant. (From Harris and Foster, 1995; data from Harris and Rose, 1990; reproduced with permission from Entomological Society of America.) However, later experiments showed the effect of colour (green) was due to its effect in increasing landing (Harris *et al.*, 1993).

In some situations it may be difficult to separate stimuli for analysis. For example, many studies showing a response of flying insects to visual targets in the presence of pheromone may at the same time be offering a barrier to the air flow, as the target is solid. In trying to separate out visual stimuli and the effects of a wind barrier on the response of insects to an odour plume, Wyatt *et al.* (1993, 1997) used a silhouette drawn on the upwind mesh of the wind tunnel to offer a visual stimulus with no effect on the air flow. A barrier could be placed upwind of the mesh when required, and when used with an unmarked mesh screen could test the effect of a barrier without a visual stimulus (Fig. 3.2). The results with male woodworm beetles, *Anobium punctatum*, responding to their sex pheromone were surprising (Wyatt *et al.*, 1997): the barrier alone was as effective as a visual stimulus with barrier, although without the barrier a visual target induced quicker landing than no visual stimulus or barrier, giving a significant interaction of visual target and hidden barrier in the analysis. The challenge for the experimenter dealing with multiple stimuli is to devise a soluble problem which nonetheless helps understanding of a more complex real world.

A notable feature of insects, and other animals, is the ability to use not only a multiplicity of stimuli but to use one sense when another is not available. Such redundancy allows an insect to function despite loss of input from one of its senses through old age, injury, or vagaries of the environment in the case of insect navigation when, for example, cloud obscures the

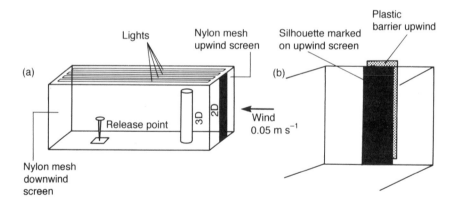

Fig. 3.2. Separating the effects of visual and barrier effects of trees on upwind orientation of insects to odour sources. (a) Diagram of the wind tunnel (not to scale) showing the position of the three-dimensional black cylinder or two-dimensional silhouette. In each experiment, the release point of single male beetles was from a take-off platform 20 cm up from the floor and 100 cm downwind of the pheromone source (also 20 cm up from the floor). (b) Separating the visual and barrier effects offered by a two way factorial experiment. The visual stimulus when present was a black shape marked on the upwind mesh and the wind barrier was a plastic shape placed immediately upwind of the mesh. (After Wyatt *et al.*, 1997.)

sun. The advantage to the insect is clear but since most experiments involve either depriving an insect of a sense at a time or providing limited controlled stimuli, drawing conclusions can be difficult. If insects can switch to another sense then the ability of the insect to continue to do the task successfully is not an indication that it did not normally use these cues. Studies on animal navigation for example, including that of bees, have been bedevilled by this problem (Dusenbery, 1992).

3.4. Investigating Behaviour

The techniques for investigating behaviour are similar whatever the system. Among the major decisions are whether to use laboratory or field experiments and observations, or often best, a combination of both.

3.4.1. Laboratory experiments

Laboratory experiments can be a very effective way of studying the behaviour of insects. When developing your set-up try to create conditions which elicit as natural behaviour as possible. Temperature, relative humidity and lighting can be important variables. For example, intriguing experiments by Kamm (1990) showed that seed chalcid wasps would not make orientated flights to host olfactory stimuli in the laboratory but would under natural light in a greenhouse (it turned out polarized light was required). Often the development of laboratory set-ups is a matter of luck. For example, use of agar as a substitute for mud for a burrowing beetle in the laboratory would not have worked if the beetle had needed to close its jaws on sand grains (Wyatt, 1986).

It is possible to observe nocturnal insects under lights with 'far-red' filters (e.g. Ilford 609, passing wavelengths light > 680 nm). This can work well. However, recent work suggests infrared light-emitting diode (LED) light sources might be better (mosquitoes can respond to backgrounds illuminated in red light (> 700 nm) but not the > 900 nm light produced by LEDs (Gibson, 1995)). Arenas lit with such LEDs can be viewed with modern solid state cameras which are very sensitive to long-wavelength light (>900 nm) (Young *et al.*, 1994).

Subtle changes in laboratory insect cultures can have important effects on insect behaviour. These often result from inadvertent artificial selection pressures. For example, Birch *et al.* (1989) were unable to demonstrate a courtship role for the male hairpencils in the cabbage moth *Mamestra brassicae*. Later it was shown that this was a laboratory artefact: the hairpencils of wild males and males from fresh cultures did have a pheromone which seemed to be lost from older laboratory cultures (Poppy and Birch, 1994).

The design of laboratory experiments aims at a happy medium. While you cannot create a complete habitat, the more it matches the natural habitat the better the experiments may predict behaviour in the wild. For example, if you are investigating responses to resource stimuli it may be worth investigating factors such as habitat structure, patch size and resource distribution (Prokopy, 1986). Ultimately, the goal is to test the predictions in the field.

3.4.2. Field observation

Laboratory experiments may allow for easier replication of experiments but they raise the question: do animals really behave in the same way in the wild? This can only be answered by observing animals in their natural environment, in the field.

Important new subjects such as the role of learning in entomophagous parasitoids (see Section 3.3.4) can grow up without a firm base of studies showing its significance in the field (Papaj, 1993). Similarly, almost all studies on the feeding patterns of herbivorous insects have been laboratory based, with a range of plant choices determined by the experimenter. Raubenheimer and Bernays (1993) heroically followed individual marked female grasshoppers (*Taenipoda eques*) for the whole day (sunrise to sunset) in Arizona. Using a hand-held electronic event recorder (see Section 3.2.3) they recorded all the feeding bouts, their duration and marked the plants for later identification as the insects moved around the habitat. For this species, perhaps reassuringly, many parameters such as the clustering of feeding events into bouts, were very similar to laboratory studies of other acridids although there were some differences which highlighted the need for equivalent laboratory studies of this species. Similar field techniques have been applied to other insects. For example, Aluja *et al.* (1989) labelled every twig on trees to allow the 3-D movement of apple maggot flies, *Rhagoletis pomonella*, to be followed. Near-focusing binoculars to follow insects may be useful in field studies. Nocturnal insects have been observed in the field by using night vision equipment, for example looking at behaviour of moths flying to pheromone traps (see Riley 1994; also this volume, Chapter 5). Opp and Prokopy (1986) discuss other aspects of field experiments. For example, you may need to mark your animals (see Chapter 5).

Much of the classic work on ant navigation has been done in the field (see e.g. Wehner *et al.*, 1996) as was Tinbergen's pioneering work with the digger wasps. In each case experimental manipulation was combined with observation. Placing models or test arenas in the field so animals encounter them naturally can be very effective (e.g. Judd and Borden's, 1991, experiments on onion fly, *Delia antiqua*).

A major advantage of field experiments is that the internal state of the animals can be assumed to be natural but with the disadvantage that it might not be known or necessarily repeatable. Another reason for studying animal behaviour in the field is when important factors, especially stimuli, might be affected by moving the experiment into laboratory. For example, cutting plants to provide leaves for feeding choices could offer wilting leaves and initiate wound responses by the leaves, all factors now known to affect insect feeding behaviour (Bernays and Chapman, 1994). When it is impossible to reproduce field conditions in the laboratory, for example the range and activity of predators, field experiments may be the only way of investigating topics such as the effectiveness of parental care. For example, by elegant *in situ* exclusion experiments Tallamy and Denno (1982) showed that eggplant lace bug, *Gargaphia solani,* females were able to successfully defend their broods against arthropod predators.

While there are many advantages in doing field experiments there are good reasons why many experimenters attempt to do at least part of their work through laboratory experiments. The principal problem in temperate regions is the unpredictability of the weather. On cool or rainy days there may be no insect activity. Low animal densities may also be a problem. With field behavioural work on some species (perhaps especially for pests – not chosen for their ease of study) the problem is not being able to identify species or their sex accurately at a distance (Finch, 1986). In these cases you will need to be able to catch the individual after the experiment.

Where field conditions are difficult or the animals are difficult to rear, a compromise may be to use field collected animals in laboratory experiments where you can better control the conditions.

It is possible to discover much about the field behaviour of insects by the use of traps as an indirect way of testing stimuli, for example odours or colours, rather than watching individual animals. Conversely, behavioural observations of individuals can be used to improve traps. Small differences in design can have large effects on trap catch (Phillips and Wyatt, 1992). Vale and co-workers used electrocuting grids to show the numbers of tsetse flies approaching and going into different trap designs (e.g. Vale, 1982).

Testing pheromone identifications based on laboratory bioassays is one important use of traps. Cardé and Elkinton (1984) discuss the design and interpretation of field trials, including trap interactions (Byers, 1992; see also this volume, Chapter 7). The design of field trials still presents a major practical problem (e.g. Sanders, 1989; Wyatt, 1997). It can be difficult to find matching control plots far enough away to be unaffected by the pheromone treatments on the experimental plots, but close enough to offer similar conditions. This reinforces the need to be able to do realistic tests of pheromone blends and release rates before field trials.

3.4.3. Visual stimuli

Visual stimuli are important in the lives of all insects. Their influence has been extensively investigated in both host-finding by phytophagous insects (see this volume, Chapter 6; Prokopy, 1986; Bernays and Chapman, 1994) and by pollinators (this volume, Chapter 9). Finch (1986) gives a good review of features to consider in bioassay of colour responses of insects. Among the most difficult problems are producing standard colours changing in hue (dominant wavelength) but not in brightness (intensity of reflected light). Colour saturation (spectral purity) is another variable. For initial experiments it is possible to use standardized colour papers to test for attraction to hues and to hues of different saturation (Finch, 1986). Neutral 'grey' papers from black to white can be used to test whether attraction apparently based on hue is really one to brightness. The shorter wavelengths have more energy (quanta) than longer ones, towards red, and thus appear brighter.

When considering brightness, an important point is that light sources, artificial and natural, do not have an even output across the spectral sensitivity of the insect eye. With a knowledge of the light source output spectrum, and the reflectance spectrophotometer readings from the pigment, it is possible to calculate the relative number of quanta reflected from a pigmented surface (Kolb and Scherer, 1982; e.g. Harris *et al.*, 1993).

Coloured lights can also be used, with the advantages that specific wavelengths can be used and one can control duration as well as intensity. For example Kolb and Scherer (1982) were able to show that the effect on drumming and egg laying behaviour for *Pieris brassicae* L. was wavelength specific rather than a response to brightness. Coloured light has been used for testing landing responses of aphids and white fly in vertical flight chambers (see Section 3.4.7).

The background on which targets are presented can also have effects. For example Harris *et al.* (1993), working with Hessian flies (*Mayetiola destructor*), found marked differences in the landing responses to targets of different reflectivities depending whether the arena walls were white or black.

3.4.4. Using physical models

Ever since Tinbergen's experiments showing stickleback male courtship responses, models have been an important way of investigating responses to stimuli. The principal advantage is the chance to allow control of the stimuli, which can be manipulated in turn, allowing experiments on the integration of behaviour (above).

A major use of models has been testing host plant selection by insects (Bernays and Chapman, 1994) (see Sections 3.3.6 and 3.4.3). Variability between plants, for example, can be eliminated by using painted models with reflectancies checked with a spectrophotometer, for example (Prokopy *et al.*, 1983) testing attraction of cabbage root flies, *Delia radicum*. A host of experiments on pollination, learning in honey bees, flower constancy, reward response have used model flowers (see this volume, Chapter 9) with or without methods for giving a nectar reward.

Visual stimuli are not limited to colour. Target shapes also have strong effects on landing. Harris *et al.* (1993) showed vertical shapes were more attractive to female Hessian flies. The 'vertical contour length' proved important, with the taller and longer the better but other things equal, the greater the area the more eggs laid (so long as the target was not horizontal), matching the predicted preferred shape as the species lays its eggs on grass stems.

Models can also be used to investigate the importance of surface texture and other tactile cues. For example, Harris and Rose (1990) used parallel vertically orientated grooves to mimic leaf venation of the host grass plants of the Hessian fly. Model eggs made from glass beads have been used to study host egg selection by parasitoid wasps (Vinson, 1985), enabling shape, size, and contact kairomones to be tested. The use of models can improve reproducibility and where the stimulus would otherwise be another animal it controls for behavioural interactions between them.

3.4.5. Chemical stimuli

Chemical stimuli are important in many aspects of insect behaviour from communication to host plant selection. This section emphasizes olfaction but techniques in contact chemoreception are very similar. While electroantennograms (EAG) or single cell recording (SCR), perhaps coupled with gas chromatography, can pinpoint potentially active compounds, behavioural analysis is still needed as EAGs or SCR cannot predict the behavioural response of the insect. Components for which there are few receptors can be missed as may synergies between components (q.v.) which are not usually observed at the peripheral level. Byers (1992) provides a concise discussion of these points and the design of bioassays.

A behavioural bioassay is a repeatable experiment for measuring response to a stimulus. In the context of semiochemicals, or equally host odours, it allows you to trace which fractions contain activity during fractionation. Baker and Cardé 's (1984) review of bioassays, although primarily for pheromones, is still useful. Depending on your objectives, simple bioassays may be the best. One of the main criteria is ease of use – in scoring as well as execution – although the crucial point is that it should be a reliable measure of the behaviour you ultimately want to assess: a laboratory bioassay, for example, should be able to identify a pheromone which works in the field (wing fanning by males has been used in this way for laboratory pheromone bioassays instead of more complicated flight orientation bioassays). You may be able to reduce the duration of the bioassay if you discover on plotting the data that the result is given more quickly than first thought.

There are many different types of walking bioassay – from still air arenas, with pheromone placed in the centre, to Y-tube olfactometers with pheromone laden air flowing down one side, clean air down the other, more elaborate four-arm olfactometers (Baker and Cardé, 1984) – and a wide variety of wind tunnels for testing in flight (Baker and Linn, 1984; Young *et al.*, 1994).

Kennedy (1977) highlights the need for discriminating bioassays as opposed to ones which do not reveal the orientation mechanisms used by responding insects. For example, animals collecting in one arm of an olfactometer could be due to insects stopping there (inverse chemoorthokinesis), or to increased or decreased random turning (direct or inverse chemoklinokinesis) or to directed turning (chemotaxis) or to odour-conditioned anemotaxis (for an explanation of terms see Table 3.1). Y-tube and four-arm olfactometers share the problem of sharp boundaries to the odour plumes which allow the insects to use chemotaxis in ways that could not be done at a distance from the source. Bioassays that allow control of the way the insect contacts the pheromone, for example by a catwalk for walking at right angles to the gradient of odour, are preferred. However, if a bioassay allows one to identify active components it can be useful even when the orientation mechanisms are not revealed.

Behavioural bioassays to assess fractions as they come direct from a gas chromatograph (GC) have been developed (e.g. Leal *et al.*, 1994; Nazzi *et al.*, 1996). A combined GC/EAG/behavioural bioassay using associative learning of a sugar reward and odour, which exploits the bee's tongue extension behaviour, has been used by Wadhams *et al.* (1994) to investigate honey bee responses to flower odours. However, searching for synergy (below) remains difficult when fractions are only tested as they come from the GC column in a linear time sequence.

Most insect pheromones appear to be composed of more than one synergistic component acting together. Each synergist alone may have little activity, but presented together at appropriate concentrations they are bioactive. This poses problems for isolating the active compounds from an insect extract as sub-fractions will not show activity unless other synergists are also present. As the number of major fractions rises the complexity of bioassaying the potential combinations increases as a power. Byers (1992) compares the efficiency of different bioassay strategies and firmly advocates 'subtractive combination' (fractions containing active synergists are revealed when removing the fraction reduces the activity of the whole extract). De Jong (1987) provides a method for efficiently determining the optimum blend if the components are known.

Apart from the design of the bioassay, additional questions for work with olfaction include the type of substrate used to release the chemicals, the strength of the stimulus, and the way in which two or more chemicals should be mixed (Baker and Cardé, 1984). Most experiments release chemical stimuli from filter paper or rubber septa. Alternative dispensers using diffusion from open tubes may be particularly appropriate for field experiments or when high release rates are required (Byers, 1988).

Bioassays for investigating glandular sources of trail substances of social insects and their chemical characterization have been critically reviewed by Traniello and Robson (1995) in particular citing the failure of many bioassays to distinguish between recruitment and orientation which are distinct behaviours in trail communication and foraging organization in both ants and termites.

The extreme sensitivities of insects to their semiochemicals dictates extreme cleanliness in the laboratory to avoid contamination (solvent controls are a safeguard). The soaking of all glassware in detergents such as Decon^R 90 between experiments is recommended.

Table 3.1. A summary of categories of odour-induced manoeuvre resulting in movement towards an odour source. (From Kennedy, 1986.)

Name	Sensory input Odour concentration differences detected by sampling:	Minimum receptors needed	Locomotory output = manoeuvre Form	Steering
Orthokinesis	Successively along path	One	Change in linear speed or frequency of locomotion	Self-steering throughout
Klinokinesis	Successively along path	One	Change in rate or frequency of turning	Self-steering throughout
Schemakinesis ('longitudal klinotaxis')*	Successively along path	One	A particular pattern	Self-steering throughout
Klinotaxis ('transverse klinotaxis')	Successively, by small swings left & right ('wig-wagging')	One	Turning to the more or less stimulated side	By the odour gradient, but 'wig-wags' probably self-steered
Tropotaxis	Simultaneously left & right	Bilateral pair	Turning to the more or less stimulated side	By the odour gradient
Menotaxis (now including 0° & 180° angles)	Wind direction may be sampled successively left & right by serial countertuning — Directional cues generated by wind	Multi-directional	Maintaining a set angle to the wind direction = *anemotaxis**	By wind-generated cues: mechanical for walkers, visual for flyers

* 'Positive' anemotaxis = upwind; 'negative' = downwind; across-wind was called anemo-*menotaxis* in the original Kühn sense.

3.4.6. *Acoustic stimuli*

The first stage is recording the sounds and analysing them. Ewing (1989) and Bailey (1991) offer practical advice on methods as well as providing good introductions to the field of insect bioacoustics. Digital recording is now possible even on portable cassette recorders. Many software packages for analysing sound are available (e.g. 'Canary' for the Mac, reviewed by Wilkinson, 1994, and 'AVISOFT-SONAGRAPH' for IBM systems, reviewed by McGregor and Holland, 1995). For up-to-date information on available software, including shareware, try the Cornell Bioacoustics Research Program (web page http://www.ornith.cornell.edu/BRP).

A playback system allows you to test your hypotheses in studies of communication. Among the many decisions to be made about playback experiments are the intensity, volume, distance between sources, original volume measurement, sound quality, distortion, fidelity, which parts of the signal to use, and the problem of between-individual variation. A particular problem with insects and other poikilotherms is the effect of temperature (see Bailey, 1991).

Some software packages can also be used for manipulating the signals for playback – or creating entirely synthetic signals for playback through a loudspeaker or other device. Synthetic signals allow much greater control over the stimulus offered (e.g. for vibratory courtship signals of lacewings (Wells and Henry, 1992)). Michelsen *et al.* (1992) used a mechanical 'robot bee' to investigate communication in the bee dance.

3.4.7. *Flight and locomotory behaviour*

Wind tunnels have become common in laboratories researching olfactory behaviour in insects but have also been used for studies of orientation, whether or not mediated by chemicals, and of migration. Very practical advice on designing and using wind tunnels is given by Baker and Linn (1984). Wind tunnels with air pushed by a fan predominate as these give fewer problems than pull-fans at the relatively low wind speeds used.

Some insects such as flies and beetles have had a reputation for being difficult to work with in the wind tunnel but much may depend on getting the conditions, such as lighting and wind speed, right. It may be worth persevering if the experiments would be useful. Prior treatment of the insects, and choice of laboratory strain for cultures, may be particularly important given the potential causes of variability (see Section 3.3).

As part of studies on flight orientation mechanisms two groups have been examining moth response to individual pheromone puffs using different ways of producing these and different but equally ingenious methods to track the passage of odour puffs down the wind tunnel (Mafra-Neto and Cardé, 1994; Vickers and Baker, 1994). This last paper describes an in-flight electroantennogram (EAG), using an antenna (from a second male) attached by Velcro across the head of the flying moth, to allow contacts with the pheromone puffs to be matched to behaviour.

Insects such as winged aphids can be kept in sustained flight in vertical wind tunnels, flight upwards towards a bright light countered by air flow from above. Such a wind tunnel has been successfully automated, with video-tracking feeding a computer controlling the air flow to keep the insect on station (see Young *et al.,* 1994). The computer also controls the intermittent flashing of lights which give visual stimuli from the side. Host plant odour cues have also been tested on aphids in this system, e.g. Nottingham and Hardie (1993). Vertical wind tunnels have also been used with beetles (Blackmer and Phelan, 1991) and whitefly (Blackmer and Byrne, 1993).

Locomotor compensators, which work like a multi-directional treadmill, have been used with success for walking insects in studies of both olfactory and auditory stimuli (see Bailey, 1991; Bell, 1991). The insect walks on a servosphere which is moved in the opposite direction by a feedback loop driven by a control system in response to movement detected by a video camera. One advantage of the system is that the orientation of the insect in relation to the odour air flow, for example, is always known. However, as only one animal can be tested at a time, use has been largely restricted to studies where the track of the insect was the focus. The cost of the hardware is also a major handicap.

3.4.8. Feeding behaviour

Feeding is a central part of any insect's behaviour and has important consequences for applied ecology including a potential role in plant and pesticide resistance. (The feeding behaviour of phytophagous insects is discussed in this volume, Chapter 6.) Feeding behaviour in blood sucking insects such as mosquitoes and tsetse flies has been studied mostly at the level of host finding (for reviews see Lehane, 1991; Colvin and Gibson, 1992; trapping techniques for mosquitoes are described in Service, 1993). However, biting behaviour on different hosts and, using artificial membranes, effects of phagostimulants have also been investigated (e.g. Moskalyk and Friend, 1994). Marked circadian rhythms of activity and thus feeding are found (e.g. Kyorku and Brady, 1994). The feeding behaviour of predators such as carabid beetles, important in biological control, has been studied partly in laboratory trials, which is not very satisfactory, but also by testing field caught animals with ELISA (Sopp *et al.,* 1992), and now using monoclonal antibodies (Symondson and Liddell, 1995; also this volume, Chapter 12), against potential prey items.

Feeding behaviour provides an excellent focus for investigating the integration of internal and external influences on behaviour, and its consequences such as life history effects, at both immediate and evolutionary time scales. Simpson and Raubenheimer (1995) offer a practical framework for studying feeding behaviour in the context of the multiple nutrient requirements of animals.

3.4.9. Reproductive behaviour

Over the last 25 years the emphasis of research on insect reproductive behaviour has been increasingly moving towards a focus on individual selection (Thornhill and Alcock, 1983; Alcock and Gwynne, 1991). The most active research areas, and methodologies used, tend to follow those of behavioural ecology more broadly, that is a recent focus on the evolution of mating systems and sexual selection. The specific insect techniques involved in this area of research are mostly those of catching and marking individuals. The practical advantages insect research has over vertebrates include the short generation time (which makes genetic studies possible) and the ability to continue studies with laboratory cultures to elucidate the fitness benefits of paternal investment for example. Work on reproductive behaviour contributes to studies on speciation (Butlin and Ritchie, 1994).

Arthropod studies have made important contributions to our understanding of sperm competition, including a recent reappraisal of two-male sperm precedence experiments using genetic markers or sterilization of one male with irradiation. Using single locus minisatellite profiling to check the paternity of offspring of multiple matings, Zeh and Zeh (1994) have shown that the strong last male-sperm precedence usually shown in laboratory two-male experiments breaks down when more than two males mate with a female.

3.5. Conclusions

Keep it simple: neat experimental design wins over high technology. Although it may sound imprecise, developing a 'feeling for the organism' through time spent watching is likely to reward you with better experiments. Relating questions to the ecology and evolutionary background of the animal may often provide the best answers.

3.6. Acknowledgements

I thank M.C. Birch, S.P. Foster, J. Gordon, M.O. Harris, D. Raubenheimer, J. Rojas, and S.J. Simpson for generously reading and commenting on drafts of this chapter.

3.7. References

Alcock, J. (1993) *Animal Behaviour. An Evolutionary Approach*, 5th edn. Sinuaer Associates, Sunderland, Massachusetts.

Alcock, J. and Gwynne, D.T. (1991) Evolution of insect mating systems: the impact of individual selectionist thinking. In: Bailey, W.J. and Ridsdill-Smith, J. (eds)

Reproductive Behaviour of Insects. Individuals and Populations. Chapman & Hall, London, pp. 10–42.

Alcock, J., Jones, C.E. and Buchmann, S.L. (1977) Male mating strategies in the bee *Centris pallida* Fox (Anthophoridae: Hymenoptera). *American Naturalist* 111, 145–155.

Aluja, M., Prokopy, R.J., Elkinton, J.S. and Laurence, F. (1989) Novel-approach for tracking and quantifying the movement patterns of insects in 3 dimensions under seminatural conditions. *Environmental Entomology* 18, 1–7.

Bailey, W.J. (1991) *Acoustic Behaviour of Insects. An Evolutionary Perspective.* Chapman & Hall, London.

Baker, T.C. and Cardé, R.T. (1984) Techniques for behavioural bioassays. In: Hummel, H.E. and Miller, T.A. (eds) *Techniques in Pheromone Research.* Springer Verlag, New York, pp. 45–73.

Baker, T.C. and Linn, C.E.J. (1984) Wind tunnels in pheromone research. In: Hummel, H.E. and Miller, T.A. (eds) *Techniques in Pheromone Research.* Springer Verlag, New York, pp. 75–110.

Barton Brown, L. (1993) Physiologically induced changes in resource-oriented behavior. *Annual Review of Entomology* 38, 1–25.

Beerwinkle, K.R., Lopez, J.J., Cheng, D., Lingren, P.D. and Meola, R.W. (1995) Flight potential of feral *Helicoverpa zea* (Lepidoptera: Noctuidae) males measured with a 32-channel, computer-monitored, flight-mill system. *Environmental Entomology* 24, 1122–1130.

Bell, W.J. (1991) *Searching Behaviour. The Behavioural Ecology of Finding Resources.* Chapman & Hall, London.

Bell, W.J., Kipp, L.R. and Collins, R.D. (1995) The role of chemo-orientation in search behavior. In: Cardé, R.T. and Bell, W.J. (eds) *Chemical Ecology of Insects vol. 2.* Chapman & Hall, London, pp. 105–153.

Bernays, E.A. (1995) Effects of experience on host-plant selection. In: Cardé, R.T. and Bell, W.J. (eds) *Chemical Ecology of Insects, vol. 2.* Chapman & Hall, London, pp. 47–64.

Bernays, E.A. and Chapman, R.F. (1994) *Host-plant Selection by Phytophagous Insects.* Chapman & Hall, London.

Birch, M.C., Lucas, D. and White, P.R. (1989) The courtship behavior of the cabbage moth, *Mamestra brassicae* (Lepidoptera: Noctuidae), and the role of male hair-pencils. *Journal of Insect Behavior* 2, 227–239.

Blackmer, J.L. and Byrne, D.N. (1993) Environmental and physiological factors influencing phototactic flight of *Bemisia tabaci. Physiological Entomology* 18, 336–342.

Blackmer, J.L. and Phelan, P.L. (1991) Behavior of *Carpophilus hemipterus* in a vertical flight chamber – transition from phototactic to vegetative orientation. *Entomologia experimentalis et applicata* 58, 137–148.

Borden, J.H., Hunt, D.W.A., Miller, D.R. and Slessor, K.N. (1986) Orientation in forest Coleoptera: an uncertain outcome of responses by individual beetles to variable stimuli. In: Payne, T.L., Birch, M.C. and Kennedy, C.E.J. (eds) *Mechanisms in Insect Olfaction.* Oxford Scientific Publications, Oxford, pp. 97–116.

Butlin, R.K. and Ritchie, M.G. (1994) Behaviour and speciation. In: Slater, P.J.B. and Halliday, T.R. (eds) *Behaviour and Evolution.* Cambridge University Press, Cambridge, pp. 43–79.

Byers, J.A. (1988) Novel diffusion–dilution method for release of semiochemicals: testing pheromone component ratios on western pine beetle. *Journal of Chemical Ecology* 14, 199–212.

Byers, J.A. (1992) Optimal fractionation and bioassay plans for isolation of synergistic chemicals: the subtractive–combination method. *Journal of Chemical Ecology* 18, 1603–1621.

Cardé, R.T. and Elkinton, J.S. (1984) Field trapping with attractants: methods and interpretation. In: Hummel, H.E. and Miller, T.A. (eds) *Techniques in Pheromone Research*. Springer Verlag, New York, pp. 111–129.

Colvin, J. and Gibson, G. (1992) Host-seeking behavior and management of tsetse. *Annual Review of Entomology* 37, 21–40.

De Jong, M.C.M. (1987) A direct search approach to characterize the sex pheromone composition giving the maximal male response. *Physiological Entomology* 12, 11–21.

Dusenbery, D.B. (1992) *Sensory Ecology. How Organisms Aquire and Respond to Information*. WH Freeman and Company, New York.

Eigenbrode, S.D., Moodie, S. and Castagnola, T. (1995) Predators mediate host-plant resistance to a phytophagous pest in cabbage with glossy leaf wax. *Entomologia experimentalis et applicata* 77, 335–342.

Ewing, A.W. (1989) *Arthropod Bioacoustics. Neurobiology and Behaviour*. Edinburgh University Press, Edinburgh.

Finch, S. (1986) Assessing host-plant finding by insects. In: Miller, J.R. and Miller, T.A. (eds) *Insect–Plant Interactions*. Springer Verlag, New York, pp. 23–63.

Gibson, G. (1995) A behavioral-test of the sensitivity of a nocturnal mosquito, *Anopheles gambiae*, to dim white, red and infrared light. *Physiological Entomology* 20, 224–228.

Gould, J.L. (1993) Ethological and comparative perspectives on honey bee learning. In: Papaj, D.R. and Lewis, A.C. (eds) *Insect Learning. Ecology and Evolutionary Perspectives*. Chapman & Hall, London, pp. 18–50.

Harris, M.O. and Foster, S.P. (1995) Behavior and integration. In: Cardé, R.T. and Bell, W.J. (eds) *Chemical Ecology of Insects, vol. 2*. Chapman & Hall, London, pp. 3–46.

Harris, M.O. and Rose, S. (1990) Chemical, color, and tactile cues influencing oviposition behavior of the Hessian fly (Diptera: Cecidomyiidae). *Environmental Entomology* 19, 303–308.

Harris, M.O., Rose, S. and Malsch, P. (1993) The role of vision in the host plant-finding behavior of the Hessian fly. *Physiological Entomology* 18, 31–42.

Heinrich, B. (1993) *The Hot-blooded Insects: Strategies and Mechanisms of Thermoregulation*. Springer Verlag, Berlin.

Hoffmann, A.A. (1994) Behaviour genetics and evolution. In: Slater, P.J.B. and Halliday, T.R. (eds) *Behaviour and Evolution*. Cambridge University Press, Cambridge, pp. 7–42.

Islam, M.S., Roessingh, P., Simpson, S.J. and McCaffery, A.R. (1994) Effects of population-density experienced by parents during mating and oviposition on the phase of hatchling desert locusts, *Schistocerca gregaria*. *Proceedings of the Royal Society of London Series B – Biological Sciences* 257, 93–98.

Jones, J.S. (1982) Genetic differences in individual behaviour associated with shell polymorphism in the snail *Cepaea nemoralis*. *Nature* 298, 749–750.

Judd, G.J.R. and Borden, J.H. (1991) Sensory interaction during trap-finding by female onion flies – implications for ovipositional host-plant finding. *Entomologia experimentalis et applicata* 58, 239–249.

Kamm, J.A. (1990) Control of olfactory-induced behavior in alfalfa seed chalcid (Hymenoptera: Eurytomidae). *Journal of Chemical Ecology* 16, 291–300.

Kennedy, J.S. (1977) Behaviorally discriminating assays of attractants and repellents. In: Shorey, H.H. and McKelvey, J.J.J. (eds) *Chemical Control of Insect Behavior. Theory and Application.* John Wiley, New York, pp. 215–230.

Kennedy, J.S. (1986) Some current issues in orientation to odour sources. In: Payne, T.L., Birch, M.C. and Kennedy, C.E.J. (eds) *Mechanisms in Insect Olfaction.* Oxford Scientific Publications, Oxford, pp. 1–25.

Kennedy, J.S. (1992) *The New Anthropomorphism.* Cambridge University Press, Cambridge.

Kolb, G. and Scherer, C. (1982) Experiments on the wavelength specific behaviour of *Pieris brassicae* L. during drumming and egglaying. *Journal of Comparative Physiology – A Sensory Neural and Behavioral Physiology* 149, 325–332.

Krebs, J.R. and Davies, N.B. (1993) *An Introduction to Behavioural Ecology*, 3rd edn. Blackwell Scientific Publications, Oxford.

Kyorku, C. and Brady, J. (1994) A free-running bimodal circadian-rhythm in the tsetse-fly *Glossina longipennis. Journal of Insect Physiology* 40, 63–67.

Leal, W.S., Hasegawa, M., Sawada, M. and Ono, M. (1994) Sex-pheromone of oriental beetle, *Exomala orientalis*: identification and field-evaluation. *Journal of Chemical Ecology* 20, 1705–1718.

Lehane, M.J. (1991) *Biology of Blood-sucking Insects.* Harper Collins Academic, London.

MacDonald, I.L. and Raubenheimer, D. (1995) Hidden Markov models and animal behaviour. *Biometrical Journal* 6, 701–712.

Mafra Neto, A. and Cardé, R.T. (1994) Fine-scale structure of pheromone plumes modulates upwind orientation of flying moths. *Nature* 369, 142–144.

Martin, P. and Bateson, P. (1993) *Measuring Behaviour. An Introductory Guide*, 2nd edn. Cambridge University Press, Cambridge.

McGregor, P.K. and Holland, J. (1995) AVISOFT-SONAGRAPH Pro – a PC-program for sonagraphic analysis – V-2.1. *Animal Behaviour* 50, 1137–1138.

McNeil, J.N. (1991) Behavioral ecology of pheromone-mediated communication in moths and its importance in the use of pheromone traps. *Annual Review of Entomology* 36, 407–430.

McNeil, J.N. (1992) Evolutionary perspectives and insect pest control: an attractive blend for the deployment of semiochemicals in management systems. In: Roitberg, B.D. and Isman, M.B. (eds) *Insect Chemical Ecology. An Evolutionary Approach.* Chapman & Hall, New York, pp. 334–352.

Michelsen, A., Andersen, B.B., Storm, J., Kirchner, W.H. and Lindauer, M. (1992) How honeybees perceive communication dances, studied by means of a mechanical model. *Behavioral Ecology and Sociobiology* 30, 143–150.

Moskalyk, L.A. and Friend, W.G. (1994) Feeding behaviour of female *Aedes aegypti*: Effects of diet temperature, bicarbonate and feeding technique on the response to ATP. *Physiological Entomology* 19, 223–229.

Mousseau, T.A. and Dingle, H. (1991) Maternal effects in insect life histories. *Annual Review of Entomology* 36, 511–534.

Nazzi, F., Powell, W., Wadhams, L.J. and Woodcock, C.M. (1996) On the sex pheromone of the aphid parasitoid *Praon volucre* (Hymenoptera: Braconidae). *Journal of Chemical Ecology* 22, 1169–1175.

Nottingham, S.F. and Hardie, J. (1993) Flight behavior of the black bean aphid, *Aphis fabae*, and the cabbage aphid, *Brevicoryne brassicae*, in host and nonhost plant odor. *Physiological Entomology* 18, 389–394.

Opp, S.B. and Prokopy, R.J. (1986) Approaches and methods for direct behavioural observation and analysis of plant–insect interactions. In: Miller, J.R. and Miller, T.A. (eds) *Insect–Plant Interactions*. Springer Verlag, New York, pp. 1–22.

Papaj, D.R. (1993) Afterword: learning, adaptation, and the lessons of *O*. In: Papaj, D.R. and Lewis, A.C. (eds) *Insect Learning. Ecology and Evolutionary Perspectives*. Chapman & Hall, London, pp. 374–386.

Papaj, D.R. and Lewis, A.C. (eds) (1993) *Insect Learning. Ecological and Evolutionary Perspectives*. Chapman & Hall, New York.

Peterson, C.H. and Renaud, P.E. (1989) Analysis of feeding preference experiments. *Oecologia* 80, 82–86.

Phelan, P.L. (1992) Evolution of sex pheromones and the role of asymmetric tracking. In: Roitberg, B.D. and Isman, M.B. (eds) *Insect Chemical Ecology. An Evolutionary Approach*. Chapman & Hall, New York, pp. 245–264.

Phillips, A.D.G. and Wyatt, T.D. (1992) Beyond origami – using behavioral observations as a strategy to improve trap design. *Entomologia experimentalis et applicata* 62, 67–74.

Poppy, G.M. and Birch, M.C. (1994) Evidence of the eversion of *Mamestra brassicae* (Lepidoptera: Noctuidae) hair-pencils during courtship. *Journal of Insect Behavior* 7, 885–889.

Prokopy, R.J. (1986) Visual and olfactory stimulus interaction in resource finding by insects. In: Payne, T.L., Birch, M.C. and Kennedy, C.E.J. (eds) *Mechanisms in Insect Olfaction*. Oxford Scientific Publications, Oxford, pp. 81–89.

Prokopy, R.J. and Lewis, W.J. (1993) Application of learning to pest management. In: Papaj, D.R. and Lewis, A.C. (eds) *Insect Learning. Ecology and Evolutionary Perspectives*. Chapman & Hall, London, pp. 308–342.

Prokopy, R.J., Collier, R.H. and Finch, S. (1983) Leaf color used by cabbage root flies to distinguish among host plants. *Science* 221, 190–192.

Raubenheimer, D. and Bernays, E.A. (1993) Patterns of feeding in the polyphagous grasshopper *Taeniopoda eques*: a field study. *Animal Behaviour* 45, 153–167.

Riley, J.R. (1994) Flying insects in the field. In: Wratten, S.D. (ed.) *Video Techniques in Animal Ecology and Behaviour*. Chapman & Hall, London, pp. 1–15.

Roessingh, P., Simpson, S.J. and James, S. (1993) Analysis of phase-related changes in behavior of desert locust nymphs. *Proceedings of the Royal Society of London Series B – Biological Sciences* 252, 43–49.

Sanders, C.J. (1989) The further understanding of pheromones: biological and chemical research for the future. In: Jutsum, A.R. and Gordon, R.F.S. (eds) *Insect Pheromones in Plant Protection*. John Wiley, Chichester, pp. 323–351.

Service, M.W. (1993) *Mosquito Ecology: Field Sampling Methods*, 2nd edn. Elsevier Applied Science, London.

Simpson, S.J. and Ludlow, A.R. (1986) Why locusts start to feed: a comparison of causal factors. *Animal Behaviour* 34, 480–496.

Simpson, S.J. and Raubenheimer, D. (1995) The geometric analysis of feeding and nutrition – a users guide. *Journal of Insect Physiology* 41, 545–553.

Simpson, S.J., Raubenheimer, D. and Chambers, P.G. (1995) The mechanisms of nutritional homeostasis. In: Chapman, R.F. and de Boer, G. (eds) *Regulatory Mechanisms in Insect Feeding*. Chapman & Hall, New York, pp. 251–277.

Singer, M.C. (1986) The definition and measurement of oviposition preference in plant-feeding insects. In: Miller, J.R. and Miller, T.A. (eds) *Insect–Plant Interactions*. Springer Verlag, New York, pp. 65–94.

Snowball, M.F. and Holmqvist, M.H. (1994) An electronic device for monitoring escape behavior in *Musca* and *Drosophila*. *Journal of Neuroscience Methods* 51, 91–94.

Sopp, P.I., Sunderland, K.D., Fenlon, J.S. and Wratten, S.D. (1992) An improved quantitative method for estimating invertebrate predation in the field using an enzyme-linked-immunosorbent-assay (ELISA). *Journal of Applied Ecology* 29, 295–302.

Stone, G.N. and Willmer, P.G. (1989) Endothermy and temperature regulation in bees – a critique of grab and stab measurement of body-temperature. *Journal of Experimental Biology* 143, 211–223.

Strickland, T.R. and Franks, N.R. (1994) Computer image analysis provides new observations of ant behaviour patterns. *Proceedings of the Royal Society of London Series B – Biological Sciences* 257, 279–286.

Symondson, W.O.C. and Liddell, J.E. (1995) Decay-rates for slug antigens within the carabid predator *Pterostichus melanarius* monitored with a monoclonal-antibody. *Entomologia experimentalis et applicata* 75, 245–250.

Tallamy, D.W. and Denno, R.F. (1982) Maternal care in *Gargaphia solani* (Hemiptera: Tingidae). *Animal Behaviour* 29, 771–778.

Thornhill, R. and Alcock, J. (1983) *The Evolution of Insect Mating Systems*. Harvard University Press, Cambridge, Massachusetts.

Tinbergen, N. (1963) On aims and methods of ethology. *Zeitschrift Fur Tierpsychologie – Journal of Comparative Ethology* 20, 410–433.

Traniello, J.F.A. and Robson, S.K. (1995) Trail and territorial communication in insects. In: Cardé, R.T. and Bell, W.J. (eds) *Chemical Ecology of Insects, vol. 2*. Chapman & Hall, London, pp. 241–286.

Turlings, T.C.J., Tumlinson, J.H., Lewis, W.J. and Vet, L.E.M. (1989) Beneficial arthropod behavior mediated by airborne semiochemicals. 8. Learning of host-related odors induced by a brief contact experience with host by-products in *Cotesia marginiventris* (Cresson), a generalist larval parasitoid. *Journal of Insect Behavior* 2, 217–225.

Turlings, T.C.J., Wäckers, F.L., Vet, L.E.M., Lewis, W.J. and Tumlinson, J.H. (1993) Learning of host-finding cues by hymenopterous parasitoids. In: Papaj, D.R. and Lewis, A.C. (eds) *Insect Learning. Ecology and Evolutionary Perspectives*. Chapman & Hall, London, pp. 51–78.

Unwin, D.M. (1980) *Microclimate Measurement for Ecologists*. Academic Press, London.

Unwin, D.M. and Corbet, S.A. (1991) *Insects, Plants and Microclimate*. Richmond Publishing, Slough.

Vaillant, J. and Derridj, S. (1992) Statistical analysis of insect preference in two-choice experiments. *Journal of Insect Behavior* 5, 773–781.

Vale, G.A. (1982) The improvement of traps for tsetse flies (Diptera: Glossinidae). *Bulletin of Entomological Research* 72, 95–106.

Varley, M.J., Copland, M.J.W., Wratten, S.D. and Bowie, M.H. (1994) Parasites and predators. In: Wratten, S.D. (ed.) *Video Techniques in Animal Ecology and Behaviour*. Chapman & Hall, London, pp. 33–63.

Vet, L.E.M., Lewis, W.J. and Cardé, R.T. (1995) Parasitoid foraging and learning. In: Cardé, R.T. and Bell, W.J. (eds) *Chemical Ecology of Insects, vol. 2*. Chapman & Hall, London, pp. 65–104.

Vickers, N.J. and Baker, T.C. (1994) Reiterative responses to single strands of odor promote sustained upwind flight and odor source location by moths. *Proceedings of the National Academy of Sciences of the United States of America* 91, 5756–5760.

Vinson, S.B. (1985) The behavior of parasitoids. In: Kerkut, G.A. and Gilbert, L.I. (eds) *Comprehensive Insect Physiology, Biochemistry and Pharmacology*, vol. 9. Pergamon Press, Oxford, pp. 417–470.

Wadhams, L.J., Blight, M.M., Kerguelen, V., Lemetayer, M., Marion Poll, F., Masson, C., Pham Delegue, M.H. and Woodcock, C.M. (1994) Discrimination of oilseed rape volatiles by honey-bee – novel combined gas-chromatographic electrophysiological behavioral assay. *Journal of Chemical Ecology* 20, 3221–3231.

Wehner, R., Michel, B. and Antonsen, P. (1996) Visual navigation in insects: coupling of egocentric and geocentric information. *Journal of Experimental Biology* 199, 129–140.

Wells, M.M. and Henry, C.S. (1992) Behavioral-responses of green lacewings (Neuroptera: Chrysopidae: Chrysoperla) to synthetic mating songs. *Animal Behaviour* 44, 641–652.

White, P.R. and Birch, M.C. (1987) Female sex-pheromone of the common furniture beetle *Anobium punctatum* (Coleoptera: Anobiidae) – extraction, identification, and bioassays. *Journal of Chemical Ecology* 13, 1695–1706.

Wilkinson, G.S. (1994) Canary 1.1: sound analysis for Macintosh computers. *Bioacoustics* 5, 227–238.

Wratten, S.D. (ed.) (1994) *Video Techniques in Animal Ecology and Behaviour*. Chapman & Hall, London.

Wyatt, T.D. (1986) How a saltmarsh beetle, *Bledius spectabilis*, prevents flooding and anoxia in its burrow. *Behavioral Ecology and Sociobiology* 19, 323–331.

Wyatt, T.D. (1997) Putting pheromones to work: paths forward for direct control. In: Cardé, R.T. and Minks, A.K. (eds) *Pheromone Research: New Directions*. Chapman & Hall, London (in press).

Wyatt, T.D., Phillips, A.D.G. and Gregoire, J.C. (1993) Turbulence, trees and semiochemicals – wind-tunnel orientation of the predator, *Rhizophagus grandis*, to its barkbeetle prey, *Dendroctonus micans. Physiological Entomology* 18, 204–210.

Wyatt, T.D., Vastiau, K. and Birch, M.C. (1997) Orientation of male *Anobium punctatum* (Coleoptera: Anobiidae): use of visual and turbulence cues to its own pheromone. *Physiological Entomology* (in press).

Young, S., Hardie, J. and Gibson, G. (1994) Flying insects in the laboratory. In: Wratten, S.D. (ed.) *Video Techniques in Animal Ecology and Behaviour*. Chapman & Hall, London, pp. 17–32.

Zeh, J.A. and Zeh, D.W. (1994) Last-male sperm precedence breaks down when females mate with 3 males. *Proceedings of the Royal Society of London Series B – Biological Sciences* 257, 287–292.

4

Quantifying Insect Populations: Estimates and Parameters

D.R. Dent
International Institute of Biological Control, Silwood Park, Buckhurst Road, Ascot, Berkshire SL5 7TA, UK

4.1. Introduction

Basic estimates of populations are required by both ecologists and applied entomologists for the purposes of understanding the population dynamics of a species and as the basis for developing management strategies to successfully combat pest infestations. Estimates of insect fecundity, development and mortality are commonly required in combination with measures of the population age distribution and the extent of migration. Dispersal and migration are dealt with in Chapter 5. The methods associated with obtaining estimates of fecundity, development and mortality are considered here, with particular emphasis placed on the role of temperature. Temperature is emphasized because it is one of the fundamental driving variables in the population dynamics of insects.

The chapter deals first with general methods common to the techniques of quantifying insect populations and is followed by sections on reproduction, development and growth, survival and mortality. A section is devoted to the summary statistic, the intrinsic rate of increase and then the final section deals with methods associated with the construction of lifetables, key factor analysis and density dependence.

4.2. General Methods

Experimental methods to determine insect population estimates tend to follow a general form involving confinement of individual insects on an individual plant or plant part, usually in the laboratory, and then exposing them to a constant set of conditions, particularly temperature, and evaluating their performance in terms of fertility, development, reproduction and survival. Monitoring of individual performance is most often measured on a daily basis (e.g. Asante *et al.*, 1991; Kaakeh and Dutcher, 1992; Omer *et al.*, 1992).

4.2.1. Confinement of the insects

Most cages consist of a transparent plastic 'Mylar' and/or a fine mesh which covers the whole or part of the plant. 'Mylar' cylinders with nylon mesh tops or windows have been used for confining insects to cereal seedlings (Wratten, 1977; Kay *et al.*, 1981; Heinrichs *et al.*, 1985) while nylon mesh sleeve cages are commonly used to confine insects to woody plants (e.g. Leather and Dixon, 1981; Asante and Danthanarayana, 1990). Clip cages (Noble, 1958) are popular for confining aphids and whiteflies to leaves (e.g. Enkegaard, 1993a; Van Helden *et al.*, 1993). Other forms of cage include plastic screwtop vials covered at one end by a mesh secured by an elastic band (Shanower *et al.*, 1993), glass topped and sided cages (Leather *et al.*, 1985), perforated plastic bags closed around the petiole of each leaf (Enkegaard, 1993b).

The use of leaf discs or excised plant tissue often produces problems associated with differences in nutrient status and physiology compared with intact plants. Despite this, leaf discs have been used in experiments to obtain population estimates (Hughes and Woolcock, 1965; Dean, 1974; Wyatt and Brown, 1977) as have excised plant tissues, e.g. excised cowpea pods, used to test resistance against *Clavigralla tomentoscollis* (Olatunde and Odebiyi, 1991). The use of leaf discs and excised plant tissue should be avoided where possible, since their use can confound experimental results and reduce the likelihood of their relating to the field situation.

4.2.2. Temperature control

The influence of temperature on insect development, fecundity and survival is commonly assessed by entomologists. Although some studies are carried out in the field, the majority are carried out in laboratory temperature controlled cabinets or rooms.

Reported variations around the required temperature for cabinets, growth chambers or controlled temperature rooms used in experiments tend to range from $\pm 0.5°C$ up to $\pm 2°C$. Certainly, the ability to control temperatures to an accuracy of better than $0.5°C$ is unlikely given diurnal variations in temperature due to radiant energy from lights and spatial variation in temperature within the cabinet (especially where heat transfer is not fan assisted) (Allsopp *et al.*, 1991). According to Browning (1952), errors of more than $0.5°C$ can spoil the fit of models and make it impossible to distinguish between them.

Temperature variation is usually reported as a range, i.e. a plus and minus value above the required temperature, which tends to be remarkably consistent across temperatures and cabinets, perhaps reflecting wishful thinking rather than actual measures of temperature. Enkegaard (1993b) takes a different approach, which may be more appropriate for temperature-development studies, of providing the standard errors for the mean temperature.

4.3. Reproduction

Populations increase in size because of natality, where natality means the production of new individuals in the broadest sense including birth, hatching, germination and fission (Krebs, 1978). Reproduction can be considered both in terms of fertility or fecundity although the vast majority of ecological and agricultural studies deal with fecundity due to the difficulties of measuring fertility (Barlow, 1961). Reproduction in insects is affected by a wide range of abiotic and biotic (intrinsic and extrinsic) factors, which are usually studied to determine the extent of their influence on the population dynamics of a species.

4.3.1. Fertility

Fertility refers to the proportion of eggs which are produced or laid that develop to viable progeny (Hyde, 1914). In species that are not parthenogenic, fertility depends in the first instance on the successful fertilization of the eggs. For the most part infertility is measured as the number of eggs which do not hatch (e.g. Serit and Keng-Hong, 1990) although in some cases insects are dissected to determine whether failure to mate was responsible for infertility (Royer and McNeil, 1993). Failure of the eggs to develop can be assessed by searching for evidence of embryonic differentiation, appearance of a head capsule or through changes in shape or colour when viewed under a light microscope (Willers *et al.*, 1987; Fox, 1993). The distinctions between physiological death (death in embryo) and unfertilization (no embryonic development) are in some cases clearly detected by using a microscope (Chang and Moritomo, 1988).

Mating experiments can provide important information on the causes of infertility. Multiple matings of females may increase fertility through replenishing sperm supply or compensating for previous matings with infertile males (Danthanarayana and Gu, 1991), while delays in matings can also reduce fertility (Leather *et al.*, 1985; Unnithan and Page, 1991) and the conditions experienced by males during pupal development can lower the fertility rate in eggs of females with which they mate (Gerber *et al.*, 1991). The absence of host odours affected fertility in *Earias insulana* (Tamhankar, 1995) while in laboratory reared *Helicoverpa* spp., fertility was influenced by a whole range of factors including cage volume, absence of natural daylight, high humidity and adult food (Teakle, 1991).

The examples above make it clear that a wide range of factors can influence insect fertility and any assessment that makes use of fertility as a comparative measure of performance should take these into account.

4.3.2. Fecundity

The term fecundity refers to an insect's reproductive output in terms of the total number of eggs produced or laid during the lifetime of the female (Jervis and Copland, 1996). In insects which mature eggs throughout their adult life fecundity is measured directly by keeping females caged under as natural conditions as possible and recording the total number of eggs laid (Southwood, 1978). In such cases potential fecundity can be predicted from relationships between fecundity and pupal weight (e.g. Gilbert, 1984a; Fig. 4.1) or head width (e.g. Chua, 1992). In insects where all eggs are mature on emergence the total potential fecundity may be estimated by examining the ovaries (Southwood, 1978). Potential fecundity in aphids can be assessed a number of different ways through relationships with adult weight, embryo and/or ovariole number depending on the species (Dixon and Wratten, 1977; Wratten, 1977; Kay *et al.*, 1981; Leather and Wellings, 1981). Embryo number in aphids may be obtained by placing an individual in a drop of glycerol on a microscope slide and squashing under a coverslip by applying gentle pressure with a mounted needle (Kay *et al.*, 1981). This aborts the embryos and those with eye pigment visible under a ×70 magnification (i.e. the largest embryos) are counted. The complete reproductive system of aphids may be obtained undamaged by gently pulling the terminal abdomen segments with a pair of fine forceps while holding the thorax with another pair (Leather and Wellings, 1981). The ovarioles are then clearly

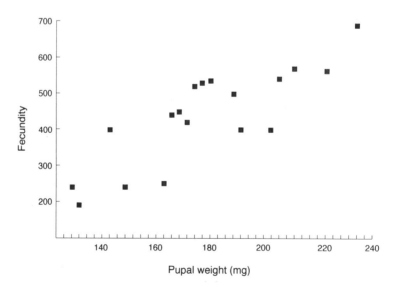

Fig. 4.1. Lifetime fecundity vs. pupal weight of *Pieris rapae*. Each point is mean of five females. (After Gilbert, 1984a.)

visible and can be counted along with the number of embryos per ovariole. Such methods are useful as a means of providing rapid assessments of reproductive potential of insects on different host plants and avoid the problems in resistance studies of changes in plant condition during a 1–2 week period when fecundity is measured (Kay *et al.*, 1981). An alternative approach is to compare reproductive performance in terms of a 7-day fecundity (Dent and Wratten, 1986) or 15-day fecundity (Bintcliffe and Wratten, 1982), etc. Such approaches are particularly relevant when a high proportion of total progeny are produced within the specified period. They have been extensively used as a means of measuring insect reproductive performance on a range of different host plants.

4.3.3. Influence of biotic factors on fecundity

Biotic factors that influence fecundity may be classified as intrinsic, e.g. insect size, morph or clone, and extrinsic such as host plant effects, which may include plant species, cultivar or growth stage differences (Leather and Dixon, 1982; Leather *et al.*, 1985; Soroka and Mackay, 1991; Kaakeh and Dutcher, 1993). Insect size is generally acknowledged to influence fecundity with larger females consistently laying more eggs than small ones (Gilbert, 1984c) and in addition, in some species they produce more eggs of a larger size (Speight, 1994). Average egg size may be determined by weight or by measuring the volume or length and breadth of the eggs. The egg volume of *Bembidion lampros* was determined by measurement of their length and breadth under stereomicroscope and then using the formula for a rotating ellipsoid

$$V = LB^2\pi/6 \qquad\qquad (4.1)$$

where L is the length , B is the breadth of the eggs (Wallin *et al.*, 1992). In trying to identify differences in fecundity between insect morphs it may be necessary to take weight differences into account. For example, individuals need to be divided into weight classes to compare fecundity between aphid alate and apterous morphs. Dixon and Wratten (1971) found that when weight was taken into account there was no difference between 5-day fecundity of apterous or alatae *Aphis fabae* whereas apterae of both *Sitobion avenae* and *Metopolophium dirhodum* were more fecund than alatae of comparable weight producing about three more nymphs on average in any 5-day period (Wratten, 1977).

4.3.4. Influence of abiotic factors on fecundity

Temperature is the main abiotic factor influencing fecundity in insects, although such effects may be modified by other abiotic factors such as light intensity (e.g. Wyatt and Brown, 1977) and biotic factors such as the host plant (e.g. Leather and Dixon, 1982; Kaakeh and Dutcher, 1993). The rate of reproduction of insects is dependent on temperature usually up to a critical maximum (Dent, 1995). The number of offspring produced increases with temperature (e.g. Enkegaard, 1993b; Vansteenis, 1993; Yang *et al.*, 1994) but high temperatures reduce fecundity. Hence, relationships tend not to be linear, which limits the usefulness of linear regression equations for analysis of these data. Artigues *et al.* (1992) used the function developed by Hilbert and Logan (1983) for egg laying and temperature in the parasitoid *Encarsia tricolor* (Fig. 4.2) while Madden *et al.* (1986) used Richards' (1959) model to represent the relationship between fecundity and temperature for *Dalbulus* leafhoppers. The latter model takes the form:

$$N_{(t)} = G(1 + B_e^{-kt})^{1/(1-m)} \qquad (4.2)$$

where $N_{(t)}$ = cumulative number of eggs laid until time t
G = the maximum number of eggs laid, set to equal the gross reproductive rate
B = a location parameter that adjusts the fecundity curve along the time axis
k = a rate parameter
m = a shape parameter

The rate parameter k provides a measure of the compactness of the fecundity data along the time axis, i.e. a low k is obtained when the insect spreads its egg laying over a long time. Madden *et al.* (1986) point out that values of k cannot be directly compared when G or m differ, therefore the weight led mean fecundity rate k^1 is used for comparisons where

$$k^1 = Gk/(2m + 2) \qquad (4.3)$$

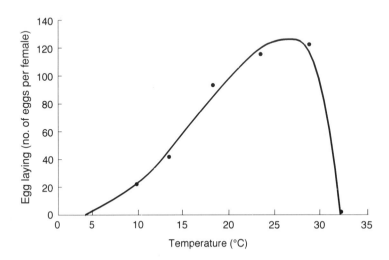

Fig. 4.2. Total number of eggs laid by *E. tricolor* females from days 3 to 20 after adult emergence at different constant temperatures. (After Artigues *et al.*, 1992.)

The location parameter B is positive when $m > 1$ and negative when $m < 1$. Hence, it is inappropriate to compare values of B of this model when m is not constant. The advantage of Richards' model is that, unlike R_0 and r_m, the model permits the analysis of fecundity independent of survival. Other bimodal fecundity curves also have been modelled (e.g. Adachi and Korenaga, 1991).

4.4. Development and Growth

Insect development is characterized by a period of time, a number of instars, and an increase in size and weight as the insect passes from its immature to adult phase. The increase in size and weight is often referred to as insect growth, as opposed to development. The development time of an insect can be defined as the period between birth and the production of the first offspring by the adult female and, hence, includes the pre-oviposition period (Bonnemaison, 1951). Development time can be determined either as a total time from birth to first offspring or as a series of times for each instar, in which case the pre-oviposition period is defined as a specific stage from the final adult moult to production of the first offspring (e.g. Dean, 1974). The number of instars which constitute the immature stage of the insect will often vary according to conditions such as host quality and temperature. The rate of development of insect eggs and pupae is primarily dependent on temperature while the development of larval and nymphal stages is dependent both on abiotic (particularly temperature) and biotic (primarily host plant influences) factors.

4.4.1. Influence of temperature

Early attempts to relate temperature to insect development go back to the work of Glen (1922) on *Cydia pomonella*. The whole approach adopted by entomologists revolves around the concept of physiological time. Physiological time is the amount of heat required over time for an insect to complete development or a stage of development. Its calculation requires knowledge of the developmental threshold temperature of the particular stage of the insect and its rate of development in relation to temperature. Both these parameters can be obtained by taking the reciprocals of development times measured over the normal range of operating temperatures for the insect, plotted as a development rate vs. temperature graph. A sigmoid curve of the form depicted in Fig. 4.3 is then obtained.

The curve has three distinct regions: the lower region A, the linear region B, and the optimal region C. The curve of the lower region A levels out because at low temperatures insects can often survive long periods with little or no development. There is also a greater degree of variation in this part of the curve created by slight differences in threshold temperatures for individual insects (Howe, 1967). As the temperature increases, development

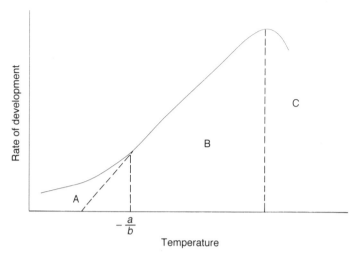

Fig. 4.3. Typical relationship between development rate and constant temperatures. The curve can be divided into three regions: Region A, where development rate asymptomatically approaches zero rather than the threshold temperature estimate ($-a/b$); Region B, the approximate linear region where the degree-day model ($y = a + bT$); Region C where development rate declines after reaching an optimum. (After Allsopp *et al.*, 1991.)

rates become proportional to temperature resulting in a linear region in the middle of the response curve (region B). Development then begins to slow up to a maximum, the so-called optimum temperature (Allsopp *et al.*, 1991). Development rates then fall off sharply with further increases in temperature. Insect mortality rates are often high in this region of the response curve.

For the greater range of temperatures the response curve is linear which has led many researchers to utilize a linear regression model as a means of determining the threshold for development and the thermal constant K; the total quantity of thermal energy required to complete development. The linear regression model takes the form:

$$y = \alpha + \beta T \tag{4.4}$$

where y = the rate of development at temperature T and α and β are constants. The threshold for development (Th) is determined by extrapolating the line to the abscissa, i.e. to the temperature at which there is no development, $Th = -\alpha/\beta$, while the thermal constant K is the reciprocal of the slope β, i.e. $1/\beta$ (standard errors can be calculated using the method of Campbell *et al.*, 1974). The values of Th and K are then used to determine the rate of development in the field using the thermal summation principle. The duration of development is calculated by adding up the number of thermal units (degree-hours or degree-days) above the threshold Th that have occurred in the field until the value *K* is reached, at which point development is complete.

The linear model is the simplest description of the development rate–temperature relationship, and it is widely used despite the inherent inaccuracies associated with the assumption of linearity in the sigmoid response curve. Over the region B, the model will provide satisfactory results but in the regions A and C the linear model will underestimate and overestimate the degree-days per unit time respectively, leading to inaccuracies in the summing procedure. A number of other models have been developed that try to describe the sigmoid response more precisely and, thus, enable the more accurate prediction of insect development. Most of these emphasize the shape of the response curve around region C, the optimum temperature. This is a particularly important region of the response curve but, in practical terms, it is difficult to define because high mortality rates often occur at high temperatures and hence results tend to be based on relatively few individual insects.

The sigmoid model developed by Stinner *et al.* (1974) has been widely used (e.g. Whalon and Smilowitz, 1979; Allsopp, 1981). The equation takes the form:

$$R_T = \frac{C}{(1 + e^{\,k_1 + k_2 T'})}. \qquad (4.5)$$

Where R_T = rate of development at temperature *T*
   C = maximum development rate x $e^{k_1 + k_2}$, i.e. the asymptote
   k_1 = intercept (constant)
   k_2 = slope (constant)
   T' = *T* (temperature) for $T < t_{opt}$
   T' = $2.t_{opt} - T'$ for $T > t_{opt}$
   t_{opt} = temperature at which maximum development rate occurs

The curve produced by this model is symmetrical around the optimum temperature, hence it tends to reflect poorly critical development rates at higher temperatures which, in practice, are not symmetrical; a problem it shares with the Pradhan (1946) exponential equation based on the form of the normal distribution. The problem of symmetry around the optimum temperature can be removed by combining two exponential equations to describe temperatures above and below the optimum (Logan *et al.*, 1976). However, such approaches become increasingly complex and the data points above the optimum are very difficult to obtain in practice. In addition, it is debatable whether the small improvement in accuracy provided by these more complex models is of any value in relation to conditions prevailing in the field (Kitching, 1977; Whalon and Smilowitz, 1979; Allsopp, 1981). For the most part the widely used linear approximation of the rate vs. temperature curve is sufficient given the range of other potential errors that are prevalent when using physiological times as a predictor of insect development.

Potential sources of error which need to be taken into account include the following: (i) constant vs. fluctuating temperatures; (ii) measurement of field temperatures; and (iii) the selection of the biofix from which the degree-day accumulation begins.

4.4.2. Constant vs. fluctuating temperatures

In a number of cases the development of insects determined under constant conditions of temperature has been shown to differ from the development of those maintained under fluctuating temperatures (Siddiqui and Barlow, 1973), usually with faster development times occurring under fluctuating conditions (Foley, 1981; Sengonca et al., 1994). It should not be assumed that rates derived under constant conditions of temperature will always be applicable. Tokeshi (1985) proposed a method of degree-day accumulation which is applicable to both laboratory and field data that avoids the above problem.

The method was developed for an aquatic insect *Ephemera danica* but it is equally valid for terrestrial insects. Random samples of *E. danica*, which has a slow growth rate, were taken twice monthly and the body length of each individual measured. These data were used to calculate a sample mean for each month. The water temperature was measured at 4 h intervals every day over the same year. Determination of temperature related growth was based on the summation of degree-days above a threshold (MT) repeated for a range of values for MT, in this case at 1°C intervals between −2 and +10°C. Subsequently this was narrowed to 0.1°C intervals. For each value of MT a linear regression analysis was carried out between mean body length (BL) and the derived number of degree-days (DD). The MT which yielded the highest r^2 (coefficient of variation) was assumed to be the minimum threshold temperature (Fig.4.4). This approach yielded estimates of DD above the MT for nymphs to complete growth and reach maturity at a range of body lengths obtained in the field. The regression line was then extrapolated back to provide an estimate of DD required for hatching.

4.4.3. Measurement of field temperatures

Field temperatures used in degree-day summations are most often measured at meteorological stations which may be some distance from the point or area of interest. Errors will inevitably arise when specific measurements are made at the sites where the insects normally live (Baker, 1980). Willmer (1982) reviewed the problem and identified three types of solution: (i) determine the development–temperature relationships under field conditions, e.g. Tokeshi (1985); (ii) relate meteorological station temperatures to microhabitat temperatures; and (iii) measure microhabitat temperatures directly.

Relating meteorological station data to microhabitat temperatures is rarely done but where attempts have been made (e.g. Bernal and González, 1993) differences were found to be small (< 1.0°C) and, hence, weather

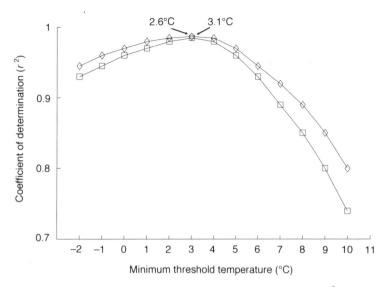

Fig. 4.4. Relationship between coefficient of determination (r^2) of body length–degree-days regression and minimum threshold temperatures (□) males; (◇) females. (After Tokeski, 1985.)

station data were considered adequate to predict insect development in the field. However, the variation experienced will be very dependent on the distance from the weather station, topography and local conditions.

Ambient air temperature is the measure most commonly used for temperature summations even though this may be inappropriate for some species such as those that are soil dwelling. Collier and Finch (1985) achieved a considerable improvement in degree-day prediction of emergence of *Delia radicum* through using soil instead of air temperatures. The use of air temperatures underestimated the number of degree-days required for emergence probably because changes in soil temperatures varied according to soil type, aspect and variable moisture content.

Modern instruments based on microprocessors can measure temperatures very accurately at microhabitat levels. They also have the advantage of having frequent measurements which can be logged and downloaded to computers. Accurate, frequent measurements (even hourly mean temperatures) will provide more accurate degree-day accumulations than those conventionally used which are based on the mean of the maximum and minimum temperature minus the threshold temperature, i.e.

$$\sum [\{(T_{max} - T_{min})/2\} - \text{threshold } T] \tag{4.6}$$

This method of summation will result in considerable inaccuracies if a temperature contributing to the mean lies outside the linear portion of the relationship (Jervis and Copland, 1996). Use of continuous measures (temperature dataloggers) or hourly mean temperatures will provide much more accurate estimates.

4.4.4. Selection of a biofix

By far the greatest inaccuracies in degree-day models occur when there is difficulty in defining a single discrete time or biological event (a biofix) that can be used to initiate an accumulation (Welch et al., 1981). This leads to arbitrary selected biofixes such as calendar dates (Baker et al., 1982; Collier and Finch, 1985) especially for insects where accumulations are used to predict development after a period of dormancy or diapause. The justification for use of an arbitrary date is that development prior to this will have been negligible, although rarely is any biological basis for such an argument put forward. Where accumulations are initiated by presence of a particular insect stage in order to predict some future stage, they are usually successful (Pruess, 1983; Gargiullo et al., 1984).

4.4.5. Relative growth rate

Most studies on temperature relationships have dealt with development but have ignored growth (Jervis and Copland, 1996). Growth is usually measured by determination of insect weight or size at birth and then subsequently again as an adult. The standard measure is the Mean Relative Growth Rate (MRGR) which is calculated by the formula (Fisher, 1921; Radford, 1967; Van Emden, 1969; Kogan, 1986; Castle and Berger, 1993)

$$\text{MRGR} = \frac{1_n(W_1) - 1_n(W_0)}{d} \tag{4.7}$$

where W_1 = adult weight, W_0 = weight at birth and d = development time from birth to adult.

The relationship between growth rate and temperature tends to be linear within the range of temperatures normally experienced by the insects in the field. Gilbert (1984b) defined the following relationship

$$\frac{1}{w} \cdot \frac{dw}{dt} = a(T - \theta) \tag{4.8}$$

where T = temperature, θ = the threshold below which no growth occurs, w is the larval weight at time t and a is a constant.

4.4.6. General experimental requirements

The majority of experiments carried out to determine the develement temperature relationships for an insect species are conducted in the laboratory using controlled temperature cabinets. There are numerous examples of such work but they are carried out with a greater or lesser regard for accuracy. The most important factors to take into account are those reviewed by Howe (1967), namely: (i) the number of temperature treatments; (ii) replication; and (iii) frequency of observation. Temperature control has been dealt with in Section 4.2.2.

NUMBER OF TEMPERATURE TREATMENTS

An approximate degree-day relationship can be obtained for the linear part of the response with as few as three or four temperature treatments (e.g. Minkenberg and Helderman, 1990; Enkegaard, 1993a; Kocourek *et al.*, 1994) although more usually five or six temperatures are used (e.g. Reissig *et al.*, 1979; Walgenbach *et al.*, 1988; Bernal and Gonzalez, 1993; Enkegaard, 1993b). Baker (1980) recommended use of temperatures every 2°C over a 20°C range while Howe (1967) suggested at least 10 temperatures separated by intervals of no more than 2.5°C in order to describe a non-linear curve. Further, to distinguish between models that are symmetrical around the optimum temperature and those that are not, requires several temperatures spaced at 1°C or 0.5°C intervals to be added 3°C either side of the optimum. As Allsopp *et al.* (1991) pointed out, few experiments meet these stringent requirements and while this may not matter for the purposes of pest management it is unwise to compare the merits of different models or to speculate on their physiological implications unless data of such quality are available.

REPLICATION

Variation in the development of insects in response to temperature may be considerable, as 30–40 insects surviving to the end point in each temperature treatment should be the aim (Allsopp *et al.*, 1991). In practice most researchers report use of higher numbers of replicates than this, e.g. 60–80 nymphs of *Schizaphis granarium* (Walgenbach *et al.*, 1988); 80 eggs of *Aproaerema modicella* (Shanower *et al.*, 1993) and 600 pupae per temperature of *Rhagoletis pomonella* (Reissig *et al.*, 1979). Decisions about replication will be constrained by the insect stage, with mobile stages presenting more of a problem than immobile, also those that need to be confined in cages or on host plants may restrict the number of replicates due to constraints of space in an incubator. Replications must always be sufficient to cope with mortality, particularly at extreme temperatures, and mortality should be reported so that the possibility of any biases can be assessed (Allsopp *et al.*, 1991).

FREQUENCY OF OBSERVATION

The frequency of observation will influence the accuracy of the development rates obtained. Howe (1967) recommends observations in a range of at least ten of the time units around each experimental end point. Fewer than five indicates that the variance is not worth estimating and the mean is inaccurate (Allsopp *et al.*, 1991). It may also be appropriate to change the frequency of observations according to the temperature with fewer observations at lower compared to higher temperatures (e.g. Kocourek *et al.*, 1994). Frequency of observation ranges from every 2–3 hs (e.g. Dean, 1974) to every 24 hs (e.g. Omer *et al.*, 1992).

4.4.7. Development and other abiotic factors

Although temperature is a major influence on development of insects, other abiotic factors also have been shown to affect development. The tendency for an inverse relationship between humidity and temperature is well known. Subramanyam and Hagstrum (1991) used the biophysical four parameter regression model of Wagner *et al.* (1984) to describe the development of *Prostephanus truncatus* at four different humidities, 40, 70, 80 and 90% humidity. Relative humidity influenced the development times of the larvae but not the eggs or the pupae. However, Subramanyam and Hagstrum were unsure from their experiments whether the humidity had directly affected the metabolism of the larvae or whether it had affected them indirectly by changing the palatability or physical condition of the diet on which they were maintained.

Brodsgaard (1994) investigated the influence of daylength on the bionomics of the glasshouse pest *Frankliniella occidentalis* reared on bean leaves at 25°C. Insects were subjected to one of three photoperiod treatments 4:20, 8:16 or 16:8 (L:D) hs and immature development time was found to significantly increase with decreasing daylength from 13.15 to 14.75 days at 16:8 to 4:20 (L:D) photoperiod respectively.

4.4.8. Influence of the host and diet

The development time of insects varies according to the condition, growth stage and type of host or diet on which the insect is feeding. Where artificial diet is used development times tend to be faster than on most host plants (Butler, 1976; Pretorius, 1976; Allsopp *et al.*, 1991). The quality of diet can be significantly affected by environmental conditions such as low temperature (Gilbert, 1988) and high humidity (Subramanyam and Hagstrum, 1991) which subsequently affects insect development times. It has been well established from studies of host plant resistance that insect development times vary both between plant species and crop cultivars (e.g. Dent and Wratten, 1986; Sekhon and Sajjan, 1987; Olatunde and Odebiyi, 1991; Yang *et al.*, 1994) as well as host growth stage (Leather and Dixon, 1982). Howe (1971) proposed an index for assessing the suitability of an environment or medium for the development of insects. The index is:

$$I = \log S / T \qquad (4.9)$$

where *S* is the percentage survival and *T* is the mean development period of the insect from egg hatch to adult. Howe's index of suitability has been used to determine the suitability of maize varieties for the development of *Plodia interpunctella* (Mbata, 1990) and *Chilo partellus* (Sekhon and Sajjan, 1987).

The MRGR has also been used in studies to determine the susceptibility of plants in host plant resistance studies (e.g. Van Emden, 1969; Leather and Dixon, 1981). As Radford (1967) points out the MRGR can always be used without involving any assumptions about the form of the growth curves and is therefore particularly valuable for comparing treatment differences. The use of MRGR also makes possible the comparison of insects of some range in initial size provided relative growth remains constant during immature development.

4.5. Survival and Mortality

Few insects die through senescence, most of them are killed by predators, disease and other hazards long before they reach old age (Krebs, 1978). The factors responsible for causing mortality in insects and the stage at which mortality occurs are of great interest to the population ecologist and pest manager. Measurement of mortality may be direct or indirect. Direct measurement is achieved through mark and recapture experiments while indirect measurements are made by estimating numbers of individuals in successive developmental stages. Mortality can be expressed in a number of different ways (Table 4.1) but one of the simplest graphical methods is that of the survivorship curve, i.e. the change in the numbers of survivors with time. The logarithm of the numbers living at a given age is plotted against age and the shape of the curve describes the distribution of mortality with age. A number of general survivorship curves have been described. Slobodkin (1962) described four survivorship curves, whereas Pearl (1928) and Deevey (1947) refer to only three (Fig. 4.5). The type I curve describes

Table 4.1. Ways of expressing population change and age-specific mortality when stages of the insect do not overlap. (From Varley *et al.*, 1973.)

Line	Eggs (N_E)	Small larvae (N_{L1})	Large larvae (N_{L2})	Pupae (N_P)	Adults (N_A)	
1 Population	1000	100	50	20	10	
2 Number dying in interval	900 +	50 +	30 +	10		Sum: 990 dead
3 % mortality	90 +	5 +	3 +	1		Sum: 99% mortality
4 Successive % mortality	90	50	60	50		
5 Successive % survival	10	50	40	50		
6 Fraction surviving	0.1 ×	0.5 ×	0.4 ×	0.5		Product = 0.01 survival
7 Log population	3.0	2.0	1.7	1.3	1.0	
8 k-value	1.0 +	0.3 +	0.4 +	0.3		Sum: $K = 2.0$

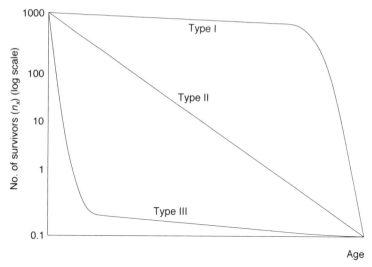

Fig. 4.5. Hypothetical survivorship curves. (After Pearl, 1928.)

mortality in populations which exhibit low initial losses that increase markedly with age. The diagonal survivorship curve (type II) implies a constant rate of mortality independent of age while the type III exhibits high initial losses followed by a period of much lower relatively constant mortality. Rockstein and Miguel (1973) provide further variation on a theme, with two main types similar to the types I and II and then a third intermediate form, between I and II, having an initial slow rate of death which gradually increases during the middle part of the lifespan and then slows down towards the end. All of these theoretical curves have been illustrated through the use of graphs but no parametric forms are given. Birley (1977) provided three relatively parametric forms based on the exponential function. However, with the exception of Birley's simplest model (the exponential model) the curves lack any theoretical justification such as might be supplied by simple probabalistic assumptions (Madden *et al.*, 1986). More recently, Gompertz and Weibull functions in conjunction with the exponential function have been used to describe survivorship curves.

4.5.1. Exponential, Gompertz and Weibull functions

Many investigators assume in their analyses of insect survival and mortality that mortality rates are independent of age (Clements and Paterson, 1981). This leads to the use of the exponential model based on the plot of the survival function $l_n(S(t))$ against age (t) which gives a straight line with a negative slope, described by:

$$S(t) = S_0 e^{-\alpha t} \tag{4.10}$$

a simple exponential curve where $S(t)$ is the proportion alive at any time t

(equivalent to (l_x)). The usefulness of such a distribution for survival times will depend in part on its having a hazard function with suitable properties. The hazard function $h(t)$ measures the instantaneous risk in that $h(t)dt$ is the probability of dying in the next small interval dt given survival to t (McCullagh and Nelder, 1983). The simplest hazard function, a constant, is associated with the exponential distribution of survival times and hence with the Poisson process (e.g. Peferoen *et al.*, 1981). The Gompertz and Weibull models both allow for a hazard rate that increases over time. The hazard function $M(t)$ for the Gompertz function increases exponentially with age thus

$$\mu(t) = \alpha e^{\beta t}$$
$$\text{or } l_n(\mu(t)) = l_n(\alpha) + \beta t \tag{4.11}$$

where α and β are constants. The Gompertz survivorship curve, has a curve which is concave below so that mortality increases with age. The Gompertz function is described by

$$S(t) = S_0 \exp\left(\frac{\alpha}{\beta}(e^{\beta t} - 1)\right) \tag{4.12}$$

where $S(t)$ is the proportion of females surviving to day t, α is the instantaneous rate of mortality at birth and β is a constant describing the exponential rate of increase in mortality rate with increasing age. The Gompertz function has been used by actuaries and has been applied to mortality rates in mosquitoes (Clements and Paterson, 1981) and in *Lucilia sericata* (Readshaw and van Gerwen, 1983).

The Weibull model is probably the most common model for representing survival distributions in both medical and engineering research (Madden and Nault, 1983), it also has applications in entomology (Madden and Nault, 1983; Madden *et al.*, 1984, 1986; Bartlett and Murray, 1986). The Weibull model can be written as:

$$S(t) = \exp\left(-(t/b)^c\right) \tag{4.13}$$

where $S(t)$ is the probability that an insect lives at least to time t, b is a scale parameter that is inversely related to the mortality rate and c is a shape parameter that allows the model to represent survival distributions of different forms (Lawless, 1982; Madden, 1985) (Fig. 4.6). The hazard function for the above equation is:

$$h(t) = (c/b)(t/b)^{c-1} \tag{4.14}$$

The Weibull model has also been used to investigate the influence of temperature on survival (e.g. Madden *et al.*, 1984).

4.5.2. Influence of abiotic factors on survival

A number of abiotic factors have been shown to influence survival in insects. Common among these are temperature and humidity but also

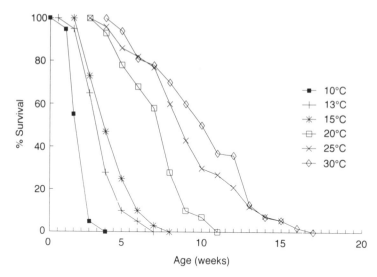

Fig. 4.6. Survival of adult apterous virginoparae of *E. lanigerum* at constant temperatures. (After Asante *et al.*, 1991.)

photoperiod and moisture, rainfall and wind speed. The influence of temperature, humidity and photoperiod on survival is usually assesssed in the laboratory. Insect survival tends to be reduced by extreme temperatures (e.g. Asante *et al.*, 1991; Moralesranos and Cate, 1992; Shanower *et al.*, 1993) (Fig. 4.6) and low humidities (e.g. Kfir, 1981) whereas photoperiod can have variable effects (e.g. Peferoen *et al.*, 1981; Brodsgaard, 1994). Rainfall can cause direct mortality through insects drowning (as larvae; Chang and Morimoto, 1988) (as pupae; Neuenschwander *et al.*, 1981) or indirectly through increasing soil moisture and hence susceptiblity to pathogen attack (Dancer and Varlez, 1992). High winds can dislodge insects from their host plants and thereby make them more susceptible to predation (Cannon, 1986). The extent to which such factors influence mortality may be best evaluated through direct field observation due to the difficulties associated with the necessary environmental effects in a laboratory situation.

4.5.3. Influence of biotic factors on survival

The influence of predation and parasitism on insect survival and the techniques associated with its quantification are largely dealt in Chapter 11. In addition to this, the following section briefly considers the effects of insect aggregation on survival and the use of mortality as an indicator of antibiotic host plant resistance.

The aggregation of insect populations is not unusual, especially for folivorous insects (Taylor *et al.*, 1978). In such aggregations, however, certain mortality factors may not act independently on individual insects but simultaneously destroy a number of individuals in a colony (Iwao, 1970). In order to know the mode of action of a mortality factor on a colony the following equation can be used (Iwao, 1970):

$$\tilde{i} = \overset{*}{m}.m_0/(\overset{*}{m}_0.m) \tag{4.15}$$

where m_0 is the mean colony size, $\overset{*}{m}_0$ is the mean crowding of the colony before the action of the mortality factor while m and $\overset{*}{m}$ are those after its action. When $\tilde{i} > 1, i = 1$ and $\tilde{i} < 1$ the mortality acts contagiously, randomly and uniformly among individuals of the colony respectively. The $\overset{*}{m}-m$ regression analysis (Iwao, 1968, e.g. Fig. 4.7) for a particular mortality factor will indicate by the intercept (α) and slope (β) the tendency for crowding ($+\alpha$) or repulsion ($-\alpha$) and extent to which colonies are contagious ($\beta > 1$; Southwood, 1978). Figure 4.7 shows the $\overset{*}{m}-m$ regression for predation of *Gastrolina depressa* by *Aiolocaria hexaspilota*. The predator tended to eat more than one individual per attack in a colony and their action seemed to be contagious in some particular colonies because $\alpha < 0$ and $\beta > 1$. Thus, the aggregation of the insects influenced their survival. One recurring explanation for aggregations of insects is that insect distribution reflects variation in nutritional or secondary chemistry among and within host plants.

Variation in host plant quality may influence clumping and insect survival (Faeth, 1990). Insect survival is often used as a measure in conjunction with others as indicators of antibiotic resistance in crop plants. In this context, however, survival is rarely used as the sole measure because host plant effects are usually not severe enough to kill insects, but impact on their rates

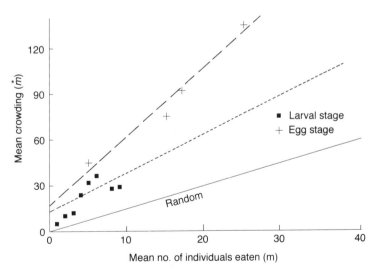

Fig. 4.7. Distribution pattern of the number of eggs and larvae eaten by *Ailocaria hexaspilota* in each colony analysed by mean crowding ($\overset{*}{m}$) to mean (m). (After Chang and Moritomo, 1988.)

of development and reproduction. Survival is usually measured as numbers or proportions surviving to adults on different host plant species or cultivars (e.g. Bintcliffe and Wratten, 1982) or more rarely as l_x values across insect stages and presented as a survivorship curve for each cultivar (Easwaramoorthy and Nandagopal, 1986).

4.5.4. Sampling and analysis of stage frequency data

Data obtained from the field on insect population size are often based on counts or estimates of numbers of insects in different stages. The data are collected from successive samples taken at different points in time at the same location. The sampling methods used might be absolute or relative measures depending on the insect stage, its habitat and mobility. The common forms of sampling methods used by entomologists are covered in Chapter 2. In addition to these, the use of mark–recapture sampling and transect sampling methods has been devised specifically for more mobile species or insect stages. These methods are briefly reviewed below.

LINE TRANSECTS AND MARK–RECAPTURE METHODS
Counts along a fixed route or line have become a standard method for evaluating the abundance of insects (e.g. Pollard, 1977, 1984; Thomas, 1983; Munguira and Thomas, 1992; Pollard *et al.*, 1995). Methods used can be classed into three categories (Eberhardt, 1978): line intercept sampling; strip transect sampling; and line transect sampling. The first considers the numbers crossing a transect line, the second the numbers of insects within a strip and the third where an observer moves in a line through the area covered by a population and counts the numbers of insects of a population that are seen. This latter approach usually involves the 'flushing' of the insects as the observer moves along the transect. If the proportion 'flushed' is constant then numbers themselves give an index of absolute population and incorporating the proportion and the area covered, an estimate of absolute population density can be obtained. If the efficiency of flushing varies then only a relative measure of abundance is obtained (Southwood, 1978). Death rates are generally obtained from mark–recapture methods but they can be estimated from transect counts using the model of Zonneveld (1991). The model (related to that of Manly, 1974, see below) describes the time course of abundance of adult insects for species emerging in discrete generations using four parameters: total population size; time of maximal appearance rate; a measure for the dispersion of the appearance rate; and the death rate. Apart from the population size, these parameters can be estimated from transect counts. The time course of abundance $x(t)$ is given by

$$x(t) = Ne^{\alpha(t-\mu)}\int_0^b \frac{r^{\alpha\beta}}{(l+r)^2} - \mathrm{d}t \qquad (4.16)$$

N is the total population size (i.e. the total number to appear), α is the death rate, μ the time of maximal appearance rate, β a measure for the dispersion of the emergence curve. The integral equals the incomplete beta function $Bb/(1+b)$ $(1 + \alpha\beta, 1-\alpha\beta)$ and has to be evaluated numerically. The shape of the equation depends only on the dimensionless parameter $\lambda = \alpha\beta$. With increasing λ a smaller fraction of the total population is present at one time.

Zonneveld (1991) tested the above model with transect data for five species of butterfly and found the model provided an excellent fit. The death rate (α) estimates provided by the model were in accordance with those obtained by mark–recapture methods for the same species (Thomas, 1983; Emmet and Heath, 1989). Zonneveld considered that this strengthened both methods and provided an opportunity to evaluate some of the assumptions of mark–recapture methods.

The methods developed for marking insect species for mark–recapture experiments have been reviewed in Chapter 5, along with the use of the methods for studying insect dispersal. The following account restricts itself to the methods of analysis used for quantifying insect abundance and mortality. The choice of methods for use, however, is not simple, since there are a number of alternatives which make different assumptions and provide different results. The subject is quite involved and has been the subject of a number of reviews (Southwood, 1978; Begon, 1979; Seber, 1982). Three of the more commonly used methods are considered here and their relative merits are compared. They are the methods of Fisher and Ford (1947), Jolly (1965), Seber (1965) and Manly and Parr (1968). The equations and assumptions made for each are considered below.

Fisher and Ford (1947):

$$N_t = \frac{n_t a_i \theta_{i-t}}{r_{ti}} \tag{4.17}$$

where N_t is the population estimated, n_t the total sample at time t, a_i the total of marked insects released at time i, θ_{i-t} the survival rate over the period $i-t$ and r_{ti} the recaptures at time t of insects marked at time i. This method assumes that sampling is random, survival rates and probabilities of capture are unaffected by marking, survival rates are independent of age and are approximately constant (Manly, 1974).

Seber (1965) and Jolly (1965) independently developed the following method to cover situations in which there is both death and emigration and births and immigration in a population. Their methods give similar solutions except that Jolly's makes allowance for any insects killed after capture and, hence, not released again (a common occurrence in entomological experiments; Southwood, 1978). The equation takes the form

$$N_i = \frac{M_i n_i}{r_i} \tag{4.18}$$

where N_i is the population estimate on day i, M_i is the estimate of the total number of marked insects on day i, r_i is the total number of marked insects recaptured on day i and n_i is the total number captured on day i. This model assumes that sampling is random, survival rates and probabilities or capture are unaffected by marking and that survival rates are independent of age.

The Manly and Parr (1968) method takes account of age-dependent mortality, but does, like the other two methods, assume random sampling and capture unaffected by marking. The marking of insects in this method should be date specific or enable identification of specific individuals. The equation is

$$N_i = \frac{an_i}{r_i} \tag{4.19}$$

where as before N_i is the population estimate on day i, n_i is the total number captured on day i and r_i the total number of marked insects recaptured on day i, a is then the total number marked.

Although Manly and Parr's (1968) method is not affected by age-dependent mortality rates it is more sensitive than the other two methods to poor data resulting from small sample sizes because it divides recapture data into smaller categories (Blower *et al.*, 1981). The method of Jolly (1965) is less affected than Ford and Fisher by age-dependent mortality although high infant mortality can result in a considerable bias in estimates (Manly, 1974). Ford and Fisher's method is particularly useful when sample sizes are small (Bishop and Sheppard, 1973). Published examples of the use of these and other mark–recapture methods can be found in Brakefield (1982) and Manguira and Thomas (1992).

TECHNIQUES FOR ANALYSING STAGE FREQUENCY DATA

Stage frequency data take the form indicated below:

Time	Stage				
	1	2	3	...	q
t_1	f_{11}	f_{12}	f_{13}	...	f_{1q}
t_2	f_{21}	f_{22}	f_{23}	...	f_{2q}
$\vdots$					
t_n	f_{n1}	f_{n2}	f_{n3}	...	f_{nq}

Such data are usually collected in order to estimate the mean durations of stages, stage specific survival, unit time survival rates and the total numbers entering different stages (Manly and Seyb, 1989). The data may be multicohort or single cohort data. In the former, individuals are entering the population for a substantial part of the sampling time whereas in the latter all individuals enter the population at the start of the sampling period. The analysis of single cohort data is much simpler than the multicohort data, and some of the same models can be used for both, but modelling is more straightforward with single cohort data because entry distributions need not be heeded. A large number of different approaches for analysing multi-cohort are available; far too many to consider here. Table 4.2 summarizes various approaches; 24 in all, but there may be others. Few comparisons have been made of the different methods (but see Manly, 1974, 1989, 1990; Munholland *et al.*, 1989) and, hence, it is difficult to advise on which are the best models for any particular situation. Two of the simpler models are described below to explain some of the basic principles.

Table 4.2. Summary of methods that have been proposed for analysing stage-frequency data. (From Manly, 1990 with additions.)

References	Parameters estimated	Comments
Richards and Waloff	Numbers entering stages; unit time survival rate	Uses linear regression to relate the number in a stage and all higher stages to time. Assumes a constant survival rate in all stages
Richards *et al.* (1960)	Unit time survival rates in stages	The number entering stage 1 and the durations of stages must be known. The area under the stage-frequency curve is equated to its expected value
Dempster (1961)	Total number entering stage 1; unit time survival rates in stages	Changes in the total stage-frequencies are related to proportions entering stage 1 between samples and survival rates by a linear regression. Entry rates to stage 1 must be known
Southwood and Jepson (1962); Sawyer and Haynes (1984)	Numbers entering stages	The mean durations of stages must be known. Based on an equation that holds only if all mortality occurs in stage transitions
Kiritani and Nakasuji (1967); Manly (1976, 1977a, 1985)	Numbers entering stages; unit time survival rate; durations of stages	Relates the area under the stage-frequency curve and the time-stage frequency curve to population parameters. Assumes a constant survival rate in all stages
Kobayashi (1968)	Numbers entering stage	Deaths are apportioned to different stages by a two-part correction process. The number entering stage 1 must be known
Read and Ashford (1968); Ashford *et al.* (1970)	Numbers entering stages; unit time survival rate; distributions of stage durations	The first model with a proper statistical model. Estimation by the method of maximum likelihood

(Continued)

Table 4.2. (*Continued*)

References	Parameters estimated	Comments
Rigler and Cooley (1974)	Numbers entering stages; durations of stages	There has been some discussion about the validity of this method (Hairston and Twombly, 1985; Aksnes and Hoisaeter, 1987; Hairston *et al.*, 1987; Saunders and Lewis, 1987)
Lakhani and Service (1974)	Survival rates in different stages	Equations relating areas under stage-frequency curves to survival rates are solved. Durations of stages and the proportion surviving to the last stage must be known
Manly (1974)	Numbers entering stages; unit time survival rate; durations of stages	Uses a non-linear regression to relate the number in a stage and all higher stages to population parameters
Ruesink (1975)	Stage-specific survival rates that vary with time	Durations of stages must be known
Birley (1977); Bellows and Birley (1981)	Survival rates in stages; numbers entering stages	The rate of entry to stage 1 must be known. Estimation by non-linear regression
Derr and Ord (1979)	Stage-specific survival rates that vary with time	Durations of stages must be known
Kempton (1979)	Distributions of durations of stages; time-dependent survival rates; numbers entering stages	Distributions of entry times to stages can take various forms. The survival function is not stage-dependent
Mills (1981a, 1981b)	Mean durations of stages; numbers entering stages	One duration, or the ratio of two durations must be known
Bellows *et al.* (1982)	Unit time survival rates that can vary with time; distributions of stage durations; numbers entering stages	Survival rates are estimated by regressing total stage-frequencies against time. The Bellows and Birley (1981) model is then used to estimate other parameters
van Straalen (1982, 1985)	Numbers entering stages; unit time survival rate; duration of stages; a growth function	Assumes that each individual has an associated measurable development variable that increases with time according to the growth function. Temperature effects are allowed for by using physiological time. Estimation by the method of maximum likelihood
Osawa *et al.* (1983); Stedinger *et al.* (1985); Dennis *et al.* (1986); Kemp *et al.* (1986); Dennis and Kemp (1988)	Proportion of the population in different stages at different times (and places)	Estimates a normal or logistic function for the distribution of a development variable at different times. Estimation by the method of maximum likelihood. Temperature effects are allowed for by using physiological time

(Continued)

Table 4.2. (*Continued*)

References	Parameters estimated	Comments
Shoemaker *et al.* (1986)	Recruitment rates for stage 1; stage-specific survival rates; a unit time survival rate; the durations of stages	A non-linear regression relating stage-frequencies to recruitment and survival is estimated. Temperature effects are allowed for by using physiological time
Manly (1987)	Numbers entering stages; a unit time survival rate in each stage; duration of stages	Uses a linear regression to relate the number of individuals in one sample, to the numbers in different stages in the previous sample
Munholland (1988); Kemp *et al.* (1989); Munholland *et al.* (1989)	Proportions of the population in different stages at different times; survival rate that can depend on time	A development of the model of Osawa and others given above that allows for mortality. Estimation is by the method of maximum likelihood. Temperature effects are allowed for by using physiological time
Braner (1988); Braner and Hairston (1989)	Numbers entering stages; stage-specific or a constant survival rate; durations of stages	A gamma model is used to model variation in sampling intensities with time. Normal distributions are assumed for stage durations. Estimation by maximum likelihood
Manly (1989)	Numbers entering stages; a unit time survival rate; distribution of a development variable; amount of development needed to enter each stage	Assumes that each individual has an unobservable value for a development variable with this being normally distributed at the time of the first sample. Estimation by maximum likelihood
Manly (1993)	Unit time survival rate; stage-specific survival rates; the duration of stages; numbers entering each stage	Can be applied with any distribution of entry times to stage 1 and any distribution of numbers in stages when sampling begins. Simulation required for estimating standard errors. Assumes survival rate per unit time is constant and transition between one stage and the next occurs in one unit of time

Richards and Waloff (1954) provided one of the earliest methods for analysing stage frequency data based on stage-specific data having a well defined peak and a constant rate of mortality. Working with British grasshoppers they assumed that once the hoppers had emerged the numbers in the population over time would be defined by

$$N_t = N_0 K^t \tag{4.20}$$

where N_t is the population occurring on day t, N_0 is the total number of hoppers which emerge and K is the fraction surviving each day (the survival rate per unit time). Hence, when the data are plotted on a log N_t vs. t graph the fall-off of numbers after the peak can be extrapolated back to the time when the first stage was found using the equation

$$\log N_t = \log N_0 + t \log K \qquad (4.21)$$

which provides an estimate of the initial number entering the stage (Dempster, 1956; Southwood, 1978). A calculation for the population size less the first instar (i.e. for successive accumulated totals of the second instar), gives an estimate of the total numbers of hoppers entering the second instar and so on through each of the instars. From these estimates of the total number of hoppers entering each stage, the mortality during each stage can be determined.

Manly (1974) modified the approach of Richards and Waloff (1954) avoiding the need for a distinct peak in the data. The method assumes that the time insects take to enter a particular stage follows a frequency distribution with a function $f_{(x)}$. The constant daily survival rate is $e^{\theta(t-x)}$. The expected number of insects in the stage (or a later stage) at time t is given by

$$N_t = M \int_{-\infty}^{t} e^{-\theta(t-x)} f_{(x)} d_x \qquad (4.22)$$

where N_t includes the insects in the stage and all later stages and M is the total number of insects entering the stage. Manly used the normal frequency distribution for the entry distribution $f_{(x)}$, although alternative models such as the Weibull and gamma distributions could be used.

The normal frequency distribution $f_{(x)}$ is given by

$(2\pi)^{-\frac{1}{2}} \exp\left(-\frac{1}{2}(x - \mu)^2/\sigma^2\right)$ and when substituted into the above equation

$$N_t = M^* e^{-\theta t} \int_{-\infty}^{(t-\mu^*)/\sigma} (2\pi)^{-\frac{1}{2}} \exp\left(-\frac{1}{2} x^2\right) d_x \qquad (4.23)$$

where $M^* = \exp\left(\theta\left(\mu + \frac{1}{2}\theta\sigma^2\right)\right) M$ and $\mu^* = \mu + \theta\sigma^2$, where μ = the mean time of entry to the stage and σ its standard deviation. The four unknowns N_0, μ, θ and σ may be obtained from a series of equations derived from at least four sampling occasions (Southwood, 1978).

Danthanaryana (1983) working with *Ephiyas postivittana* used Manly's method to determine the number of larvae recruited to higher instars and when populations were very low approximate estimates of recruits to stages were obtained by the method of Richards and Waloff (1954). Manly's method is illustrated in Fig. 4.8 which shows the weekly and observed and calculated values for instar VI larvae of *E. postivittana* in the summer generation of 1974.

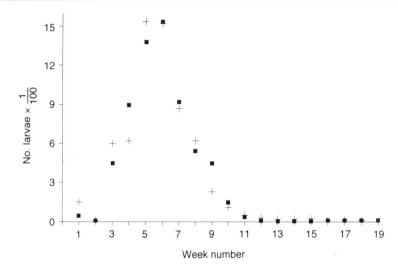

Fig. 4.8. Weekly estimates of instar VI larvae for *E. postivittana* based on Manly (1974) and Danthanaraya (1983).

4.6. Intrinsic Rate of Natural Increase

The intrinsic rate of natural increase (r) provides an effective summary of an insect's life history traits (Dixon, 1987). It is the rate of increase per individual female under specified physical conditions in an unlimited environment where the effects of increased density do not need to be considered (Birch, 1948). The r value is derived from the birth rate minus the death rate under known conditions and is often used by ecologists and pest management scientists as a comparative statistic for revealing the impact of a parameter (e.g. temperature, host plant) on insect demographic potential.

The intrinsic rate of increase is best defined as the constant r in the logistic equation:

$$N_t = N_0 e^{rt} \tag{4.24}$$

where N_0 is the number of insects at time zero
N_t is the number of insects at time t
r is the infinite rate of increase (the finite rate of increase λ is the natural antilog of r).

For a given species r can take a number of values. Theoretically, a species has an optimum natural environment in which r will attain its maximum possible value, r_m, with a stable age distribution (Jervis and Copland, 1996). A stable age distribution comes about when a population with constant age schedules of fecundity and mortality multiplies in an unlimited environment (Birch, 1948). The intrinsic rate of increase of a population may be calculated from the age-specific fecundity and survival rates observed under defined environmental conditions.

The experimental data required for the calculation of r are based on the female population and consist of a cohort of individuals monitored under specified conditions to determine their age-specific fecundity and survival which is then presented as a life and fecundity table (Table 4.3). The probability at birth of being alive at age x is d_x (l_0-1) while the mean number of female offspring produced in a unit of time by a female aged x is m_x. The product $l_x m_x$ is obtained for each age group and the sum of these products is the net reproductive rate R_0 (Table 4.3) defined by the equation:

$$R_0 = \int_0^\infty l_x m_x d_x$$

$$R_0 = \sum l_x m_x$$

(4.25)

Alternatively R_0 may be expressed as the ratio of individuals in a population at the start of one generation to the numbers at the beginning of the previous generation.

$$R_0 = \frac{N_t + T}{N_t}$$

(4.26)

where T is the generation time. Values of R_0 between populations should only be compared when generation times are the same. Two or more populations may have the same net reproductive rate but their intrinsic rates of increase may be quite different because of the different lengths of their generation times (Birch, 1948).

The mean generation time can be determined from the equation:

$$T = \frac{\log_e R_0}{r}$$

(4.27)

Table 4.3. Lifetable, age specific fecundity (m_x), and net reproduction rate (R_0) of a 600-µg apterous and alate individual of A. fabae kept at 20°C. (From Dixon and Wratten, 1971.)

Pivotal age 5-day unit (x)	Apterae			Alatae		
	Survival rate (l_x)	Fecundity (m_x)	$l_x m_x$	Survival rate (l_x)	Fecundity (m_x)	$l_x m_x$
0.5*						
1.5						
2.5	0.97	13.4	13.0	0.99	13.4	13.27
3.5	0.83	17.3	14.36	0.83	11.9	9.88
4.5	0.54	12.2	6.59	0.51	4.6	2.35
5.5	0.39	12.2	4.76	0.36	4.6	1.66
			R_0 38.71			R_0 27.16

Immature stages, mortality assumed to be 0.

but this requires knowledge of the value for r. Where r is not known an approximate value for T can be obtained from the equation:

$$\frac{T - \sum_x l_x m_x}{\sum l_x m_x} \tag{4.28}$$

A number of other useful population statistics can be calculated as soon as values for $l_x m_x$ and r are known (Messenger, 1964). These include the gross reproduction (GRR) (the mean total number of eggs produced by females over their lifetime; measured in female/female/generation) given by

$$GRR = \sum m_x \tag{4.29}$$

The doubling time of a population, i.e. the time required for a population to double its numbers, is given by the equation:

$$DT = \log_e 2 / r \tag{4.30}$$

Finally, the finite capacity for increase (λ), which is the number of times the population will multiply itself per unit time (measured in units of female/female/day), is obtained from

$$\lambda - e^r \equiv \text{antilog}_e r \equiv \frac{N_t + 1}{N_t} \tag{4.31}$$

for a population that is increasing exponentially.

4.6.1. Calculation of the intrinsic rate of increase

The procedure for determining the intrinsic rate of increase for insect pests first proposed by Birch (1948) has had widespread application through ecology and agricultural entomology, despite the experimental difficulties associated with compiling lifetables and the rather complex iterative calculations required to determine a value for r_m. Although the process is now made much easier with the use of computers, simpler methods of calculation are available, although inevitably they tend to compromise accuracy for ease of determination. The methods of Birch (1948), Howe (1953), Laughlin (1965), Wyatt and White (1977), and Lewontin (1965) are considered below.

THE BIRCH METHOD

The Birch (1948) method for calculating r_m is based on the work of Lotka and the application of the principle to human populations (e.g. Dublin and Lotka, 1925). The method assumes a stable age distribution and is based on age-specific fecundity and survival data (Table 4.3). With a stable age distribution the value of r can be calculated from the equation:

$$\int_0^\infty e^{-rx} l_x m_x d_x = 1 \tag{4.32}$$

r can be determined by iteratively solving the equation

$$\sum_{x=0}^{n} e^{-rx} l_x m_x = 1 \qquad (4.33)$$

where x is the midpoint of age intervals in days and m_x and l_x are defined above. A trial number of values for r are substituted into the equation until the left hand side of the equation tends towards one, at which point r_{max} is identified.

It is possible to estimate a standard error for the r values (Meyer *et al.,* 1986) using the Jackknife procedure. This is a computer-intensive technique based on: (i) recombining the original data; (ii) calculating pseudo values of the parameter of interest for each combination of the original data; and (iii) estimating the standard error from the resulting frequency distributions of pseudo values. Given the r value the Jackknife procedure omits one of the n replicate insects (the ith insect, $i = 1, 2, \dots n$) from the original data set and r is recalculated as r' using the data from the remaining $n-1$ insects. The Jackknife pseudo value $\hat{r}$ is then calculated for this subset according to:

$$\bar{r}_j = \frac{1}{n}\sum_{i=1}^{n} \tilde{r}_i$$
$$\qquad (4.34)$$
$$\tilde{SE}(\bar{r}_j) = \sqrt{s^2_{\tilde{r}} / n}$$

where $s^2_{\tilde{r}}$ is the variance of the pseudo values.

THE HOWE METHOD

Howe (1953) proposed a method that applies only to insects that produce all their offspring over a short period of time. The equation for calculating r is as follows:

$$r = (\log_e k)/(d + \tfrac{1}{2}l) \qquad (4.35)$$

where k = the number of female offspring
d = the development time
l = the period over which the offspring are produced.

The method assumes that the mean generation time lies halfway through the reproductive period which is seldom true, especially for insects such as aphids that have a relatively short development time and a long reproductive period (Wyatt and White, 1977). The method has been used to determine r for *Encarsia tricolor* a parasitoid of *Trialeurodes vaporariorum* (Artigues *et al.,* 1992). Howe produced two modifications of the above equation which involved weighting coefficients and tables that are mainly applicable to insects with long development times and reproductive periods.

THE LAUGHLIN METHOD

Laughlin (1965) proposed r_c, the capacity for increase as an approximation for r_m based on the equation:

$$r_c - \log_e R_0 / T \tag{4.36}$$

where R_0 = net reproductive rate $(\sum_x l_x m_x)$ $\tag{4.37}$

  T = cohort generation time $(\sum_x l_x m_x / \sum_x l_x m_x)$ $\tag{4.38}$

hence

$$rc = \frac{(\sum_x l_x m_x)(\log_e(\sum_x l_x m_x))}{(\sum_x x l_x m_x)} \tag{4.39}$$

The general relationship between r_c and r_m has been discussed by May (1976); r_c tends to provide an underestimate of r_m (Pielou, 1974). r_c can be used as an initial r value for the iterative calculation of r_m (see above Birch, 1948) or as an approximation of r_m in its own right (e.g. Yu *et al.*, 1990).

THE WYATT AND WHITE METHOD

DeLoach (1974) calculated the reproductive time required to contribute 95% to the r_m for three aphid species. This work led to the development of a simplified method of calculating r_m for aphids and mites by Wyatt and White (1977). Their approach is based on the assumption that if the development time is 'd' then 95% of the r_m will be achieved in about 2d, measured from birth. The effective fecundity can then be regarded as the number of offspring (Md) produced in a reproductive period equal to d. Provided the form of the distribution of progeny over time remains fairly constant under all conditions then Md may be assumed to act as a single effective date (Td), equivalent to a generation time.

The equation for the calculation of r_m is:

$$r_m = 0.738 \, (\log_e Md)/d \tag{4.40}$$

where 0.738 is a correcting constant based on calculations of 45 sets of data for four different species of aphid: *Myzus persicae*, *Brachycaudus helichrysi*, *Macrosiphoniella sanborni* and *Aphis gossypii*.

The method has been widely used for a range of aphid species (e.g. Soroka and Mackay, 1991; Aldyhim and Khalil, 1993; Castle and Berger, 1993; van Helden *et al.*, 1993) but Southwood (1978) warns that when a whole series of determinations are to be made under different climatic or biotic conditions then the value of the constant should be confirmed since it is very dependent on the form of the $\sum_x l_x m_x$ curve.

LEWONTIN'S REPRODUCTIVE FUNCTION

The model developed by Lewontin (1965) for calculating r_m follows a different approach to those mentioned above. Often referred to as Lewontin's reproductive triangle, the model describes reproduction during the period of exponential population increase as a simple function V_1 of age (x) where A, T and W represent the times at which reproduction starts, peaks and ends respectively. The area S of the triangle (Fig. 4.9) represents overall reproduction and from these values r_m can be calculated from the equation:

$$\frac{r_m^2(W-A)}{2S} = \frac{\exp(-r_m A) - \exp(-r_m T)}{T-A} + \frac{\exp(-r_m W) - \exp(-r_m T)}{W-T} \quad (4.41)$$

Lewontin's triangular reproductive function provides a basic tool for comparing population development in terms of r_m and is especially useful in comparing insect performance on different host plants where there are overlapping insect generations (Caswell and Hastings, 1980; Panda and Khush, 1995), for instance whitefly populations (e.g. Romanow et al., 1991).

4.6.2. Applications of r_m

The intrinsic rate of increase is determined under standardized conditions which provide the opportunity for the comparison of organisms or strains under different treatments (e.g. temperature and host plant), and for revealing the impact of these on the expression of the insect demographic potential (Hance et al., 1994).

The intrinsic rate of increase has been used as a means of comparing the dynamics of a number of pest species on a crop host (e.g. Kaakeh and Dutcher, 1992) and as a means of assessing the intrinsic factors such as

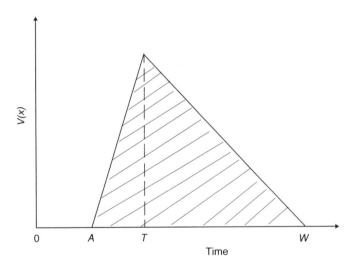

Fig. 4.9. The Lewontin reproductive function model showing an exponential population increase as a simple function, V, of age (x) where A = start of reproduction, T = peak reproduction, W = end of reproduction, and S (hatched area) = total reproduction. (After Lewontin, 1965.)

clones and morphs within a species (e.g. Dixon and Wratten, 1971; Simon *et al.*, 1991). More commonly, the effects of host plant and temperature on the parameters of the intrinsic rate of natural increase are the subject of investigation. The impact of temperature on pararmeters of r_m has been considered for insect pests (e.g. Wyatt and Brown, 1977; Zhou and Carter, 1992; Asante *et al.*, 1991; Aldyhim and Khalil, 1993; Enkegaard, 1993a; Kocourek *et al.*, 1994) and for their parasitoids (e.g. Moralesranos and Cate, 1992; Enkegaard, 1993b; Vansteenis, 1993), with constant and fluctuating temperatures (e.g. Elliott and Keickhefer, 1989) and in conjunction with evaluation of plants as hosts (e.g. Kaakeh and Dutcher, 1993). The majority of such studies in the latter case are to evaluate, through use of *r*, the antibiotic resistance of potential crop cultivars (e.g. Soroka and Mackay, 1991; van Helden *et al.*, 1993).

4.6.3. Lifetables, key factors and density dependence

The intrinsic rate of increase is a description of a lifetable by a single parameter. Where there is a need to determine the role of each factor the lifetable can be taken apart to reveal the impact of various factors on age-specific mortalities and population densities. The understanding gained by such analyses can provide a more rational basis for devising pest management programmes by identifying vulnerable life stages, as well as providing an ecological understanding of factors influencing a species population dynamics.

There are two types of lifetable: the age specific (syn = cohort, horizontal) and the time specific (syn = static, vertical). The time-specific lifetable considers census data taken on a single occasion when it is assumed that all generations are completely overlapping and, hence, all age classes are present simultaneously. By contrast, age-specific lifetables involve repeated counting of a single cohort of similar aged individuals over time. Age-specific lifetables are more commonly used in entomology than time-specific lifetables (Dent, 1991). Room *et al.* (1991) reported a third type of lifetable, a 'rolling lifetable' that combines elements of both age- and time-specific tables. Counts of individuals in all age classes were made daily in the manner of a repeating time-specific lifetable. These data combined with temperature measurements and known development rates were used to calculate the proportion of cohorts each day that should have been present in each age class on each census day. Subtraction of observed numbers from expected in each age class provided a daily estimate of age-specific mortality. However, the method is labour intensive and has not been widely attempted (Room, 1983). The methods associated with collecting data, building lifetables and their analyses have been reviewed in a number of ecological texts (Varley *et al.*, 1973; Southwood, 1978; Begon and Mortimer, 1981; Horn, 1988).

CENSUS DATA FOR AGE-SPECIFIC LIFETABLES

The census required to obtain data for a lifetable is simplified if the species of interest has a single generation per year, because age-specific mortality is then seasonal. The insect counts are most usefully expressed as a number per metre squared or some other unit measure, since this allows direct comparisons to be made between samples taken at different times. However, this is not always possible, especially where relative methods are required to estimate numbers of a mobile adult stage. Often different methods will be required for determining densities of different insect stages, depending on the insect's biology. For each sample, the age class or insect stage must be identified (by size, presence of exuviae or distinguishing feature). Sampling must take place at intervals that are at least no longer than the duration of a developmental stage. Since the census involves following the fate of individual insects it may be necessary to mark the insect (Room, 1977), plant part or plant (e.g. Easwaramoorthy and Nandagopal, 1986) on which the insect is found, depending on the mobility of the insect and stage concerned. Considerable problems arise in obtaining appropriate data if individual insects are mobile and cannot be located and 'identified' on each sampling occasion.

On each sampling occasion identifiable causes of mortality must be recorded. These causes fall into a number of general classes including weather, predation, parasitism, diseases, host nutrition or a resistance mechanism and insecticides (Room et al., 1991). Evidence of mortality is often difficult to obtain, more often than not insects just disappear leaving no record of their fate. Exceptions to this are corpses attacked by fungi, bacteria or viruses, and when a predator or parasitoid kills but remains in the host long enough to be counted. In the latter case, samples can be removed to the laboratory where they are reared through and the number and species of parasitoids that emerge recorded, providing the required estimates of mortality due to parasitism.

The data collected by census must conform to a number of rules if they are to be included in a lifetable (Varley et al., 1973). These include the following:

1.   Mortality events which are separated by time should be treated as entirely separate, whereas events which overlap may be more conveniently considered to act contemporaneously
2.   Each insect sampled must be classified as either dead or alive; a parasitized insect although alive should be considered as certain to die.
3.   No individual can be killed more than once, for instance, where a pest is attacked by two parasitoids then the death of the pest must be accredited to the first parasitoid species. If the second parasitoid species eventually kills the pest it must be accredited with the death of the first parasitoid. The second attack is entered into the lifetable of the parasitoid but not the pest.

In addition to the insect census data, measurements of other potential causes of mortality such as temperature, rainfall and host plant condition and growth stage must be made and recorded for the period of the study.

CONSTRUCTION OF THE LIFETABLE

The method most widely used for the construction of a lifetable follows the form developed by Varley and Gradwell (1960, 1963, 1965) and Varley *et al.* (1973). Their approach requires data for a number of generations from adult to adult in a series of successive lifetables. Key factors and density-dependent relationships can be determined provided data are collected over a sufficient number of generations.

Generally tables are constructed utilizing the headings given in Table 4.4 and described below. Although the degree of detail provided will vary markedly according to the biology of a pest and the way data are collected (Begon and Mortimer, 1981).

The term x is normally used to denote the development stage or age class. The number entering the age class from the previous class is denoted as lx and is sometimes transformed so that the initial cohort size is 1000 which permits comparison between different studies. dxF is the factor responsible for the mortality which is quantified as dx, the number dying in age class x. The apparent mortality 100 qx is the numbers dying as a percentage of the numbers entering the stage and its main value is for simultaneous comparison either with independent factors or with the same factor in different parts of the habitat (Southwood, 1978). The survival rate within x is denoted by Sx. The real mortality 100 rx is dx as a percentage of the population density at the start of the generation and it is the only percentage figure that is additive; useful for comparing the population factors within the same generation. The indispensable mortality 100 lx, is that mortality that would not occur should the dx under consideration be removed, as a percentage of the initial number (Room *et al.*, 1991). kx is the difference between successive values of log lx and is referred to as the k-value. A number of these k-values are used in key factor analysis.

KEY FACTOR ANALYSIS

Morris (1959) was the first to introduce the term 'key factor' to describe the predictive value of the effect of parasitism in determining the population density in the next generation of a pest species. Morris used regression and correlation methods to identify key factors, however, it is the graphical method of Varley and Gradwell (1960) which has had most influence and has been adopted widely for key factor analysis.

Key factor analysis is used to estimate the contribution of each separate mortality to the overall generation mortality (Varley and Gradwell, 1960). The separate mortalities are calculated as k-values ($k_1 \ldots k_n$) and K the total mortality:

$$K = \log_{10}N - \log_{10}N_s = k_1 + k_2 + k_3 \ldots + k_n = \sum_{i=1}^{n} k_i \qquad (4.42)$$

Table 4.4. Lifetable of *Chilo sacchariphagus indicus*. (From Easwaramoorthy and Nandagopal, 1986.)

Age interval	No. living at beginning of x (*lx*)	Factors responsible for dx (*dxf*)	Number dying during x (*dx*)	*dx* as a % of *lx* (100 *qx*)	*kx*
Egg	7556	Infertility	601	7.95	0.036
		Parasitism	322	4.26	0.021
		Arthropod predation	1718	22.74	0.130
		Dessication	364	4.82	0.033
		Blown by winds	215	2.85	0.021
		Sub-total	3220	42.62	0.241
Larva I instar	4336	Arthropod predation + inability to establish	2256	58.95	0.387
		Diseased	36	0.83	0.009
		Sub-total	2592	59.78	0.396
II instar	1774	Migration + unknown	426	24.43	0.121
		Diseased	150	8.60	0.053
		Sub-total	576	33.03	0.174
III instar	1168	Migration + unknown	504	43.15	0.245
		Diseased	23	1.97	0.016
		Parasitism	9	0.77	0.006
		Sub-total	536	45.89	0.267
IV instar	632	Migration + unknown	307	48.58	0.289
		Diseased	3	0.47	0.004
		Sub-total	310	49.05	0.293
V instar	322	Migration + unknown	197	61.18	0.411
		Sub-total	197	61.18	0.411
Pupa	125	Parasitisms	20	16.00	0.076
		Diseased	2	1.60	0.008
		Failure to emerge	15	12.00	0.068
		Sub-total	37	29.60	0.152
Adult	88				1.934K

Number of female moths emerged: 45
Number of eggs produced: 9370
Trend index (*N2/N1*): 1.24

where N = the number of individuals before mortality, N_s = the number surviving the mortality, K = the total generation mortality comprising the sum of a number of mortalities K_1-k_n, are then plotted against generation number (Fig. 4.10). These plots are examined to determine which k-value graph most closely follows the form of the variation in total K. The k-value which most closely resembles K is the key factor, i.e. it is the key to changes in the population density. Although this method is simple to use it suffers from two important drawbacks: (i) sometimes there is no obvious key factor; and (ii) the relative importance of the other key factors is often ignored.

Podoler and Rogers (1975) proposed a modification to the Varley and Gradwell (1960) approach which is less arbitrary in its identification of the key factors. The method is based on the use of regression analysis of k-values against total mortality K and the comparison of regression coefficients as a means of identifying the key factor(s). The key factor is the k-value for which the regression coefficient has a value closest to 1.0. This k-value will accurately reflect the changes in total K (Fig. 4.11).

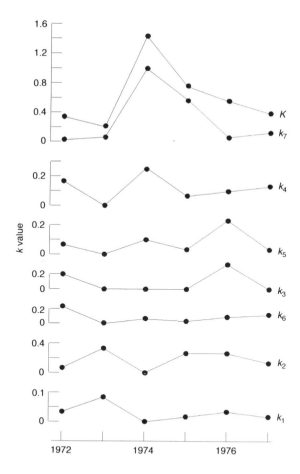

Fig. 4.10. k-Factor analysis of mortalities of *E. postvittana* larval instar VI at La Trobe in six summer generations. (After Danthanarayana, 1983.)

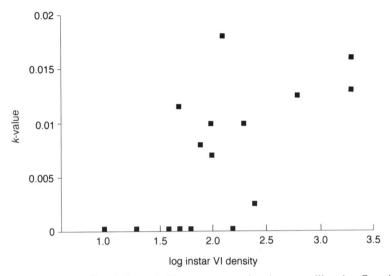

Fig. 4.11. Mortality of *E. postvittana* pupae due to parasitism by *Exochus* sp. (*k*), plotted on sixth instar larval density for 16 generations at La Trobe indicates a density-dependent relationship (*r* = 0.615; *P* < 0.01). (After Danthanarayana, 1983.)

Smith (1973) extended the key factor analysis to ascertain the relative importance of the remaining *k*-values once the key factor had been identified. The method is essentially the same as the graphical approach except when having found the key factor (*kp*) the following equation:

$$\sum_{i=1}^{n} k_i$$

is rearranged in terms of the residual killing power *K'*

$$K' = K - kp = \sum_{i=1}^{n} k_i (i \neq p) \tag{4.43}$$

The contribution of the remaining *k*-values after the removal of *kp* can then be assessed through regression analysis until all the *k*-values in turn have been removed in order of importance. Calculations can be done either in terms of covariances or as regression coefficients of *k*-values on residual killing powers. The process is analogous to a stepwise multiple regression analysis but without an obvious stopping rule.

A more sophisticated approach to key factor analysis was developed by Manly (1977b). This approach is based on an equation that partitions the variance of the number alive at the end of the generation into components associated with variation in the numbers entering the first stage, variation in survival rates and also density-dependent aspects of survival. The population model used by Manly (1977b) involves estimating parameters that allow for density-dependent mortality and from these estimate the contribution of each life stage to the variation in the numbers entering the final stage. A key factor can then be defined as a life stage that substantially either increases or reduces this variation (Manly, 1990).

The logarithm of the population size at the end of the *j*th stage in a particular generation will be:

$$R_j = S_0 + S_1 + S_2 + \dots \; S_j \tag{4.44}$$

where S_0 is the logarithm of the initial number entering the stage and $S_j \equiv -k_j$ values. S_j is used here to avoid complications with signs in the model, but apart from this the S_j values are the same as the *k*-values used above.

To allow for density-dependent mortality the standard regression model is used which assumes that

$$S_j = \alpha_j + \beta_j \, (R_j - 1) + e_j \tag{4.45}$$

where α_j and β_j are regression constants and e_j is a random variable, independent of $R_j - 1$ (the *x* variable) with a zero mean and constant variance $Var(e_j)$. It then follows that

$$R_j = R_j - 1 + S_j \tag{4.46}$$

substituting S_j from Eqn 4.45 gives

$$R_j = \alpha_j + (1 + \beta_j) \, R_j - 1 + e_j \tag{4.47}$$

$$\text{with Var} \quad (R_j) = (1 + \beta_j)^2 \, \text{Var} \, (R_j - 1) + \text{Var} \, (e_j) \tag{4.48}$$

which with repeated use gives

$$\begin{aligned}
\text{Var} \, (R_n) = {} & (1 + \beta_n)^2 \, (1 + \beta_{n-1})^2 \dots (1 + \beta_1)^2 \, \text{Var} \, (S_0) \\
& + (1 + \beta_n)^2 \, (1 + \beta_{n-1})^2 \dots (1 + \beta_1)^2 \, \text{Var} \, (e_1) \\
& + \dots + (1 + \beta_n)^2 \, \text{Var} \, (e_{n-1}) + \text{Var} \, (e_n)
\end{aligned} \tag{4.49}$$

where Var (R_n) is the variation in the population size at the end of the life cycle. This equation shows how Var (R_n) depends upon the variation in the initial size of the population (S_0), random variation in the S_j values and the density dependence of the S_j values (i.e. the value of β_j; for an S_j value that is not density dependent $\beta_j = 0$).

The terms in Eqn 4.49 can be written in terms of variances and covariances of $S_0, S_1, S_2 \dots S_n$.

$$\beta_j = \text{Cov} \, (S_{jI} \, R_{j-1}) / \text{Var} \, (R_{j-1}) \tag{4.50}$$

where Cov $(S_{jI} \, R_{j-1})$ indicates the covariance between S_j and R_{j-1}

$$\text{Var} \, (e_j) = \text{Var} \, (S_j) - \beta_j^2 \, \text{Var} \, (R_{j-1}) \tag{4.51}$$

If estimates of these variances are available then estimates of the regression coefficients, the β_j values, can be obtained using Eqn 4.50 and estimates of Var (e_j) values from Eqn 4.51. The variance of the population size (Var (R_n)) at the end of the life cycle can then be partitioned according to Eqn 4.49 which will indicate the key factor(s) contributing to the total variance.

Manly (1990) compared the methods of Varley and Gradwell (1960), Podoler and Rogers (1975), Smith (1973) and Manly (1977b) and found that they all identified the same key factors, although the latter two methods were useful in clarifying the importance of correlations between k-values, and in addition, the Manly (1977b) model incorporates density-dependence aspects of survival.

DENSITY DEPENDENCE

One of the major uses for key factor analysis is the search for density dependence in mortality factors since such factors may be responsible for the regulation of population density. There are four classes of such density-relatedness survival:

1.   Density dependence where mortality increases as density increases which has a stabilizing effect on the population dynamics.
2.   Inverse density-dependent mortality where mortality decreases with increasing population density which has a destabilizing effect on populations.
3.   Delayed density-dependent mortality where the effect on the population density may take effect in subsequent generations.
4.   Density-independent mortality where the mortality factor acts irrespective of population density, e.g. adverse weather conditions (Varley et al., 1973).

Following the determination of the key factors, the k-values are plotted against the logarithm of the stage population density on which they act in order to see if there is a density-dependent relationship (Fig. 4.11). If a regression analysis is significant then density dependence is suspected (Southwood, 1978).

Evidence of delayed density dependence may be obtained from a k-value vs. density graph if when the points for successive generations are linked together an anticlockwise spiral is formed (Varley and Gradwell, 1970) (Fig. 4.12).

A number of drawbacks exist with the regression technique for density dependence. The lack of independence of the variables could produce a spurious regression result, due to sampling errors (Varley and Gradwell, 1960; Ito, 1972; Bulmer, 1975) and even if there are no sampling errors the standard regression model may be inappropriate (Ito, 1972; Manly, 1990).

Further spurious mortality–density relationships may be obtained when k-values are considered in the case of contemporaneous mortality factors. Varley et al. (1973) suggested that contemporaneous mortality factors, e.g. attacks by two parasitoids in one stage interval, be treated as if they had acted in a sequential manner or by grouping them in a single category

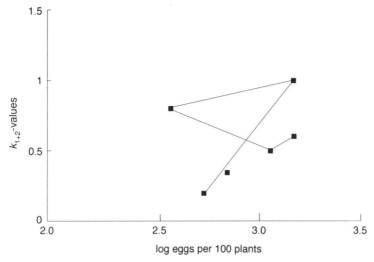

Fig. 4.12. Density relationships of egg plus early larval mortality (k_{1+2}). (After Benson, 1973.)

which is left unseparated in the lifetable. This approach has been shown to be inadequate in many cases for contemporaneous factors. Elkinton *et al.* (1992) show how marginal attack rates and their corresponding *k*-values can be calculated from data typically available during the construction of a lifetable and demonstrate how their use alters the type of density-dependent relationship identified.

The demonstration of density dependence from census data is fraught with difficulties: in particular failure to detect it in no way proves its absence (Southwood, 1978). Precise studies on individual cohorts and experimental work with particular components of the population system would seem to be profitable ways of investigating its role in population dynamics.

4.7. Conclusions

The ability to estimate the size of an insect population and the parameters that define the population lies at the very heart of ecological entomology and pest management. The work covered in this chapter spans five decades. In the 1990s inclusion of, and reference to, some of the earlier techniques that were developed in the 1940s–1960s is a testament to their continued value to ecology and pest management. Many of the experimental techniques have altered very little in this time, but significant advances have been made in the techniques of analysis used, especially where sophisticated models have required computer-aided analysis. For instance, the analysis of population data in terms of the intrinsic rate of natural increase has now become widely utilized. It is the development of techniques of analysis that has been emphasized more than the experimental methods throughout this chapter, because these represent the greatest advances in the subject of methods for determining population estimates. It is likely that this trend will continue in the near future.

4.8. References

Adachi, I. and Korenaga, R. (1991) Fertility schedules of *Unaspis yanonensis* (Hemiptera: Diaspididae) in relation to daily temperature. *Researches on Population Ecology* 33(1), 57–68.

Aldyhim, Y.N. and Khalil, A.F. (1993) Influence of temperature and daylength on population development of *Aphis gossypii* on *Cucurbita pepo*. *Entomologia Experimentalis et Applicata* 67, 167–172.

Allsopp, P.G. (1981) Development, longevity and fecundity of the false wireworms *Pterohelaeus darlingensis* and *P. alternatus* I. Effect of constant temperatures. *Australian Journal of Zoology* 29, 605–619.

Allsopp, P.G., Daglish, G.J., Taylor, M.F.J. and Gregg, P.C. (1991) Measuring development of *Heliothis* species. In: Zalucki, M.P. (ed.) *Heliothis: Research Methods and Prospects*. Springer Verlag, New York, pp. 90–108.

Artigues, M., Avilla, J., Sarasua, M.J. and Albajes, R. (1992) Egg laying and host stage preference at constant temperatures in *Encarsia tricolor* [Hym.: Aphelinidae]. *Entomophaga* 37, 45–53.

Asante, S.K. and Danthanarayana, W. (1990) Laboratory rearing of woolly aphid, *Eriosoma lanigerum*. *Plant Protection Quarterly* 5(2), 52–54.

Asante, S.K., Danthanarayana, W. and Heatwole, H. (1991) Bionomics and population growth statistics of apterous virginoparae of woolly apple aphid, *Eriosoma lanigerum*, at constant temperatures. *Entomologia Experimentalis et Applicata* 60, 261–270.

Ashford, J.R., Read, K.L.Q. and Vickers, G.G. (1970) A system of stochastic models applicable to animal population dynamics. *Journal of Animal Ecology* 39, 29–50.

Asknes, D.L. and Hoisaeter, T.J. (1987) Obtaining life table data from stage-freqency distributional statistics. *Limnology and Oceanography* 32, 514–517.

Baker, C.R.B. (1980) Some problems in using meteorological data to forecast the timing of insect lifecycles. *Bull EOPP: Organ Eur Et Medterr Pour La Prot TX Des Plant (France)* 10(2), 83–91.

Baker, P.B., Selton, A.M. and Andaloro, J.T. (1982) Monitoring of diamondback moth (Lepidoptera: Yponomeutidae) in cabbage with pheromones. *Journal of Economic Entomology* 75, 1025–1028.

Barlow, C.A. (1961) On the biology and reproductive capacity of *Syrphus corollae* in the laboratory. *Entomologia Experimentalis et Applicata* 4, 91–100.

Bartlett, P.W. and Murray, A.W.A. (1986) Modelling adult survival in the laboratory of diapause and non-diapause Colorado beetle *Leptinotarsa decemlineata* (Coleoptera: Chrysomelidae) from Normandy, France. *Annals of Applied Biology* 108(3), 487–501.

Begon, F.J. (1979) *Investigating Animal Abundance*. Edward Arnold, London.

Begon, M. and Mortimer, M. (1981) *Population Ecology*. Blackwell Scientific Publications, Oxford.

Bellows, T.S. and Birley, M.H. (1981) Estimating developmental and mortality rates and stage recruitment from insect stage-frequency data. *Researches on Population Ecology* 23, 232–244.

Bellows, T.S., Ortiz, M., Owens, J.C. and Huddleston, E.W. (1982) A model for analysing insect stage-frequency data when mortality varies with time. *Researches on Population Ecology* 24, 142–156.

Benson, J.F. (1973) Population dynamics of cabbage root fly in Canada and England. *Journal of Applied Ecology* 10, 437–446.

Bernal, J. and González, D. (1993) Experimental assessment of a degree-day model for predicting the development of parasites in the field. *Journal of Applied Entomology* 116(5), 459–466.

Bintcliffe, E.J.B. and Wratten, S.D. (1982) Antibiotic resistance in potato cultivars to the aphid *Myzus persicae. Annals of Applied Biology* 100, 383–391.

Birch, L.C. (1948) The intrinsic rate of natural increase of an insect population. *Journal of Animal Ecology* 17, 15–26.

Birley, M. (1977) The estimation of insect density and instar survivorship functions from census data. *Journal of Animal Ecology* 46, 497–510.

Bishop, J.A. and Sheppard, P.M. (1973) An evaluation of two capture–recapture models using the technique of computer simulation. In: Bartlett, M.S. and Harris, R.W. (eds) *The Mathematical Theory of the Dynamics of Biological Populations.* Academic Press, London, pp. 235–252.

Blower, J.G., Cook, L.M. and Muggleton, J. (1981) *Estimating the Size of Animal Populations.* George Allen & Unwin, London.

Bonnemaison, L. (1951) Contribution a l'etude des facteurs provoquant l'apparition des formes ailees et sexuees chez les Aphidinae. *Annals Epiphyties* 2, 1–380.

Brakefield, P.M. (1982) Ecological studies on the butterfly *Maniola jurtina* in Britain. II. Population dynamics: the present position. *Journal of Animal Ecology* 51(3), 727–738.

Braner, M. (1988) Dormancy, dispersal and staged development: ecological and evolutionary aspects of structured populations in random environments. PhD Dissertation, Cornell University, New York.

Braner, M. and Hairston, N.G. (1989) From cohort data to life table parameters via stochastic modelling. In: McDonald, L.L., Mauly, B.F.J., Lockwood, J.A. and Logan, J.A. (eds) *Estimation and Analysis of Insect Populations.* Springer-Verlag Lecture Notes in Statistics 55. Springer-Verlag, Berlin, pp. 81–92.

Brodsgaard, H.F. (1994) Effect of photoperiod on the bionomics of *Frankliniella occidentalis. Zeitschrift fur Angewandte Entomologie* 117(5), 498–507.

Browning, T.O. (1952) The influence of temperature on the rate of development of insects, with special reference to the eggs of *Gryllas commodus. Australian Journal of Science Research B.* 5, 96–111.

Bulmer, M.G. (1975) The statistical analysis of density dependence. *Biometrics* 31, 901–911.

Butler, G.D. (1976) Bollworm development in relation to temperature and larval food. *Environmental Entomology* 5(3), 520–522.

Campbell, A., Frazer, B.D., Gilbert, N., Gutierrez, A.P. and Mackauer, M. (1974) Temperature requirements of some aphids and their parasites. *Journal of Applied Ecology* 11, 431–438.

Cannon, R.J.C. (1986) Summer populations of the cereal aphid *Metopolophium dirhodium* (Walker) on winter wheat: three contrasting years. *Journal of Applied Ecology* 23(1), 101–114.

Castle, S.J. and Berger, P.H. (1993) Rates of growth and increase of *Myzus persicae* on virus-infected potatoes according to type of virus–vector relationship. *Entomologia Experimentalis et Applicata* 69, 51–60.

Caswell, H. and Hastings, A. (1980) Fecundity, developmental time, and population growth rate: an analytical solution. *Theoretical Population Biology* 17, 71–79.

Chang, Kwang-S. and Moritomo, N. (1988) Life table studies of the walnut leaf beetle, *Gastrolina depressa* (Coleoptera: Chrysomelidae), with special attention to aggregation. *Research in Population Ecology* 30, 297–313.

Chua, T.H. (1992) Relationship between female body size and demographic parameters in *Bactrocera* Malaysian A (Diptera: Tephritidae). *Research in Population Ecology* 34, 285–292.

Clements, A.N. and Paterson, G.D. (1981) The analysis of mortality and survival rates in wild populations of mosquitoes. *Journal of Applied Ecology* 18, 373–399.

Collier, R.H. and Finch, S. (1985) Accumulated temperatures for predicting the time of emergence in the spring of the cabbage root fly, *Delia radicum* (L.) (Diptera: Anthomyiidae). *Bulletin of Entomological Research* 75, 395–404.

Dancer, B.N. and Varlez, S. (1992) Applications of microbial pesticides in integrated management of pests of olives. In: Haskell, P.T. (ed.) *Monograph No.52 Research Collaboration in European IPM Systems*. BCPC, Brighton, pp. 13–21.

Danthanarayana, W. (1983) Population ecology of the light brown apple moth, *Epiphyas postvittana* (Lepidoptera: Tortricidae). *Journal of Animal Ecology* 52(1), 1–33.

Danthanarayana, W. and Gu, H. (1991) Multiple mating and its effect on the reproductive success of female *Epiphyas postvittana* (Lepidoptera: Tortricidae). *Ecological Entomology* 16, 169–175.

Dean, G.J. (1974) Effect of temperature on the cereal aphids *Metopolophium dirhodum* (Wlk.), *Rhopalosiphum padi* (L.) and *Macrosiphum avenue* (F.) (Hem., Aphididae). *Bulletin of Entomological Research* 63, 401–409.

Deevey, E.S. (1947) Lifetables for natural populations of animals. *Quarterly Review of Biology* 22, 283–314.

DeLoach, C.J. (1974) Rate of increase of populations of cabbage, green peach and turnip aphids at constant temperatures. *Annals of the Entomological Society of America* 67, 332–361.

Dempster, J.P. (1956) The estimation of the number of individuals entering each stage during the development of one generation of an insect population. *Journal of Animal Ecology* 25, 1–5.

Dempster, J.P. (1961) The analysis of data obtained by regular sampling of an insect population. *Journal of Animal Ecology* 30, 429–432.

Dennis, B. and Kemp, W.P. (1988) Further statistical inference methods for a stochastic model of insect phenology. *Environmental Entomology* 17, 887–893.

Dennis, B., Kemp, W.P. and Beckwith, R.C. (1986) Stochastic model of insect phenology: estimation and testing. *Environmental Entomology* 15, 540–546.

Dent, D. (1995) *Integrated Pest Management*. Chapman & Hall, London.

Dent, D.R. (1991) *Insect Pest Management*, CAB International, Wallingford, Oxon.

Dent, D.R. and Wratten, S.D. (1986) The host-plant relationships of apterous virginoparae of the grass aphid *Metopolophium festucae cerealium*. *Annals of Applied Biology* 108, 1–10.

Derr, J.A. and Ord, K. (1979) Field estimates of insect colonization. *Journal of Animal Ecology* 48, 521–534.

Dewar, A.M. (1977) Assessment of methods for testing varietal resistance to aphids in cereals. *Annals of Applied Biology* 87, 183–190.

Dixon, A.F.G. (1987) Parthogenetic reproduction and the rate of increase in aphids. In: Minks, A.K. and Harrewijn, P. (eds) *Aphids, their Biology, Natural Enemies and Control,*. Elsevier, Amsterdam, pp. 269–287.

Dixon, A.F.G. and Wratten, S.D. (1971) Laboratory studies on aggregration, size and fecundity in the black bean aphid, *Aphis fabae* Scop. *Bulletin of Entomological Research* 61, 97–111.

Dublin, L.I. and Lotka, A.J. (1925) On the time rate of natural increase as exemplified by the population of the United States, 1920. *Journal of American Statistical Association* 20, 305–339.

Easwaramoorthy, S. and Nandagopal, V. (1986) Life tables of internode borer, *Chilo sacchariphagus indicus* (K.), on resistant and susceptible varieties of sugarcane. *Tropical Pest Management* 32(3), 221–228.

Eberhardt, L.L. (1978) Transect methods for population studies. *Journal of Wildlife Management* 42, 1–31.

Elkinton, J.S., Buonaccorsi, J.P., Bellows, T.S. and Van Driesche, R.G. (1992) Marginal attack rate, *k*-values and density-dependence in the analysis of contemporaneous mortality factors. *Research in Population Ecology* 34, 29–44.

Elliott, N.C. and Kieckhefer, R.W. (1989) Effects of constant and fluctuating temperatures on immature development and age-specific life tables of *Rhopalosiphum padi* (L.) (Homoptera: Aphididae). *The Canadian Entomologist* 121, 131–140.

Emmet, A.M. and Heath, J. (1989) *The Moths and Butterflies of Great Britain and Ireland*. Harley Books, Colchester.

Enkegaard, A. (1993a) *Encarsia formosa* parasitizing the Poinsettia-strain of the cotton whitefly, *Bemisia tabaci*, on Poinsettia: bionomics in relation to temperature. *Entomologia Experimentalis et Applicata* 69, 251–261.

Enkegaard, A. (1993b) The poinsettia strain of the cotton whitefly, *Bemisia tabaci* (Homoptera: Aleyrodidae), biological and demographic parameters on poinsettia (*Euphorbia pulcherrima*) in relation to temperature. *Bulletin of Entomological Research* 83, 535–546.

Faeth, S.H. (1990) Aggregation of a leafminer, *Cameraria* sp. Nov. (Davis): Consequences and causes. *Journal of Animal Ecology* 59, 569–586.

Fisher, R.A. (1921) Some remarks on the methods formulated in a recent article on 'The quantitative analysis of plant growth'. *Annals of Applied Biology* 7, 367–372.

Fisher, R.A. and Ford, E.B. (1947) The spread of a gene in natural conditions in a colony of the moth *Panaxia dominula* L. *Heredity* 1(II), 143–174.

Foley, D.H. (1981) Pupal development rate of *Heliothis armigera* under constant and fluctuating temperatures. *Journal of Australian Entomological Society* 20, 13–20.

Fox, C.W. (1993) Multiple mating, lifetime fecundity and female mortality of the bruchid beetle, *Callosobruchus maculatus* (Coleopetera: Bruchidae). *Functional Ecology* 7, 203–208.

Gargiullo, P.M., Berisford, C.W., Canalos, C.G., Richmond, J.A. and Cade, S.C. (1984) Mathematical descriptions of *Rhyacionia frustrana* (Lepidoptera: Tortricidae) cumulative catches in pheromone traps, cumulative eggs hatching and their use in timing of chemical control. *Environmental Entomology* 13(6), 1681–1685.

Gerber, G.H., Bodnaryk, R.P. and Walkof, J. (1991) Effects of elevated, sub-lethal temperatures during pupal–adult development of male Bertha armyworm *Manestia configurata* on fertility, mating and egg laying. *Invertebrate Reproduction and Development* 19, 213–224.

Gilbert, N. (1984a) Control of fecundity in *Pieris rapae* I. The problem. *Journal of Animal Ecology* 53, 581–588.

Gilbert, N. (1984b) Control of fecundity in *Pieris rapae* II. Differential effects of temperature. *Journal of Animal Ecology* 53, 589–597.

Gilbert, N. (1984c) Control of fecundity in *Pieris rapae* III. Synthesis. *Journal of Animal Ecology* 53, 599–609.

Gilbert, N. (1988) Control of fecundity in *Pieris rapae* V. Comparisons between populations. *Journal of Animal Ecology* 57, 395–410.

Glen, P.A. (1922) Relation of temperature to development of the codling moth. *Journal of Economic Entomology* 15, 193–199.

Hairston, N.G. and Twombly, S. (1985) Obtaining life table data from cohort analysis – a critique of current methods. *Limnology and Oceanography* 30, 886–893.

Hairston, N.G., Braner, M. and Twombly, S. (1987) Perspective on prospective methods for obtaining life table data. *Limnology and Oceanography* 32, 517–520.

Hance, T., Nibelle, D., Lebrun, P., Van Impe, G. and Van Hove, C. (1994) Selection of *Azolla* forms resistant to the water lily aphid, *Rhopalosiphum nymphaceae*. *Entomologia Experimentalis et Applicata* 70, 11–17.

Heinrichs, E.A., Medrano, F.G. and Rapusas, H.R. (1985) *Genetic Evaluation for Insect Resistance in Rice*. International Rice Research Institute, Los Banos, Philippines.

Hilbert, D.W. and Logan, J.A. (1983) Empirical model of nymphal development for the migratory grasshopper *Melanopus sanguinipes*. *Environmental Entomology* 12, 1–5.

Horn, D.J. (1988) *Ecological Approach to Pest Management*. Elsevier Applied Science Publishers, London.

Howe, R.W. (1953) The rapid determination of the intrinsic rate of increase of an insect population. *Annals of Applied Biology* 40, 134–151.

Howe, R.W. (1967) Temperature effects on embryonic development in insects. *Annual Review of Entomology* 12, 15–42.

Howe, R.W. (1971) A parameter for expressing the suitability of environment for insect development. *Journal of Stored Products Research* 7, 63–65.

Hughes, R.D. and Woolcock, L.T. (1965) A modification of Johnson's method of rearing aphids for ecological studies. *New Zealand Journal of Agricultural Research* 8, 728–736.

Hyde, R.R. (1914) *Journal of Experimental Zoology* 17, 141–172.

Ito, Y. (1972) On the methods for determining density-dependence by means of regression. *Oecologia (Berl.)* 10, 347–372.

Iwao, S. (1968) A new regression method for analyzing the aggregation pattern of animal populations. *Research Population Ecology* 10, 1–20.

Iwao, S. (1970) Analysis of contagiousness in the action of mortality factors on the western tent caterpillar by using the m–m relationship. *Research in Population Ecology* 12, 100–110.

Jervis, M.A. and Copland, M.J.W. (1996) The lifecycle. In: *Insect Natural Enemies: Practical approaches to their study and evaluation.* Chapman & Hall, London, pp. 63–160.

Johnson, K.B., Teng, P.S. and Radcliffe, E.B. (1987) Coupling feeding effects on potato leafhopper, *Empoasca fabae* (Homoptera: Cicadellidae), nymphs to a model of potato growth. *Environmental Entomology* 16, 250–258.

Jolly, G.M. (1965) Explicit estimates from capture–recapture data with both death and immigration – stochastic model. *Biometrika* 52, 225–247.

Kaakeh, W. and Dutcher, J.D. (1992) Estimation of life parameters of *Monelliopsis pecanis, Monellia caryella* and *Melanocallis caryaefolicae* on single pecan leaflets. *Environmental Entomology* 21, 632–639.

Kaakeh, W. and Dutcher, J.D. (1993) Rates of increase and probing behaviour of *Acyrthosiphon pisum. Environmental Entomology* 22, 1016–1021.

Kay, D.J., Wratten, S.D. and Stokes, S. (1981) Effects of vernalisation and aphid culture history on the relative susceptibilities of wheat cultivars to aphids. *Annals of Applied Biology* 99, 71–75.

Kemp, W.P., Dennis, B. and Beckwith, R.C. (1986) Stochastic phenology model for the western spruce budworm (Lepidoptera: Tortricidae). *Environmental Entomology* 15, 547–554.

Kemp, W.P., Dennis, B. and Munholland, P.L. (1989) Modelling grasshopper phenology with diffusion processes. In: McDonald, L.L., Manly, B.F.J., Lockwood, J.A. and Logan, J.A. (eds) *Estimation and Analysis of Insect Populations*. Springer-Verlag Lecture Notes in Statistics 55. Springer-Verlag, Berlin, pp. 118–127.

Kempton, R.A. (1979) Statistical analysis of frequency data obtained from sampling an insect population grouped by stages. In: Ord, J.K., Patil, G.P. and Taillie, C. (eds) *Statistical Distributions in Scientific Work*. International Cooperative Publishing House, Maryland, pp. 401–418.

Kfir, R. (1981) Fertility of the polyembryonic parasite *Copidosoma koehleri*, effect of humidities on life length and relative abundance as compared with that of *Apanteles subandinus* in potato tuber moth. *Annals of Applied Biology* 99, 225–230.

Kiritani, K. and Nakasuji, F. (1967) Estimation of the stage-specific survival rate in the insect population with overlapping stages. *Researches on Population Ecology* 9, 143–152.

Kitching, R.L. (1977) Time, resources and population dynamics in insects. *Australian Journal of Ecology* 2, 31–42.

Kobayashi, S. (1968) Estimation of the individual number entering each development stage in an insect population. *Researches on Population Ecology* 10, 40–44.

Kocourek, F., Havelka, J., Berankova, J. and Jarosik, V. (1994) Effect of temperature on development rate and intrinsic rate of increase of *Aphis gossypii* reared on greenhouse cucumbers. *Entomologia Experimentalis et Applicata* 71, 59–64.

Kogan, M. (1986) Bioassays for measuring quality of insect food. In: Miller, J.R. and Miller, T.A. (eds) *Insect Plant Interactions*. Springer Series in Experimental Entomology. Springer-Verlag, New York, pp. 155–189.

Krebs, C.J. (1978) *Ecology: The Experimental Analaysis of Distribution and Abundance*. Harper Row, New York, 678pp.

Lakhani, K.H. and Service, M.W. (1974) Estimating mortalities of the immature stages of *Aedes cantans* (Mg.) (Diptera: Culicidae) in a natural habitat. *Bulletin of Entomological Research* 64, 265–276.

Laughlin, R. (1965) Capacity for increase: a useful population statistic. *Journal of Animal Ecology* 34, 77–91.

Lawless, J.F. (1982) *Statistical Models and Methods for Lifetime Data*. John Wiley & Sons, New York.

Leather, S.R. and Dixon, A.F.G. (1981) Growth, survival and reproduction of the bird-cherry aphid, *Rhopalosiphum padi*, on its primary host. *Annals of Applied Biology* 99, 115–118.

Leather, S.R. and Dixon, A.F.G. (1982) Secondary host preference and reproductive activity of the bird cherry-oat aphid, *Rhopalosiphum padi. Annals of Applied Biology* 101, 219–228.

Leather, S.R. and Wellings, P.W. (1981) Ovariole number and fecundity in aphids. *Entomologia Experimentalis et Applicata* 30, 128–133.

Leather, S.R., Watt, A.D. and Barbour, D.A. (1985) The effect of host-plant and delayed mating on the fecundity and lifespan of the pine beauty moth, *Panolis flammea* (Denis and Shiffermuller) (Lepidoptera: Noctuidae): their influence on population dynamics and relevance to pest management. *Bulletin of Entomological Research* 75, 641–651.

Lewontin, R.C. (1965) Selection for colonizing ability. In: Baker, H.G. and Stebbins, G.L. (eds) *The Genetics of Colonizing Species.* Academic Press, New York, pp. 77–94.

Logan, J.A., Wollking, D.J., Hoyt, S.C. and Tanigoshi, L.K. (1976) An analytical model for description of temperature dependent rate phenomena in arthropods. *Environmental Entomology* 5, 1133–1140.

Madden, L.V. (1985) Modelling the population dynamics of leafhoppers. In: Nault, L.R. and Rodriquez, J.G. (eds) *The Leafhopper and Planthopper.* John Wiley & Sons, New York, pp. 235–238.

Madden, L.V. and Nault, L.R. (1983) Differential pathogenicity of corn stunting mollicutes to leafhopper vectors in *Dalbulus* and *Baldulus* species. *Phytopathology* 73, 1608–1614.

Madden, L.V., Nault, L.R., Heady, S.E. and Styer, W.E. (1984) Effect of maize stunting mollicutes on survival and fecundity of *Dalbulus* leafhopper vectors. *Annals of Applied Biology* 105, 431–441.

Madden, L.V., Nault, L.R., Heady, S.E. and Styer, W.E. (1986) Effect of temperature on the population dynamics of three *Dalbulus* leafhopper species. *Annals of Applied Biology* 108, 475–485.

Manly, B.F.J. (1974) Estimation of stage-specific survival rates and other parameters for insect populations passing through stages. *Oecologia (Berl.)* 15, 277–285.

Manly, B.F.J. (1976) Extensions to Kiritani and Nakasuji's method for the analysis of stage frequency data. *Researches on Population Ecology* 17, 191–199.

Manly, B.F.J. (1977a) The determination of key factors from life table data. *Oecologia (Berl.)* 31, 111–117.

Manly, B.F.J. (1977b) A further note on Kiritani and Nakasuji's model for stage-frequency data including comments on Tukey's jackknife technique for estimating variances. *Researches on Population Ecology* 18, 177–186.

Manly, B.F.J. (1985) *The Statistics of Natural Selection on Animal Populations.* Chapman & Hall, London.

Manly, B.F.J. (1987) A regression method for analysing stage-frequency data when survival rates vary from stage to stage. *Researches on Population Ecology* 29, 119–127.

Manly, B.F.J. (1989) A review of methods for the analysis of stage-frequency data. In: McDonald, L.L., Manly, B.F.J., Lockwood, J.A. and Logan, J.A. (eds) *Estimation and Analysis of Insect Populations.* Springer-Verlag Lecture Notes in Statistics 55. Springer-Verlag, Berlin, pp. 3–69.

Manly, B.F.J. (1990) *Stage Structured Populations: Sampling Analysis and Simulation.* Chapman & Hall, London.

Manly, B.F.J. (1993) Note on a method for analysing stage-frequency data. *Research in Population Ecology* 35, 215–222.

Manly, B.F.J. and Parr, M.J. (1968) A new method for estimating population size, survivorship and birth rate from capture–recapture data. *Transactions of the Society for British Entomology* 18, 81–89.

Manly, B.F. and Seyb, A. (1989) A comparison of three maximum likelihood models for stage-frequency data. *Researches on Population Ecology* 31(2), 367–380.

May, R.M. (1976) Estimating *r*: a pedagogical note. *The American Naturalist* 110, 496–499.

Mbata, G.N. (1990) Suitability of maize varieties for the oviposition and development of *Plodia interpunctella* (Hubner) (Lepidoptera: Pyralidae). *Tropical Pest Management* 36(2), 122–127.

McCullagh, P. and Nelder, J.A. (1983) *Generalized Linear Models*. Chapman & Hall, London.

Messenger, P.S. (1964) Use of life tables in a bioclimatic study of an experimental aphid–braconid wasp host–parasite system. *Ecology* 45, 119–131.

Meyer, J.S., Ingersoll, C.G., McDonald, L.L. and Boyce, M.S. (1986) Estimating uncertainty in population growth rates: jackknife vs. bootstrap techniques. *Ecology* 67(5), 1156–1166.

Mills, N.J. (1981a) The estimation of mean duration from stage frequency data. *Oecologia (Berl.)* 51, 206–211.

Mills, N.J. (1981b) The estimation of recruitment from stage frequency data. *Oecologia (Berl.)* 51, 212–216.

Minkenberg, O.P.J.M. and Helderman, C.A.J. (1990) Effects of temperature on the life history of *Liriomyza bryoniae* (Diptera: Agromyzidae) on tomato. *Journal of Economic Entomology* 83(1), 117–125.

Moralesranos, J.A. and Cate, J.R. (1992) Rate of increase and adult longevity of *Catolaccus grandis* in the laboratory at four temperatures. *Environmental Entomology* 21, 620–627.

Morris, R.F. (1959) Single factor analysis in population dynamics. *Ecology* 40, 580–588.

Munguira, M.L. and Thomas, J.A. (1992) Use of road verges by butterfly and burnet populations and the effects of roads on adult dispersal and mortality. *Journal of Applied Ecology* 29(2), 316–329.

Munholland, P.L. (1988). Statistical aspects of field studies on insect populations. PhD Dissertation, University of Waterloo, Ontario, Canada.

Munholland, P.L., Brennan, L.A. and Block, W.M. (1989) Arthropod sampling methods in ornithology: goals and pitfalls. In: McDonald, L.L., Manly, B.F.J., Lockwood, J.A. and Logan, J.A. (eds) *Estimation and Analysis of Insect Populations*. Springer-Verlag Lecture Notes in Statistics 55. Springer-Verlag, Berlin, pp. 484–492.

Neuenschwander, P., Michelakis, S. and Bigler, F. (1981) Abiotic factors affecting mortality of *Dacus oleae* larvae and pupae in the soil. *Entomologia Experimentalis et Applicata* 30, 1–9.

Noble, M.D. (1958) A simplified dip cage for aphid investigations. *The Canadian Entomologist* 90, 760.

Olatunde, G.O. and Odebiyi, J.A. (1991) Some aspects of antibiosis in cowpeas resistant to *Clavigralla tomentosicollis* Stal. (Hemiptera: Coreidae) in Nigeria. *Tropical Pest Management* 37(3), 273–276.

Omer, A.D., Leigh, T.F., Carey, J.R. and Granett, J. (1992) Demographic analyses of organophosphate-resistant and susceptible strains of greenhouse whitefly, *Trialeurodes vaporariourum*, on three cotton cultivars. *Entomologia Experimentalis et Applicata* 65, 21–30.

Osawa, A., Shoemaker, C.A. and Stedinger, J.R. (1983) A stochastic model of balsam fir bud phenology utilizing maximum likelihood parameter estimation. *Forestry Science* 29, 478–490.

Panda, N. and Khush, G.S. (1995) *Host Plant Resistance to Insects*. CAB International, Wallingford, 431pp.

Pearl, R. (1928) *The Rate of Living*. Knopf, New York.

Peferoen, M., Huybrechts, R. and De Loof, A. (1981) Longevity and fecundity in the Colorado potato beetle, *Leptinotarsa decemlineata*. *Entomologia Experimentalis et Applicata* 29, 321–329.

Pielou, E.C. (1974) *Population and Community Ecology: Principles and Methods*. Gordon and Breach Science Publishers, New York, 424pp.

Podoler, H. and Rogers, D. (1975) A new method for the identification of key factors from life-table data. *Journal of Animal Ecology* 44, 85–114.

Pollard, E. (1984) Synoptic studies on butterly abundance. In: Vane-Wright, R.I. and Ackery, P.R. (eds) *The Biology of Butterflies, Symposium of the Royal Entomological Society of London, no.11*. Academic Press, London, pp. 59–61.

Pollard, E. (1977) A method of assessing changes in the abundance of butterflies. *Biological Conservation* 12, 115–134.

Pollard, E., Moss, D. and Yates, T.J. (1995) Population trends of common British butterflies at monitored sites. *Journal of Applied Ecology* 32(1), 9–16.

Pradhan, S. (1946) Insect population studies IV. Dynamics of temperature effect on insect development. *Proceedings of the National Institute of Science, India* 12, 385–404.

Pretorius, L.M. (1976) Laboratory studies on the development and reproductive performance of *Heliothis armigera* on various host plants. *Journal of Entomological Society of South Africa* 39, 337–343.

Pruess, K.P. (1983) Day-degree methods for pest management. *Environmental Entomology* 12(3), 613–619.

Radford, P.J. (1967) Growth analysis formulae – their use and abuse. *Crop Science* 7, 171–175.

Read, L.K.Q. and Ashford, J.R. (1968) A system of models for the life cycle of a biological organism. *Biometrika* 55, 211–221.

Readshaw, J.L. and Van Gerwen, A.C.M. (1983) Age-specific survival, fecundity and fertility of the adult blowfly, *Lucilia cuprina*, in relation to crowding, protein food and population cycles. *Journal of Animal Ecology* 52(3), 879–887.

Reissig, W.H., Barnard, J., Weires, R.W., Glass, E.H. and Dean, R.W. (1979) Prediction of apple maggot fly emergence from thermal unit accumulation. *Environmental Entomology* 8(1), 51–54.

Richards, F.J. (1959) A flexible growth function for empirical use. *Journal of Experimental Botany* 10(29), 290–300.

Richards, O.W. and Waloff, N. (1954) Studies on the biology and population dynamics of British grasshoppers. *Anti-Locust Bulletin* 17, 1–182.

Richards, O.W., Waloff, N. and Spradbery, J.P. (1960) The measurement of mortality in an insect population in which recruitment and mortality widely overlap. *OIKOS* 11, 306–310.

Rigler, F.H. and Cooley, J.M. (1974) The use of field data to derive population statistics of mulitvoltine copepods. *Limnology and Oceanography* 19, 636–655.

Rockstein, M. and Miguel, J. (1973) Ageing in insects. In: Rockstein, M. (ed.) *The Physiology of Insects*. Academic Press, London, pp. 371–471.

Romanow, L.R., de Ponti, O.M.B. and Mollema, C. (1991) Resistance in tomato to the greenhouse whitefly: analysis of population dynamics. *Entomologia Experimentalis et Applicata* 60, 247–259.

Room, P.M. (1977) 32P-labelling of immature stages of *Heliothis armigera* and *H. punctigoa*: relationships of dose to radioactivity, mortality and label half-life. *Journal of Australian Entomological Society* 16, 245–251.

Room, P.M. (1983) Calculations of temperature-driven development by *Heliothis* spp. in Nairobi Valley, New South Wales. *Journal of Australian Entomological Society* 22, 211–215.

Room, P.M., Titmarsh, I.J. and Zalucki, M.P. (1991) Life tables. In: Zalucki, M.P. (ed.) *Heliothis: Research Methods and Prospects*. Springer-Verlag, Berlin, pp. 69–80.

Royer, L. and McNeil, J.N. (1993) Male investment in the European corn borer, *Ostrinia nubilalis* (Lepidoptera: Pyralidae): impact on female longevity and reproductive performance. *Functional Ecology* 7, 209–215.

Ruesink, W.G. (1975) Estimating time-varying survival of arthropod life stages from population density. *Ecology* 56, 244–247.

Saunders, J.F. and Lewis, W.M. (1987) A perspective on the use of cohort analysis to obtain demographic data for copepods. *Limnology and Oceanography* 32, 511–513.

Sawyer, A.J. and Haynes, D.L. (1984) On the nature of errors involved in estimating stage-specific survival rates by Southwood's method for a population with overlapping stages. *Researches on Population Ecology* 26, 331–351.

Seber, G.A.F. (1965) A note on the multiple-recapture census. *Biometrika* 52, 249–259.

Seber, G.A.F. (1982) *Estimation of Animal Abundance and Related Parameters*, 2nd edn. Griffin, London.

Sekhon, S.S. and Sajjan, S.S. (1987) Antibiosis in maize (*Zea mays*) (L.) to maize borer, *Chilo partellus* (Swinhoe) (Pyralidae: Lepidoptera) in India. *Tropical Pest Management* 33(1), 55–60.

Sengonca, C., Hoffman, A. and Kleinhenz, B. (1994) Investigations on development, survival and fertility of the cereal aphids *Sitobion avenae* and *Rhopalosiphum padi* at different low temperatures. *Zeitschrift fur Angewandte Entomologie* 117, 224–233.

Serit, M. and Keng-Hong, T. (1990) Immature life table of a natural population of *Dacus dorsalis* in a village ecosystem. *Tropical Pest Management* 36(3), 305–309.

Shanower, T.G., Gutierrez, A.P. and Wightman, J.A. (1993) Effect of temperature on development rates, fecundity and longevity of the groundnut leaf miner, *Aproaerema modicella* (Lepidoptera: Gelechiidae), in India. *Bulletin of Entomological Research* 83, 413–419.

Shoemaker, C.A., Smith, G.E. and Helgesen, R.G. (1986) Estimation of recruitment rates and survival from field census data with application to poikilotherm populations. *Agricultural Systems* 22, 1–21.

Siddiqui, W.H. and Barlow, C.A. (1973) Effects of some constant and alternating temperatures on population growth of the pea aphid *Acyrthosiphon pisum* (Homoptera: Aphididae). *The Canadian Entomologist* 105, 145–156.

Siddiqui, W.H., Barlow, C.A. and Randolph, P.A. (1973) Effects of some constant and alternating temperatures on population growth of the pea aphid *Acyrthosiphon pisum* (Homoptera: Aphididae). *Canadian Entomologist* 105, 145–156.

Simon, J.C., Dedryver, C.A., Pierre, J.S., Tanguy, S. and Wegorek, P. (1991) The influence of clone and morph on the parameters of intrinsic rate of increase in the cereal aphids *Sitobion avenae* and *Rhopalosiphum padi*. *Entomologia Experimentalis et Applicata* 58, 211–220.

Slobodkin, L.B. (1962) *Growth and Regulation of Animal Populations*. Holt, Rinehart & Winston, New York.

Smith, R.H. (1973) The analysis of intra-generation change in animal populations *Journal of Animal Ecology* 42, 611–622.

Soroka, J.J. and Mackay, P.A. (1991) Antibiosis and antixenosis to pea aphid (Homoptera: Aphididae) in cultivars of field peas. *Journal of Economic Entomology* 84(6), 1951–1956.

Southwood, T.R.E. (1978) *Ecological Methods with Particular Reference to the Study of Insect Populations*, 2nd edn. University Printing House, Cambridge.

Southwood, T.R.E. and Jepson, W.F. (1962) Studies on the populations of *Oscinella frit* L. (Dipt.: Chloropidae) in the oat crop. *Journal of Animal Ecology* 31, 481–495.

Speight, M.R. (1994) Reproductive capacity of the horse chestnut scale insect *Pulvinaria regalis*. *Zeitschrift fur Angewandte* 118, 59–67.

Stedinger, J.R., Shoemaker, C. and Tenga, R.F. (1985) A stochastic model of insect phenology for a population with spatially variable development rates. *Biometrics* 41, 691–701.

Stinner, R.E. Gutierrez, A.P. and Butler, G.P. (1974) An algorithm for temperature-dependent growth rate simulation. *Canadian Entomologist* 106, 519–524.

Subramanyam, B. and Hagstrum, D.W. (1991) Quantitative analysis of temperature, relative humidity, and diet influencing development of the larger grain borer, *Prostephanus truncatus* (Horn) (Coleoptera: Bostrichidae). *Tropical Pest Management* 37(3), 195–202.

Tamhankar, A.J. (1995) Host influence on mating-behaviour and spermatophore reception correlated with reproductive output and longevity of female *Earias insulana*. *Journal of Insect Behaviour* 8, 499–511.

Taylor, L.R., Woiwod, I.P. and Perry, J.N. (1978) The density-dependence of spatial behaviour and the variety of randomness. *Journal of Animal Ecology* 47, 383–406.

Teakle, R.E. (1991) Laboratory culture of *Heliothis* species and identification of disease. In: Zalucki, M.P. (ed.) *Heliothis: Research Methods and Prospects*. Springer-Verlag, New York, 234pp.

Thomas, J.A. (1983) A quick method for estimating butterfly numbers during surveys. *Biological Conservation* 27, 195–211.

Tokeski, M. (1985) Life-cycle and production of the burrowing mayfly, *Ephemera danica*: a new method for estimating degree-days required for growth. *Journal of Animal Ecology* 54, 919–930.

Unnithan, G.C. and Page, S.O. (1991) Mating, longevity, fecundity and egg fertility of *Chilo partellus* – effects of delayed or successive matings and their relevance to pheromonal control methods. *Environmental Entomology* 20, 150–155.

Van Emden, H.F. (1969) Plant resistance to *Myzus persicae* induced by a plant regulator and measured by aphid relative growth rate. *Entomologia Experimentalis et Applicata* 12, 125–131.

van Helden, M., Tjallingii, W.F. and Dieleman, F.L. (1993) The resistance of lettuce (*Lactuca sativa* L.) to *Nasonovia ribisnigri*: bionomics of *N. ribisnigri* on near isogenic lettuce lines. *Entomologia Experimentalis et Applicata* 66, 53–58.

Vansteenis, M.J. (1993) Intrinsic rate of increase of *Aphidius colemani*, a parasitoid of *Aphis gossypii* at different temperatures. *Zeitschrift fur Angenwandte Entomologie* 116, 192–198.

van Straalen, N.M. (1982) Demographic analysis of arthropod populations using a continuous stage-variable. *Journal of Animal Ecology* 51, 769–783.

van Straalen, N.M. (1985) Comparative demography of forest floor *Collembola* populations. *OIKOS* 45, 253–265.

Varley, G.C. and Gradwell, G.R. (1960) Key factors in population ecology. *Journal of Animal Ecology* 29, 399–401.

Varley, G.C. and Gradwell, G.R. (1963) Predatory insects as density-dependent mortality factors. In: *Proceedings of 16th International Congress of Zoology*, p. 240.

Varley, G.C. and Gradwell, G.R. (1965) Interpreting winter moth population changes. In: *Proceedings of XII International Congress of Zoology*, pp.377–378.

Varley, G.C. and Gradwell, G.R. (1970) Recent advances in insect population dynamics. *Annual Review of Entomology* 15, 1–24.

Varley, G.C., Gradwell, G.R. and Hassell, M.P. (1973) *Insect Population Ecology: An Analytical Approach.* Blackwell Scientific Publications, Oxford.

Wagner, T.L., Wu, H., Sharpe, P.J.H., Schoolfield, R.M. and Coulson, R.N. (1984) Modelling insect development rates: A literature review and application of a biophysical model. *Annals of the Entomological Society of America* 77, 208–225.

Walgenbach, D.D., Elliott, N.C. and Kieckhefer, R.W. (1988) Constant and fluctuating temperature effects on developmental rates and life table statistics of the greenbug (Homoptera: Aphididae). *Journal of Economic Entomology* 81(2), 501–507.

Wallin, H., Chiverton, P.A., Ekbom, B.S. and Borg, A. (1992) Diet, fecundity and egg size in some polyphagous predatory carabid beetles. *Entomologia Experimentalis et Applicata* 65, 129–140.

Welch, S.M., Croft, B.A. and Michels, M.F. (1981) Validation of pest management models. *Environmental Entomology* 10, 425–432.

Whalon, M.E. and Smilowitz, Z. (1979) The interaction of temperature and biotype on development of the green peach aphid *Myzus persicae*. *American Potato Journal* 56, 591–596.

Willers, J.L., Schneider, J.C. and Ramaswamy, S.B. (1987) Fecundity, longevity and caloric patterns in female *Heliothis virescens*: Changes with age due to flight and supplemental carbohydrate. *Journal of Insect Physiology* 33(11), 803–808.

Willmer, P.G. (1982) Microclimate and the environmental physiology of insects. *Advances in Insect Physiology* 16, 1–57.

Wratten, S.D. (1977) Reproductive strategy of winged and wingless morphs of the aphids *Sitobion avenae* and *Metopolophium dirhodum*. *Annals of Applied Biology* 85, 319–331.

Wyatt, I.J. and Brown, S.J. (1977) The influence of light intensity, daylength and temperature on increased rates of glasshouse aphids. *Journal of Applied Ecology* 14, 391–399.

Wyatt, I.J. and White, P.F. (1977) Simple estimation of intrinsic increase rates for aphids and tetranychid mites. *Journal of Applied Ecology* 14, 757–766.

Yang, P.J., Carey, J.R. and Dowell, R.V. (1994) Temperature influences on the development and demography of *Bactrocera dorsalis* in China. *Environmental Entomology* 23(4), 971–974.

Yu, D.S., Luck, R.F. and Murdoch, W.W. (1990) Competition, resource partitioning and coexistence of an endoparasitoid *Encarsia perniciosi* and an ectoparasitoid *Aphytis melinus* of the California red scale. *Ecological Entomology* 15, 469–480.

Zhou, X. and Carter, N. (1992) Effects of temperature, feeding position and crop growth stage on the population dynamics of the rose grain aphid *Metopolophum dirhodum*. *Annals of Applied Biology* 121, 27–37.

Zonneveld, C. (1991) Estimating death rates from transect counts. *Ecological Entomology* 16, 115–121.

5 Techniques for Quantifying Insect Migration

D.R. Reynolds, J.R. Riley, N.J. Armes, R.J. Cooter, M.R. Tucker and J. Colvin
Natural Resources Institute, Central Avenue, Chatham Maritime, Chatham, Kent ME4 4TB, UK

5.1. Introduction

The spectacular nature of mass migrations of day-flying insects such as butterflies, locusts and dragonflies has amazed and even awed man since early times (Williams, 1965), and insect migration continues to fascinate both the amateur and professional entomologist today. A year seldom passes without the appearance of a learned review on some aspect of the phenomenon. Since 1980, for example, review papers by Taylor and Taylor (1983), Rankin and Singer (1984), Dingle (1985), Farrow (1990), Rankin and Burchsted (1992), Roderick and Caldwell (1992), and Pedgley (1993) have been published, and there have been chapters in edited volumes such as Gauthreaux (1980), MacKenzie *et al.* (1985), Rankin (1985), Danthanarayana (1986), Goldsworthy and Wheeler (1989) and Drake and Gatehouse (1995), as well as several reviews dealing with the effect of weather on migration (see below).

5.1.1. Definition of migration and dispersal

Like other recent authors (e.g. Gatehouse, 1987; Farrow, 1990) we believe that migration is best viewed as a 'behavioural process with ecological consequences'. On-going *vegetative* movements (e.g. those concerned with feeding and with reproduction), or even accidental displacements, can cause the birthplace of offspring to be remote from that of their parents, and these movements therefore equate to migration in the *ecological* sense (Taylor and Taylor, 1983). However, most of the population redistribution between the birthplaces of successive generations (or between breeding and diapause/dormancy sites within a generation) results from specialized *migratory behaviour* usually occurring at a particular stage of the insect's development. Kennedy (1985) defines this behaviour as **'persistent and**

straightened-out movement effected by the animal's own locomotory exertions or by its active embarkation on a vehicle. It depends on some temporary inhibition of station-keeping responses, but promotes their eventual disinhibition and recurrence'. This definition seems to be widely accepted by entomologists, although Dingle (1989) notes that it is not clear whether the complex system of antagonistic inhibitory and excitatory responses, found by Kennedy for aphids (e.g. Kennedy, 1966; Kennedy and Ludlow, 1974), occurs generally in migratory insects.

Modern behavioural definitions have made the term 'dispersal' redundant as a synonym for migration, and we agree with Kennedy (1985) that its use in this sense should be discouraged. It is perhaps best reserved to describe the increase in separation between members of a population which may occur as an incidental result of their migratory and vegetative movements. Nonetheless, we have sometimes retained the term 'dispersal' when it has been used by authors quoted here, because it seems wiser not to 'second guess' the precise behavioural reason for the observed movements.

The sustained climb to high altitude by masses of insects at dusk, which has been graphically revealed by entomological radars (e.g. Schaefer, 1976), and the persistent uni-directional daytime flight of some butterflies (Baker, 1978; Walker, 1991) are unmistakably migratory. In other cases, however, it may be difficult to determine whether flights are migratory in the strict behavioural sense (Hardie, 1993), or are examples of *extended foraging* which would be promptly terminated whenever the sought-after resource were encountered (e.g. Blackmer and Phelan, 1991). For most practical purposes, however, these academic distinctions are of little consequence; a farmer whose crops have been invaded is unlikely to be much interested in whether extended foraging or migration had brought about the infestation!

5.2. Methods for Studying Migration

In this chapter we provide an introduction to the recent literature on methods for studying long-range, airborne insect movement, with particular emphasis on techniques for quantifying high-altitude migrations. Migration over very short distances, by either pedestrian, waterborne or airborne movements, is amenable to observation by the methods for investigating *non*-migratory behaviours (e.g. see Chapter 3).

We first consider methods for measuring insect movement 'in the wild' and have grouped these into five categories following Drake (1990): *mark and capture; capture of naturally marked specimens; presence of individuals which could not have been produced locally; capture during movement;* and *observations of movement in progress.*

5.2.1. Mark and capture

The marking and later recapture of insects provides a means of answering specific questions about their movements, such as: how does an emerging moth population disperse in the local environment? (Van Steenwyk *et al.*, 1978; King *et al.*, 1990) or: what is the effect of host density on parasitoid dispersal? (Hopper, 1991). Here we concentrate on the use of mark and capture to study movement (its use in estimating population *size* is discussed in Chapter 4). Models describing the movement of marked insects from release experiments (e.g. density as a function of distance) have been discussed by Southwood (1978), Rudd and Gandour (1985), Hopper (1991) and Service (1993), among others.

A multitude of marking techniques have been described in the literature (e.g. Southwood, 1978; Begon 1979; Eddlestone *et al.*, 1984; Akey, 1991; Service, 1993), but all aim to fulfil the following requirements: the marks should not impose a disadvantage (or advantage) on the marked population; they should be easily recognizable; should last for the duration of the experiment; and should be easy to apply with minimal interference to the test insects. Tests should be carried out to ensure that the collection, handling and release of insects are all performed in ways which minimize effects on the insects' behaviour and longevity. There are two distinct methods for using artificial markers in migration and dispersal studies, viz. capture–mark–recapture and mark without capture, but the same marking materials can often be used for either method.

CAPTURE–MARK–RECAPTURE

In this method, insects are caught in the field (or sometimes bred in the laboratory), marked, and released into the field. The field population is then re-sampled at various times after release. Because marking individual insects is very time-consuming, the technique is often not practicable where a large number of insects need to be marked (i.e. for studies of long-distance movement where the chances of recapture are very small). The method has the advantage that the time of release and the number of marked insects is known, so both these parameters can be fitted into movement models in order to test them against the experimental results. Where pheromone or light traps are used both to capture specimens for marking, and subsequently to re-sample the field population, there is always an unavoidable risk that the response of marked insects to the traps during the re-sampling exercise may be modified by their experience of previously being caught.

Paints, dyes, dusts, radioisotopes, labels and physical mutilation are the principal markers used in capture–mark–recapture studies.

Paints and inks. These are usually applied externally as discrete marks. Their use is restricted to studies of one life stage, because larvae and nymphs in particular are likely to lose the mark on moulting. They have been used to investigate the movement of species from many insect orders including dragonflies (Garrison, 1978), noctuid bollworm moths (López, 1979) and coccinelid beetles (Ives, 1981). Various application methods (brushes, pins and pens) have been reviewed by Gangwere *et al.* (1964), Southwood (1978) and Service (1993), and a large variety of coding schemes are possible using different colours and positions of marks. Over-conspicuous marks should be avoided as these may modify both mating success and the risk of predation.

Dusts and dyes. Insects can also be marked externally by covering them with micronized dusts (e.g. Day-Glo^R pigments, zinc sulphide), or with dyes (e.g. Erythrosin, Rhodamine B) either in powder form, or dissolved in alcohol or acetone (see references in Southwood, 1978, and Service, 1993). It is more difficult to control dosage in the case of dusts, as particle size and insect surface texture critically affect adherence. A number of adjuvants have therefore been used to give better control of dust adhesion (e.g. Raulston, 1979). A major drawback of dusts is that they may be inadvertently transferred to unmarked insects, particularly during the process of recapture when many individuals are crowded together in the re-sampling traps. This problem of contamination is precluded by *internal* marking where a dye is added to the food source. Oil soluble dyes such as Calco Red^{TM}, Calco Blue^{TM} and Sudan Red have found the greatest utility because these marks also offer persistence across life stages. Typically, immature stages are reared in the laboratory on diet containing 0.01–0.1% dye which is readily taken up into the fat reserves and transferred into the adult stage and sometimes even to the resultant eggs (Burton and Wendell-Snow, 1970). The presence of the dye in recaptures can be determined by crushing the insect or by elution with solvent, followed by visual or spectroscopic inspection (Argauer and Cantelo, 1972), or by paper chromatography (Showers *et al.*, 1989). In the case of fluorescent dusts and dyes, the presence of the marker can best be detected with a UV illuminator. Multiple marking is possible using coded combinations of fluorescent powders (Bennet *et al.*, 1981).

Dyes and dusts may be toxic or may induce changes in behaviour, and it is important to check for these effects before implementing release programmes. In addition, insects reared in the laboratory and then released may behave differently from wild populations, so dispersal experiments using this technique need to interpreted with some caution. Laboratory breeding does not, however, necessarily impair performance in the field, e.g. Showers *et al.* (1989) have shown that laboratory cultured *Agrotis ipsilon* (Hufnagal) moths were capable of migrations in excess of 1200 km.

A number of ingenious methods have been developed to make adult insects mark themselves with dusts and dyes as they emerge from aquatic environments (Service, 1993), or enter pheromone traps (Gentry and Blythe, 1978) or bait traps (Harlan and Roberts, 1976).

Radioisotopes. Marking with radioisotopes was particularly popular during the 1950s to 1970s (see reviews in Southwood, 1978, and Service, 1993), but has lost favour in recent years, largely because of stricter environmental protection regulations throughout much of the world. Partly for reasons of safety, isotope marking has been mainly confined to laboratory reared insects, where the active agents are applied topically, by injection, insertion or the attachment of tags, or by incorporation into larval or adult diets. There have been just a few studies where insects and plants were labelled directly in the field (i.e. mark without capture), by spraying or injecting solutions containing radioisotopes (e.g. Snow *et al.*, 1969; Neilson, 1971). Phosphorus-32 is the most common isotope used, because with a half-life of 15 days, it can be detected in recaptured insects by means of Geiger–Mueller or scintillation counters, or by autoradiography (Service, 1993) for a period long enough for most dispersal studies, but without creating a long-term environmental hazard. Where more persistent marking is required, isotopes such as zinc-65 or iridium-192 have been used (Service, 1993).

Labels. Coloured plastic tags have been attached to the wings of butterflies, to study their migration (Urquhart and Urquhart, 1979), but tagging methods have been otherwise limited to the study of short-range movement; for example, Gary (1971) stuck ferrous metal labels to the abdomens of foraging bees which were subsequently recovered on magnets at the hive entrance.

Mutilation. This method is appropriate for insects robust enough to withstand procedures such as notching the pronotum or removing wing tegmina (Gangwere *et al.*, 1964), etching (Best *et al.*, 1981) or punching (Unruh and Chauvin, 1993) marks on the elytra, and is useful only for small-scale studies of dispersal because of the time taken to mark individual insects.

MARK WITHOUT CAPTURE

These methods allow the study of insect population movement without the risk of perturbing individuals' behaviour by capturing them. Insects are marked in the field directly by spraying or by the provision of larval feeding baits containing trace elements, dyes or radioisotopes. The major advantage of these techniques is that they allow large numbers of insects to be marked with minimum disturbance to their natural behaviour; the principal disadvantage is uncertainty about the number of insects which become marked, and additionally in the case of larval baits, about the time at which marked adults emerge into the population. These uncertainties can be reduced in some experiments if the field population is sampled with emergence cages so that estimates of the number and the time of emergence of marked insects can be made (King *et al.*, 1990).

Trace elements. Insects can be effectively marked by producing an increased concentration in their bodies of a normally rare element. This is achieved by simply adding the selected element to their habitat so that it is naturally assimilated into their body tissues. Rubidium (Rb), caesium (Cs) and strontium (Sr) are the most popular trace elements; they act as chemical surrogates for physiologically common elements (Rb and Cs for potassium; Sr for calcium and magnesium). The marking elements are usually detected by atomic absorption spectrophotometry (AAS) following acid digestion of individual insects (Akey and Burns, 1991). Before trace element marking is attempted, the natural *background* concentration of the selected element in insects from the experimental area must be determined on a large enough sample to ensure that they can be unequivocally distinguished from marked specimens; statistical criteria for this process are given by Hopper (1991). A limiting factor of the trace element technique is that insects tend to lose the marker by excretion, oviposition and by mating (Van Steenwyk, 1991). Rates of loss are generally higher for Rb than Sr (Armes *et al.*, 1989) and higher in sucking insects than in chewing ones (Fleischer *et al.*, 1986).

The first reported use of trace elements for marking in the field (for dispersal studies) was that of Stimmann (1974) to mark *Pieris rapae* (L.), and the methods have remained essentially unchanged. Aqueous solutions of trace element salts (usually as chlorides, e.g. RbCl, $SrCl_2$) are sprayed on to host plants at rates which are calculated from the anticipated rate of insect uptake, the toxicity of the element, and the natural concentration of trace element in the environment. Rates of 1–10 kg ha^{-1} are typical. Application is extended in time for as long as is necessary to ensure efficient marking of the immature stage and carry-over into the emerging adult population. Rb and Sr have been used in this way to study the movement of noctuid moths (Van Steenwyk *et al.*, 1978; King *et al.*, 1990; Stadelbacher, 1991), tortricid moths (Knight *et al.*, 1990), boll weevils (Wolfenbarger *et al.*, 1982), leafhoppers (Fleischer *et al.*, 1988) and planthoppers (Padgham *et al.*, 1984). A more direct method of marking adult insects in the field is to dispense honey, which has had trace elements added to it, from feeding stations or artificial nectaries. This technique has proved successful with *Helicoverpa zea* (Boddie) (Culin and Alverson, 1986). Laboratory reared insects and their eggs have been uniquely marked using coded combinations of trace elements (Hayes, 1989), but this has not been attempted under field conditions.

Field application of dyes. There are only a few reports of dyes being used in mark without capture studies. Rose *et al.* (1985) sprayed trees in Kenya with a molasses solution containing 0.5% neutral red dye, to mark teneral moths of the African Armyworm, *Spodoptera exempta* (Walker): this exploited the tendency of the moths to congregate in trees before undertaking migratory flights. The dye accumulated in the gut and fat body, and marked moths were caught in pheromone traps up to 147 km from the emergence site. In an experimental trial, Bell (1988), sprayed Calco Red™ dye mixture on to geranium plants and weeds to mark *Heliothis virescens* (F.) and *H. zea* caterpillars. Emerging moths were sampled with pheromone traps sited close to the sprayed areas, but captures of marked moths turned out to be low.

5.2.2. Capture of naturally marked specimens

The origin of migratory insects can sometimes be deduced by comparing naturally-occurring marks or features on suspected immigrants with similar marks on specimens acquired in possible source areas. This method conveniently avoids the drawbacks associated with artificial marking, but demands a thorough investigation of the phenology and distribution of the selected mark in geographically diverse populations of the species of interest. Natural markers used in dispersal studies can include: elemental composition, genetic markers, phenotypic variation (e.g. size, colour, etc.), age or stage classes, gut contents, and presence of particular parasites, pollen grains, algae and phoretic mites (Pedgley, 1982; Drake, 1990)

ELEMENTAL COMPOSITION

The elemental composition of an insect body (its *chemoprint*) is determined in part by the environment from which it comes, and this leads to the possibility of identifying the geographical origins of captured specimens. Quantitative determinations of insect elemental composition can be made either by atomic absorption spectrophotometry (e.g. Levy and Cromroy, 1973) or by X-ray spectrometry (e.g. Turner and Bowden, 1983), but the latter method is more powerful because it can provide data on several elements from a single small insect. The chemoprinting method assumes that a distribution of elements accumulates in the body of an insect during its immature growth stage, and that this distribution is characteristic of the geographical area in which the insect develops. It also assumes that the distribution persists into the adult and remains substantially unaffected by adult feeding and ageing. Dempster *et al.* (1986) showed that this assumption is not always valid; they found that the chemoprints of the long-lived Brimstone butterfly, *Gonepteryx rhamni* L. changed significantly during its adult life. Another, and very important requirement in the chemoprint method is that the differences in elemental composition *between* populations from different regions should be large relative to those *within* those populations (Dempster *et al.*, 1986). The elemental composition of herbivorous insects is affected by the host plant and the soil type in which it is growing, and hence source discrimination tends to be better for monophagous (Turner and Bowden, 1983) than polyphagous species (Bowden *et al.*, 1984).

GENETIC AND PHENOTYPIC MARKERS

Genetically-determined features such as eye and body colour mutations, as long as they carry little or no fitness cost, have potential as markers because of their ease of identification. They are also likely to persist over time and between generations, allowing long-term determination of migratory behaviour and population structure. In the Pink Bollworm, *Pectinophora gossypiella* (Saunders), Bartlett and Lingren (1984) used the genetic marker 'sooty' which produces dark body phenotypes, to monitor movement in fields neighbouring the site at which the phenotypes were released. The disadvantage of using such genetic markers is that strains have to be selected in the laboratory, and in essence the technique becomes a capture–mark–recapture method with the inherent drawbacks described earlier. Also, it has been noted (Service, 1993) that not all colour strains of the mosquito *Aedes aegyptii* (L.) moved similar distances when released into the field.

Environmentally-induced colour changes can also be used as evidence of migration. In some noctuid species for example, high temperatures during the pupal period produce pale-coloured moths, and low temperatures produce dark ones. Thus, the sudden appearance of pale *Helicoverpa punctigera* (Wallengren) moths in pheromone traps in Tasmania at times in the year when the locally emerging moth population was mainly dark coloured, clearly suggests immigration from the warmer regions of mainland Australia (Hill, 1993). Smith and MacKay (1989) used the less obvious characteristic of photoperiodic responses of the Pea Aphid, *Acyrthosiphon pisum* (Harris) as a physiological marker to identify the source of immigrant populations.

Biochemical markers resulting from genetic variation within a species can be used to provide evidence of gene flow between populations. Significant genetic differences between populations imply that little or no mixing occurs between them, and that they are isolated rather than contiguous. Daly and Gregg (1985) used allozyme electrophoresis to deduce that in both *Helicoverpa armigera* (Hübner) and *H. punctigera*, population mixing was occurring over wide regions in Australia. Conversely, Korman *et al.* (1993) used electrophoresis to show that discrete populations of *H. virescens* exist in southern USA in certain seasons. Electrophoretic studies of the enzymes of European aphids have indicated that in some species (e.g. *Sitobion fragariae* (Walker)) movement between local populations was restricted, while in others (e.g. *Sitobion avenae* (F.)) there appeared to be enough migration to prevent genetic divergence even between geographically separated populations (Loxdale *et al.*, 1993).

The appearance of strains of insects resistant to specific groups of insecticides, in areas where these chemicals are little used, provides strong circumstantial evidence of immigration. This phenomenon has been used in conjunction with synoptic weather patterns, to elucidate the migration of *H. virescens* (Wolfenbarger *et al.*, 1973) and *Spodoptera frugiperda* (Smith) (Young, 1979) moths in the USA, and of *H. armigera* in India (McCaffery *et al.*, 1989) and in Australia (Gunning and Easton, 1989). Wada *et al.* (1994) investigated possible source areas of the planthoppers which invade Japan each year, by comparing 'biotypes' of the Brown Planthopper *Nilaparvata lugens* (Stål), distinguishable by their differing ability to feed on various rice plant varieties.

POLLEN, ALGAE AND PHORETIC MITES

Insect pollinated plants have evolved pollen which readily adheres to insect cuticles, and which has obvious potential as a natural marker

(Mikkola, 1971; Turnock *et al.*, 1971). To be effective, the method requires that sources of identifiable pollen be geographically distinct, and remote from the areas in which pollen-bearing migrants are caught, but the general lack of accurate data on flora and flowering phenology means that these criteria are not easy to verify. Despite this difficulty, analysis of pollen found on *H. zea* and *A. ipsilon* moths caught in pheromone traps has been used in conjunction with synoptic weather patterns to confirm migrations of 750–1500 km in southern USA (Hendrix *et al.*, 1987; Hendrix and Showers, 1992; Lingren *et al.*, 1994), and long-range movements of *H. armigera* and *H. punctigera* from inland deserts to coastal cropping regions in eastern Australia (Gregg, 1993).

Other organic material carried by insects may give clues to their origin; for example, algae on the wings of Desert Locusts, *Schistocerca gregaria* (Forskål) (Pedgley, 1982). Lastly, Treat (1979) has suggested that phoretic mites may form a natural marker for certain moth species.

5.2.3. Evidence of migration from changes in the spatial distribution of the species of interest

There have been many instances where the migration of insects has been made obvious by their conspicuous appearance in places where they could not be accounted for by local breeding (Williams, 1930, p. 10). Examples include insects found in geographical areas far outside their normal breeding range, such as the species which arrive in northwest Europe from the south and east (Pedgley *et al.*, 1995). Also included are investigations of migratory capability where insects have been trapped or observed in inhospitable habitats far from possible sources, e.g. over the sea (Johnson, 1969; Bowden and Johnson, 1976; Hardy and Cheng, 1986; Peck, 1994) or on remote islands (Farrow, 1984; Ferguson, 1991), in deserts (Dickson, 1959; Pedgley and Yathom, 1993) or stranded on snow-fields and volcanic debris (Edwards, 1987; Ashmole and Ashmole, 1988).

Inferences that migration has occurred are commonly based on observations of temporal changes in the distribution of the species of interest. Data on the distribution of conspicuous or noteworthy insects such as macrolepidoptera are frequently available in areas where there are networks of knowledgeable amateur observers and special recording schemes (e.g. Eitschberger *et al.*, 1991; Pollard and Yates, 1993). In the case of pest species, changes in distribution are often detected by monitoring networks set up specifically for forecasting and management purposes. Dent (1991), discussing pest monitoring strategies in general, distinguishes between surveys, field-based monitoring and fixed-position monitoring.

Methods for the operational monitoring of migratory pests, particularly those in tropical and subtropical regions, have been recently reviewed by Riley and Reynolds (1995). They discuss the specialized regional surveys required for monitoring the redistribution of populations of locusts, migratory grasshoppers, armyworm moths and blackflies (*Simulium* spp.) (see also Rainey *et al.*, 1990; Meinzingen, 1993; Pedgley, 1993). Field-based monitoring is usually undertaken to help farmers and pest control operatives to decide when and where control measures are needed. However, the detection of pest immigration by visual inspection of crops (scouting) or by in-field trapping may also form part of a large-area monitoring process. For example, field scouting is included in surveillance schemes set up in East Asia to detect immigration of the rice planthoppers *N. lugens* and *Sogatella furcifera* (Horváth), so that outbreaks can be forecast (Hirao, 1985).

In fixed-position monitoring, data are usually collected over several years from a network of traps (e.g. suction, light, pheromone and pan traps), so that seasonal and year-on-year population trends are elucidated (Dent, 1991). The classic example is that of the Rothamsted Insect Survey (Woiwod and Harrington, 1994, and references therein). Many of the insects caught at the 12.2 m sampling altitude of the Rothamsted suction traps are probably in the process of migrating (see below). More commonly, trap networks catch post-migratory individuals, so techniques which work for insects in this phase (e.g. pheromone trapping) can be utilized. Hartstack *et al.* (1986), for example, were able to use a pheromone trap network to detect immigration of *H. zea* moths, because this began before the emergence of locally diapausing individuals. Tucker *et al.* (1982) investigated the windborne migration of *S. exempta* moths using data from light traps: their criterion for an immigration *event* was a fivefold increase in catch over that from the previous night, or a catch of at least five moths following a nil catch. Movements of *S. exempta* moths are now routinely monitored in eastern Africa using a pheromone trap network, supplemented by a few light traps (Rose *et al.*, 1995). In some cases, monitoring systems produce enough data to allow the redistribution of migrant pest populations to be followed over large geographical areas by *dynamic mapping*, i.e. a series of frequently updated contour maps of pest density (e.g. migrant aphid species in northwest Europe: Taylor, 1985). The migration patterns of *S. gregaria* locusts can be inferred from mapping of even fragmentary population data from over the whole of the species' range (Pedgley, 1981).

The efficiency of many trapping methods varies according to the physiological state of the target insects, and this may sometimes be turned to advantage to gain circumstantial evidence for migration. For example, if approximately equal numbers of male and female *S. exempta* moths are caught in a light trap, this is taken to indicate that migration is occurring. On the other hand, a preponderance of males suggests that migration has ended and the females are occupied laying in the vicinity (Rose and Khasimuddin, 1979). Information relating to a captured insect's migratory status can often be inferred from its physiological condition (e.g. age, mating status, maturity of the reproductive system). Generally, females migrate before their ovaries mature and often before mating (Johnson, 1969). Thus, in the Rice Leaf-roller moth, *Cnaphalocrocis medinalis* Guenée, for example, a high proportion of immature, unmated females is indicative of an *emigratory* population (Wada *et al.*, 1988). It might be expected that sexually immature females would also predominate in areas that had been recently invaded, but because *C. medinalis* immigrants mate immediately after landing and their ovaries develop quickly, areas where immigration has occurred are characterized by mated moths with developed ovaries (Wada *et al.*, 1988).

Lastly, migration may be inferred from the *effects* caused by the migrants, even though the insects themselves may not have been seen. For example, the windborne movement of animal virus vectors may be deduced from records of disease outbreaks or a rise in antibody levels in hosts (e.g. Sellers, 1980).

5.2.4. Capture during movement

CAPTURE AT LOW ALTITUDE

Absolute samples of insect populations are much easier to obtain when the insects are flying near the ground than when they occupy other, less accessible, habitats (Southwood, 1978). One of the simplest methods is to sweep a net through the air by mounting it on a moving vehicle. Service (1993) describes examples of nets mounted on trucks and other vehicles, including powerboats and even bicycles.

Fan-powered suction traps are frequently used for long-term aerial sampling (Southwood, 1978; Muirhead-Thomson, 1991; Service, 1993) because of their convenience, and empirical corrections for the reduction in trapping effectiveness caused by increases in wind speed are available for some types (e.g. the 'Johnson–Taylor trap'). Proprietary suction traps are suitable for sampling a wide variety of small-sized insects, but larger species (including most moths) which usually occur at much lower aerial densities are seldom caught.

An alternative method is to rely on the wind and/or the insect's flight to carry it into a stationary trap. A large range of interception traps of this type, including variants of the Malaise trap, window traps, ramp traps for catching mosquitoes, 'wind-sock' nets, electric grids and sticky traps are described in Southwood (1978), Muirhead-Thomson (1991) and Service (1993). Drake (1990) has used a large aperture 'goal net trap' for sampling moths flying at heights of 3–6 m above ground, and Aubert (1969) describes a very large (24 m diameter) semi-circular trap designed to capture insects flying through mountain passes. Most stationary net traps are considered to be neither visually attractive nor repellent, but (except at night) there must always be a possibility that flying insects might tend to either avoid nets or fly towards them (Service, 1993). Sticky traps can be made non-attractive or highly attractive, depending on the choice of colour.

Many insects caught flying near the ground will probably be engaged in vegetative rather than migratory flight, and so most of the above-mentioned traps will not discriminate between these two classes of behaviour. An exception may be the large interception traps used to monitor the low altitude movement of butterflies (e.g. Walker, 1985, 1991). Here, the undeviating flight direction of migrants tends to selectively trap them in such nets while non-migrant butterflies are able to fly out (E.T. Neilsen, quoted in Southwood, 1978).

CAPTURE AT HIGHER ALTITUDES

Insects flying some distance from the ground, above their *flight boundary layer* (see p. 131) are likely to be engaged in migration. Indeed, their very presence at altitude is usually the result of specialized behaviour which is highly indicative of migration, and this applies equally to many wingless arthropods such as coccid and adelgid 'crawlers', ballooning spiders and lepidopteran larvae (McManus, 1988). Thus, sampling at altitude has the great advantage that the catch will often consist largely of migrants.

A standard tool for monitoring small insects flying well above ground level is the suction trap developed by the Rothamsted Insect Survey. In this device, airborne insects are drawn into the top of a 12.2 m (40 foot) chimney by a powerful fan. A network of these traps has been operated for many years in the UK and in several European countries (Woiwod and Harrington, 1994). Another approach is to place smaller suction traps on towers or masts (e.g. Bowden and Gibbs, 1973). The traps should be of the 'enclosed cone' type so as to reduce the adverse effect of high wind speeds on the trap performance (Southwood, 1978). Wainhouse (1980) used small, portable suction traps powered by rechargeable batteries to sample the airborne first instar larvae of the scale insect *Cryptococcus fagisuga* Lindinger at various heights from the ground. These 'elevated trap' methods are suited to sampling small, day-flying insects undergoing *cumuliform downwind migration* (Taylor, 1986) where atmospheric mixing tends to ensure that individuals are circulated through a range of altitudes, including that of the trap. They are much less suitable for investigating *nocturnal* migration where, for much of the time, the insects may be flying in layers at altitudes of several hundred metres.

Routine sampling of migrants at altitudes above a few tens of metres is usually difficult and costly to carry out. Callahan *et al.* (1972) and Gregg *et al.* (1993) were nevertheless able to routinely sample higher than this by fitting upwardly-directed light traps on tall TV or communications towers. With this technique, care has to be taken to ensure that the trap lights (or any other lights on the tower) do not draw insects from below (Gregg *et al.*, 1993) particularly in calm weather when insects could vector their climbing flight towards the lights. The elevated light-trap method allows routine, qualitative sampling over long periods, but it suffers from the disadvantages that it is behaviour dependent, and trap efficiency is adversely affected by moonlight and strong winds.

Sampling at altitudes of hundreds of metres above ground level normally requires the use of tethered aerodynamically-shaped balloons (kytoons), kites or aircraft, and is practicable for only limited periods. Johnson (1969) was able to sample insects quantitatively at heights up to about 300 m by using Second World War barrage balloons to lift large suction traps, but balloons of this type are now not so readily available, and are expensive to fill.

Very much smaller (e.g. 11 m^3) helium-filled kytoons equipped with tow nets provide a more practicable alternative (Riley *et al.*, 1991), provided that winds at the sampling altitude are strong enough (i.e. >3 ms^{-1}) to ensure that the nets trap effectively. This method has been used to sample up to about 500 m, and with a lightweight line the sampling height could perhaps be increased by 100 or 200 m. In situations where the wind at altitude can be expected to remain fairly strong for long periods, parafoil kites provide a very inexpensive means of suspending aerial nets (Farrow and Dowse, 1984).

Flying insects have also been sampled with nets attached to radio-controlled planes and model aircraft (Gottwald and Tedders, 1986; Tedders and Gottwald, 1986), to piloted fixed-wing aircraft (Spillman, 1980; Reling and Taylor, 1984; Greenstone *et al.*, 1991) and to helicopters (Beerwinkle *et al.*, 1989; Hollinger *et al.*, 1991). These nets are generally of more elaborate design than those used on kites or balloons, because of the need to reduce impact and abrasion damage to captured insects, and to allow samples to be segregated or collected during the aircraft's flight. One of the main problems in sampling at altitude is that many species, particularly larger insects, are usually present at only very low aerial densities, and so catching rates are low. Beerwinkle *et al.* (1989) attempted to remedy this situation by developing a net with a very large aperture (5 m^2) for use under a helicopter.

Aerial netting is especially effective when used in conjunction with an entomological radar (see below) which provides information on the altitudinal distribution of insect migrants, and thus allows the net to be positioned at the best height for sampling, for example, within dense layers of insects.

5.2.5. Observation of movement in progress

VISUAL OBSERVATIONS

Large, conspicuous day-flying insects, especially species of butterflies which frequently migrate close to the ground, can often be easily observed by eye or with binoculars, particularly during seasons or at places where their migrations are fairly intense (Williams, 1930, 1965; Baker, 1978; Gibo, 1986). Movements in locations such as high mountain passes can be especially spectacular (Johnson, 1969, p. 574), but caution must be exercised so as not to give undue weight to such impressive phenomena when assessing general migration patterns (Johnson, 1969). Careful observations at several sites and for several seasons are needed to build up a convincing picture (see, for example, the 'vanishing bearing' studies of Schmidt-Koenig (1993) on the Monarch butterfly, *Danaus plexippus* (L.)). The range over which the movement of conspicuous groups of insects can be observed may sometimes be increased by following them in vehicles (e.g. Neilsen, 1961) or, in the case of *S. gregaria* swarms, in aircraft (Rainey, 1963).

Baker (1978, p. 361) attempted detailed assessments of the displacement pattern of the Small White butterfly, *Pieris rapae* (L.), by following an individual on foot until it became lost from view. He then waited until another appeared flying on the same track, and followed that one, and so on. In a more elaborate study, Turchin *et al.* (1991) used a team of two to quantify the movements of butterflies: one person followed the subject insect, marked its landing positions with numbered flags, and called out any significant behavioural activity to an assistant who recorded the data on a computer. The successive positions of the butterfly were later registered as spatial coordinates, from which various statistics (e.g. the mean length of each 'move', and the mean turning angle) could be extracted. Weins *et al.* (1993) suggest that flag positions in such an experiment could be conveniently recorded using an electronic distance-measuring theodolite, and they also discuss the quantification and analysis of the fine-scale movements of insects. Although detailed visual studies of this type have been most frequently used to study *non*-migratory behaviour, they may find application in the examination of pedestrian migration or of migratory flight near the ground.

Unaided visual observation is obviously of little use when movement takes place at night. Species which emigrate just before dusk can sometimes be glimpsed in silhouette against the western sky (Schaefer, 1976), but radar observations have shown that the peak emigration of larger-sized insects (moths and grasshoppers) usually occurs when it is too dark for humans to see effectively (Schaefer, 1976; Farrow, 1990). Several authors have observed night flight near the ground by using searchlights (Roffey, 1963; Brown, 1970), but this technique is obviously prone to modify the insects' behaviour, and the use of image-intensifier, night-viewing devices, with or without supplementary infrared illumination, is a preferable option nowadays (Lingren *et al.*, 1986).

Zalucki *et al.* (1980) recorded short segments of butterfly flight trajectories in three dimensions by synchronously tracking them with two separated viewing devices; the elevation and bearing angles of each of the viewing tubes were registered on a strip chart. Insect flight in three dimensions can also be recorded by stereoscopic film (Dahmen and Zeil, 1984) or video techniques (see below).

VIDEO TECHNIQUES

The use of video equipment to observe flying insects in the field has been described in detail in a recent review (Riley, 1993), and so only remarks relevant to its use in migration studies are made here. Video systems provide coverage of flying insects over a range of only a few metres, and so their use in field migration studies is rather limited. However, video recording can be a valuable supplement to radar studies of migration, by providing observations of flight below the minimum altitude normally accessible to radar, i.e. typically below 20–30 m (Riley *et al.*, 1990). In order to attain a useful range of detection, it is essential that airborne insects form high contrast images, and this is most easily achieved by viewing them against the night sky, using some form of artificial illumination. Illuminators in the near infrared region (750–900 nm) are most suitable because they produce little perturbation of insect behaviour (Riley, 1993). This procedure does not work well in twi-light, and not at all in daylight, but an alternative approach of using a very bright xenon flash lamp, working in the near infrared, and a video camera equipped with a gated image intensifier has provided high-contrast images of small flying insects against the midday sky (Schaefer and Bent, 1984).

Stereographic records can be readily obtained by the use of a pair of synchronized video cameras, and reconstruction of three-dimensional flight trajectories is relatively straightforward (Riley *et al.*, 1990).

RADAR METHODS

Radar is the *only* method of making direct quantitative measurements of the undisturbed nocturnal flight behaviour of insects at altitude, so it is not sur-prising that the bulk of our current knowledge about the magnitude and altitudinal distribution of insect migratory flight at night stems directly from the application of this powerful technique (Riley, 1974; Schaefer, 1976; Riley, 1989). By producing unequivocal evidence that *mass* movement of insect populations regularly takes place at altitude, radar has done much to validate biogeographical estimates of insect aerial movement which would otherwise have remained largely speculative. Radar has also provided a unique insight into the role of atmospheric processes in the dispersal and concentration of airborne insect populations (Schaefer, 1976; Pedgley *et al.*, 1982; Drake and Farrow, 1988; Riley and Reynolds, 1990).

Most entomological radars are based on inexpensive, 3 cm wavelength marine radar transceivers, and produce a narrow (1–2°) conical beam which sweeps in a circle about a vertical axis, and can be adjusted in elevation angle to give the desired coverage in altitude. Individual insects intercepted by the rotating beam are displayed as bright dots at the appropriate azimuth and range on a circular screen called the plan position indicator (PPI). A fading trail of dots is produced on the PPI as the rotating beam re-intercepts each flying insect, and this trail defines the insect's horizontal *displacement speed* and *direction* (Riley, 1974). The density of the primary dots on the screen can be interpreted in terms of the aerial density of the airborne insects (Riley, 1979), and a method of estimating the aerial density when dots merge into continuous echo has been described by Drake (1981a). Density measurements can be integrated over the whole flight altitude range to estimate the total number of insects flying above a unit area of the earth's surface (i.e. the *area* density) (Drake and Farrow, 1983). Density val-ues combined with the displacement speed measurements provide a quantitative measure of the *migration flux* (Drake and Farrow, 1983). The principles of operation, and analysis procedures for this type of radar have been described by Riley (1974, 1979), Schaefer (1976) and particularly by Drake (1981b). The effective range for the detection of individual, medium-sized species is usually 1.5–2 km, but concentrations of airborne insects can be detected up to several tens of kilometres. Individual *small* insects, with masses of a few milligrams, are best detected with millimetric wavelength radars because the shorter wavelength increases the strength of the insect's radar echo (Riley, 1992). To date, however, radars of this type have been used in only a few entomological studies (Riley *et al.*, 1991). If the wind velocity at the same altitude as the insects is available (from radar or theodolite measurements of free-flying balloons), insect *air* or *flying speed*, and *heading* or *orientation* can be calculated.

Radar measurements of the *start* of migratory flights rely on the detection of insect take-off and climb to altitude. This usually occurs near dusk for nocturnal migrants, and forms a spectacular and distinctive radar display. Determinations of the *duration* of flight usually depend on interpretation of the subsequent variation (usually decline) of aerial density at altitude (Schaefer, 1976; Riley and Reynolds, 1990), and are much less exact. This is because aerial density is determined not only by the insects' flight duration, but also by the geographical distribution and population density of the source areas from which the migrants have come, and these factors are often imprecisely known. In cases where a concentration of insects overflies two radars separated by a substantial distance, sequential detection of the concentration provides a less ambiguous estimate of flight duration (Riley and Reynolds, 1983). Airborne entomological radar (Schaefer, 1979; Hobbs and Wolf, 1989; Wolf *et al.*, 1990) probably offers the best method of flight duration measurement, but aircraft operations are relatively expensive to support. The displacement of airborne insect migrants is usually dominated by the wind velocity at their flight altitude, so estimates of the *range* and direction of migration are made by combining deductions about flight duration with synoptic scale wind data to compute back-tracks and forward trajectories (see p. 132).

A major constraint in radar entomology is that imposed by the difficulty of identifying the insects which are detected. This has tended to limit successful applications of the technique to situations where the species of interest is either known to be the dominant component of the aerial fauna, or where it stands out because it has a distinctive size or wing-beat frequency (Schaefer, 1976; Riley and Reynolds, 1979). Identification problems can sometimes be resolved by observing the localized takeoff of insects leaving an area known to be infested with an identified species (Riley *et al.*, 1983), or by the supplementary use of nets suspended from balloons, kites or aircraft (see above).

Although scanning entomological radars have been very successful in the investigation of insect migratory behaviour, they are unsuitable for long-term monitoring tasks. This is because both their application and the associated data analysis are very labour intensive (Drake, 1993), and because their capacity to identify the insects which they detect is limited. However, the recent development of an inexpensive vertical-looking radar (VLR) system means that routine monitoring of migration is now a practical possibility (Riley and Reynolds, 1993; Smith *et al.*, 1993). This results from the fact that the radar is controlled by a personal computer so long-term routine operation is feasible, but more importantly because the formidable, time-consuming tasks of signal analysis and data handling have now been fully automated. A further advantage of VLR is that the careful siting required for scanning radars is not necessary – virtually any small open area (e.g. the flat roof of a building) will do.

In addition to measurements of the altitude, orientation, speed and direction of individual migrating insects, VLR provides estimates of their mass, shape and wing beat frequency, so species of significantly different size and shape may be discriminated from one another. The possible applications of the VLR technique to monitoring locust migration have been outlined by Riley and Reynolds (1993) and by Drake (1993). The feasibility of operating VLR hardware for extended periods has been demonstrated by the experiments carried out over two years by Beerwinkle *et al.* (1994) in the USA, but the full potential of VLR, equipped with the powerful new analytical software, remains to be exploited.

Radar does not work well for insects flying near the ground, because the strong reflections from ground features tend to obscure those from the intended targets. The problem of ground reflections can be resolved by tagging insects with an electrically non-linear conductor (a diode) which, unlike the ground features, will generate signals at harmonics of the illuminating frequency. This method has been used to follow (from a range of a few metres) the walking movements of insects of moderate size and robust enough to carry a diode fitted with a wire antenna, such as carabid beetles (Mascanzoni and Wallin, 1986; Wallin and Ekbom, 1988) and may soon be successfully applied to airborne insects.

5.3. Laboratory Studies

All laboratory methods of measuring the flight performance of insects introduce artificial constraints to their movement, and most require that the insect be tethered so that it is either held static, or allowed a limited degree of movement on pendulum or balance systems, or on flight mills (Cooter, 1993). Great caution must therefore be used when interpreting the measurement data in terms of free flight behaviour. Perturbation effects may be subtle, and not necessarily directly related to the measurement apparatus itself, e.g. Baker *et al.* (1980) showed that there was a dramatic decline in flight performance of the Brown Planthopper, *Nilaparvata lugens*, in insects reared through even a small number of generations in laboratory cultures. Nevertheless, there is evidence that at least in some cases (Dingle, 1985), insect migratory capacity can be satisfactorily indexed by laboratory flight studies.

5.3.1. Static tethering

In the simplest static experiments, insects are attached to a fixed support and stimulated to fly, and the duration of wing beating is recorded (Dingle, 1985). Most winged insects spontaneously initiate flight (or at least wing flapping) if tarsal contact with the substrate is suddenly lost, and some may continue to fly without further stimulus. Others require an air flow similar to that which would be experienced in free flight, and this can be conveniently provided by an electric fan. Although static tethering seems rather crude, it has been successfully used to index migratory capability in studies of the genetics of migration syndromes in the milkweed bug *Oncopeltus fasciatus* (Dallas) (Dingle, 1985, 1994). The technique has also been used to estimate fuel consumption in aphids (Cockbain, 1961) and the maximum flight endurance of *N. lugens* (Padgham, 1983). More sophisticated static mounts sometimes incorporate thrust, lift and torque sensors, and use wind tunnels to more accurately simulate in-flight air flow (e.g. Preiss and Spork, 1993).

5.3.2. Pendulum and balance systems

While an insect is suspended from a tether, it does not need to support its body weight to remain 'airborne', and so the power it expends may be much less than that required to sustain free flight. A number of balance or pendulum devices have therefore been developed to indicate the lift generated during periods of wing beating. Baker *et al.* (1980) used a simple pivoted balance arm to monitor lift production in *N. lugens*, and found that those insects which made the longer flights also produced higher lift. This observation strongly suggests that flight duration is an adequate indicator of the *relative* migratory potential of insects tested on flight mills.

More elaborate devices allow insects freedom to take off and land. In the *pendulum* system, when an insect launches itself from the substrate and begins to fly, its thrust propels the tether forward and upward so that the substrate is left behind. As thrust declines towards the end of a flight, the insect swings back until it encounters the landing substrate (such as a continuously-moving paper strip) on to which it can settle, and where female moths may deposit eggs (Barfield *et al.*, 1988). Using this system, Wales *et al.* (1985) studied the temporal pattern of flight and oviposition of the Velvet Bean Caterpillar moth, *Anticarsia gemmatalis* Hübner. However, because the insect lands unnaturally (backwards) and then has to walk continuously with the landing strip, interpretation of the data from this kind of experiment requires some caution (Gatehouse and Woodrow, 1987). In the *balance system* (Gatehouse and Hackett, 1980; Woodrow *et al.*, 1987), lift generated by the test insect is detected and if this rises above a pre-set threshold, the substrate is automatically withdrawn. When the lift falls below this threshold, the substrate (a lightweight cylinder on which the insect can walk) is returned to a position in front of and beneath the insect so that it can land, rest, walk or take off again. During extended flights, any body weight lost by the insect reduces the aerodynamic lift necessary to hold the balance above the landing threshold, so the device tends to automatically simulate the changes in flight power requirements which occur during long migrations. In the balance system used by Gatehouse and his colleagues, the flying insects are also able to adjust their pitch independently of thrust as well as having a degree of freedom in yaw and roll.

5.3.3. Flight mills

In flight mills, the test insects are attached to a pivoted radius arm so that their flying action propels them round in a circle. They thus experience air flow and visual sensations of motion rather similar to those they would experience in free flight. The potentially adverse effects of perpetual circling are reduced by ensuring that the radius arm is long relative to the size of the insect. With this technique, the distance 'flown' in still air is determined by the number of revolutions of the mill, and flight power output can be related to the speed of rotation. Radius arms are designed so that their pivot friction, moments of inertia and wind resistance are all very low, and mills have been successfully used to monitor the flight performance of a wide range of different insects, some of them very small (e.g.: Diptera – Cooter, 1982, 1983; Coleoptera – McKibben 1985; Coats *et al.*, 1986; Forsse, 1987; Orthoptera – McAnelly and Rankin, 1986; Hemiptera – Moriya 1987; Lepidoptera – Noda and Kamano, 1988; Resurreccion *et al.*, 1988; Nakasuji and Nakano, 1990). A highly sophisticated, servo-controlled mill has been developed by Preiss and Kramer (1984) which actively compensates for the aerodynamic drag and inertia of the suspension arm. This apparatus also presents the test insects (moths) with a moving visual environment of the type which they might experience in free flight, and allows them to adjust their flight speeds. It has been used for investigating the sensorimotor mechanisms underlying wind-drift compensation and the control of ground speed in flying Gypsy Moths, *Lymantria dyspar* (L.).

As is the case with simple static experiments, flight initiation on mills is often prompted by the experimenter and is terminated by the insect. The flight activity data recorded may then consist of estimates of flight willingness, the durations of individual flights, or the total time spent in flight after a set number of stimulations or during a standard test period.

A few attempts have been made to design mill systems which permit take off and landing under the control of the tethered test insect and while it is exposed to natural diurnal stimuli (e.g. Ruzicka 1984; Cooter and Armes, 1993). In computer-monitored versions of these mills (and similar pendulum and balance systems), the temporal pattern of flight activity as well as duration is automatically recorded. Software routines have been developed to analyse these data, with criteria set to partition the flights into different categories, e.g. short and long, vegetative and migratory, etc. (Gatehouse and Hackett, 1980; Barfield *et al.*, 1988).

5.3.4. Vertical windtunnels

Some of the problems associated with tethered flight were elegantly overcome by Kennedy and Booth (1963), who studied the free flight of aphids in a specially-designed vertical flight chamber. The aphid's altitude in the chamber was controlled by adjusting the (downwards) air flow until it just counteracted the aphid's rate of climb towards a light source. This type of flight chamber has been further developed and automated by David and Hardie (1988), and is suitable for studies on other small, slow flying insects which have a phototactic flight phase, e.g. the small nitidulid beetle, *Carpophilus hemipterus* (L.) (Blackmer and Phelan, 1991).

5.3.5. Pre-reproductive period (PRP) assessment as a component of migratory potential

The concept of an oogenesis-flight syndrome arose out of the observation that in several insect orders, mainly the Lepidoptera, Orthoptera and Hemiptera, female migratory flight is rarely initiated while the ovaries are mature (Johnson, 1969; Gatehouse, 1995). For species where field evidence of this syndrome exists, the period when the ovaries are either immature, or have been resorbed, is of interest because it approximates to the time available for migratory flight. Consequently, it can represent one of the two main components of an insect's *migratory potential*, the other being the capacity for prolonged flight (as assessed, for example, by the laboratory techniques described above).

The pre-reproductive period (PRP, the time between emergence and the onset of reproductive maturity) of female insects can be quantified in the laboratory in several ways, including dissection to determine the extent of ovarian development (Rankin *et al.*, 1986). Non-destructive techniques involve recording the time between emergence and the first release of pheromone (pre-calling period) or first oviposition (pre-oviposition period) (Hegmann and Dingle, 1982; Colvin, 1995; Gatehouse, 1995).

Significantly fewer data are currently available on the reproductive status of migrating male insects and a syndrome integrating migration and reproduction may not necessarily be a general phenomenon (Johnson, 1969; Gatehouse, 1995). However, where it can be justified from field evidence, laboratory data on male PRP variation and the factors influencing it can provide valuable insights into both migration strategies and migratory potential. Non-destructive assessment of male PRPs has mainly been attempted on species of noctuid moth, and has involved exposing males to actively calling females on consecutive nights following emergence. The number of nights between emergence and a male's first mating attempt was recorded as its PRP (Hill, 1992; Colvin and Gatehouse, 1993a, b; Wilson and Gatehouse, 1993).

5.4. Weather and Insect Movement

Weather has a very strong influence on insect migration, as it does on most other aspects of their ecology and behaviour. In particular, temperature and moisture will determine (directly or indirectly), the *number* of potential migrants, *when* they will go, and their subsequent survival and reproduction in their new location. Beyond this, the airborne nature of most insect migration ensures that it is especially sensitive to atmospheric processes. The effects of weather on insect migration have been discussed in Johnson (1969), Rainey (1974, 1976), Pedgley (1982), McManus (1988), Drake and Farrow (1988) and Drake and Gatehouse (1995), and a further review is outside the scope of this chapter. Here we restrict our discussion to the atmospheric features which most influence insect migratory flight, and outline some biometeorological techniques for the investigation of aerial migration.

Wind speed decreases close to the ground because of friction with the earth's surface. As a result, there is usually a layer (known as the *flight boundary layer* (Taylor, 1974)) where the wind speed is lower than insects' flight speeds, and where they can therefore control their displacement direction (track): the depth of this region obviously varies with species and with the vertical gradient of wind strength. The decrease in wind speed at low altitude is especially marked during clear nights, when radiative cooling of the ground may reduce air temperatures near the surface to values below those higher up. Within this *surface temperature inversion* (Pedgley, 1982) the air is hydrostatically stable, vertical mixing is therefore inhibited and winds can become very light. At the top of the inversion, however, there is frequently a region where the vertical profile of wind speed has a local maximum and forms a type of *low-level jet* (Drake and Farrow, 1988).

Once above their flight boundary layer, insects will be carried more or less downwind, no matter in which direction they attempt to fly. On the other hand, the stronger winds ensure that their displacement speeds will usually be much higher than they could achieve in still air, and this is always the case if their orientation is in the general downwind direction (which it often is (Riley and Reynolds, 1986)) so that wind and insect air speeds are additive. It has been found that many insects, not only small, slow-flying ones, but also many large, fast flyers such as moths and grasshoppers (Drake and Farrow, 1988; Drake and Gatehouse, 1995) achieve long-range displacement at minimum energy cost by purposefully flying at high altitude. Radar observations of nocturnally-migrating insects have revealed that they are often concentrated in layers (Schaefer, 1976; Drake and Farrow, 1988; Riley *et al.*, 1991), sometimes near to the top of the nighttime surface temperature inversion where their speed tends to be maximized by the strong winds of the low-level jet. The combination of automatic monitoring of wind and temperature profiles using several types of acoustic and electromagnetic sounder (Drake *et al.*, 1994), with measurements of insect density profiles from vertical-looking entomological radars, seems likely to lead to a better understanding of the phenomenon of insect layering.

Another atmospheric feature that can exert a profound influence on the movements of airborne insects is *wind convergence* (Drake and Farrow, 1988;

Pedgley, 1990). Zones of strong horizontal convergence, e.g. storm outflows and sea-breeze fronts, can increase aerial densities of insects by two orders of magnitude. Wind convergence zones can be investigated with appropriately-instrumented aircraft; Rainey (1976) used a Doppler navigation system, but modern Global Positioning Systems (GPS) now provide a much less expensive alternative. Simultaneous measurements of insect aerial density can be made if the aircraft is also equipped with a downward-looking entomological radar (Schaefer, 1979).

5.4.1. Trajectory analysis

It is possible to use standard maps of windfields to estimate the path taken by a hypothetical, non-dispersing parcel of air embedded in the wind flow. These paths or *trajectories* have been used for many years to describe the probable flight paths, and thus the likely sources and destinations, of migrant insects (Pedgley, 1982; Scott and Achtemeier, 1987). The assumption implicit in these descriptions is that the movement of airborne insects is essentially the same as that of air parcels at their flight height, and that in this sense they may be considered to be *windborne*. Pedgley (1982) describes how to construct trajectories from the *streamlines* of wind direction and the *isotachs* of wind speed plotted on windfield maps. He also describes some of the errors associated with trajectory construction.

Windfield maps are derived directly from wind observations, and/or indirectly from analyses of barometric pressure. Such maps can be obtained from national meteorological services, but since they are usually drawn for routine forecasting use, careful re-mapping may be needed to define the windfields with enough precision for insect trajectory work. For example, winds are often calculated from a map of barometric pressure, using the *geostrophic approximation* that they will be largely unaffected by friction with the earth's surface, and so will flow parallel to the isobars. While these calculations are usually adequate to represent larger-scale wind flows, they are inappropriate near to the equator where the geostrophic approximation does *not* hold and windfields should be estimated directly from wind observations. In tropical areas where upper air data are too sparse for adequate trajectory calculation, it may be possible to use daytime *surface* windfields which are often available with a resolution of tens of kilometres. This is because vertical mixing of the atmosphere during the daytime usually ensures that the winds at altitude are at least similar to those at the surface. In relatively undisturbed windfields, night-time winds at altitude are likely to be similar to those of the preceding and following days, so *daytime surface* winds can give an indirect guide to *night-time* winds *aloft* (Tucker, 1994).

Unlike inert particles such as pollen or fungal spores, migrating insects actively control their height of flight, and the discovery by radar that migrants commonly concentrate in well defined shallow layers indicates that they have *preferred* flight altitudes. Without information about the altitudinal distribution of insect migrants, and about the corresponding winds, it is obviously difficult to calculate meaningful migration trajectories. The lowest altitude for which upper-air wind data are widely available is that of the 850 hPa (= 850 mbar) pressure surface, corresponding to a height of approximately 1500 m above sea level, and this is unfortunately higher than the altitudes at which insects are most commonly observed to fly. However, trajectories computed from the winds at this level and from those at the surface do at least provide estimates of likely extremes in windborne displacement (e.g. Rosenberg and Magor, 1983).

Recent work on insect migration in the USA has made use of air parcel trajectories that are calculated on a regular basis from National Weather Service computer models (Scott and Achtemeier, 1987; McCorcle and Fast, 1989; Showers *et al.*, 1989). These trajectories were regularly available for surface winds and for 850 hPa pressure surfaces, and comprised two 12-hour segments ending at 0600 and 1800 hours local time for designated end points in the USA. Sources of inaccuracy in trajectory construction using these methods have been investigated by Scott and Achtemeier (1987) using wind data at different spatial and temporal resolutions. In general, potential inaccuracies in calculated trajectories accumulate rapidly with their length, so plots for flights of more than a day or two are rarely worthwhile. Trajectories for windborne insects have been measured directly by tracking specially designed balloons (tetroons) ballasted to drift at a representative flight height, and for the estimated flight period, of the species in question (Westbrook *et al.*, 1995).

5.5. References

Akey, D.H. (1991) A review of marking techniques in arthropods and an introduction to elemental marking. *Southwestern Entomologist* Suppl. 14, 1–8.

Akey, D.H. and Burns, D.W. (1991) Analytical consideration and methodologies for elemental determinations in biological samples. *Southwestern Entomologist* Suppl. 14, 25–36.

Argauer, R.J. and Cantelo, W.W. (1972) Spectrofluorometric determination of fluorescein-tagged tobacco hornworms. *Journal of Economic Entomology* 65, 539–542.

Armes, N.J., King, A.B.S., Carlaw, P.M. and Gadsden, H. (1989) Evaluation of strontium as a trace-element marker for dispersal studies on *Heliothis armigera*. *Entomologia Experimentalis et Applicata* 51, 5–10.

Ashmole, N.P. and Ashmole, M.J. (1988) Insect dispersal on Tenerife, Canary Islands: high altitude fallout and seaward drift. *Arctic and Alpine Research* 20, 1–12.

Aubert, J. (1969) Un appareil de capture de grandes dimensions destiné au marquage d'insectes migrateurs. *Mitteilungen der Schweizerischen Entomologischen Gesellschaft* 42, 135–139.

Baker, P.S., Cooter, R.J., Chang, P.M. and Hashim, H.B. (1980) The flight capabilities of laboratory and tropical field populations of the brown planthopper, *Nilaparvata lugens* (Stål) (Hemiptera: Delphacidae). *Bulletin of Entomological Research* 70, 589–600.

Baker, R.R. (1978) *The Evolutionary Ecology of Animal Migration*. Hodder & Stoughton, London.

Barfield, C.S., Waters, D.J. and Beck, H.W. (1988) Flight device and database management system for quantifying insect flight and oviposition. *Journal of Economic Entomology* 81, 1506–1509.

Bartlett, A.C. and Lingren, P.D. (1984) Monitoring pink bollworm (Lepidoptera: Gelechiidae) populations, using the genetic marker sooty. *Environmental Entomology* 13, 543–550.

Beerwinkle, K.R., Lopez, J.D., Bouse, L.F. and Brusse, J.C. (1989) Large helicopter-towed net for sampling airborne arthropods. *Transactions of the American Society for Agricultural Engineers* 32, 1847–1852.

Beerwinkle, K.R., Lopez, J.D., Schleider, P.G. and Lingren, P.D. (1994) Annual patterns of aerial insect densities at altitudes from 500 to 2300 meters in east-central Texas indicated by continuously-operating vertical-looking radar. In *Proceedings of 21st Conference on Agricultural and Forest Meteorology – 11th Conference on Biometeorology, 7–11 March 1994, San Diego, California.* American Meteorological Society, Boston, pp. 415–418.

Begon, M. (1979) *Investigating Animal Abundance: Capture–Recapture for Biologists.* Edward Arnold, London.

Bell, M.R. (1988) *Heliothis virescens* and *H. zea* (Lepidoptera: Noctuidae) feasibility of using oil-soluble dye to mark populations developing on early-season host plants. *Journal of Entomological Science* 23, 223–228.

Bennet, S.R., McClelland, G.A.H. and Smilanick, J.M. (1981) A versatile system of fluorescent marks for studies of large populations of mosquitoes (Diptera: Culicidae). *Journal of Medical Entomology* 18, 173–174.

Best, R.L., Beegle, C.C., Owens, J.C. and Oritz, M. (1981) Population density, dispersion and dispersal estimates for *Scarites substriatus, Pterostichus chalcites* and *Harpalus pennsylvanicus* (Carabidae) in an Iowa cornfield. *Environmental Entomology* 10, 847–856.

Blackmer, J.L. and Phelan, P.L. (1991) Behavior of *Carpophilus hemipterus* in a vertical flight chamber: transition from phototactic to vegetative orientation. *Entomologia Experimentalis et Applicata* 58, 137–148.

Bowden, J. and Gibbs, D.G. (1973) Light-trap and suction-trap catches of insects in the northern Gezira, Sudan, in the season of the southward movement of the Inter-Tropical Front. *Bulletin of Entomological Research* 62, 571–596.

Bowden, J. and Johnson, C.G. (1976) Migrating and terrestrial insects at sea. In: Cheng, L. (ed.) *Marine Insects.* North-Holland Publishing Company, Amsterdam, pp. 97–117.

Bowden, J., Digby, P.G.N. and Sherlock, P.L. (1984) Studies of elemental composition as a biological marker in insects. I. The influence of soil type and host-plant on elemental composition of *Noctua pronuba* (L.) (Lepidoptera: Noctuidae). *Bulletin of Entomological Research* 74, 207–225.

Brown, E.S. (1970) Nocturnal insect flight direction in relation to the wind. *Proceedings of the Royal Entomological Society of London (A)* 45, 39–43.

Burton, R.L. and Wendell-Snow, J. (1970) A marker dye for the corn earworm. *Journal of Economic Entomology* 63, 1976–1977.

Callahan, P.S., Sparks, A.N., Snow, J.W. and Copeland, W.W. (1972) Corn earworm moth: vertical distribution in nocturnal flight. *Environmental Entomology* 1, 497–503.

Coats, S.A., Tollefson, J.J. and Mutchmor, J.A. (1986) Study of migratory flight in the western corn rootworm (Coleoptera: Chrysomelidae). *Environmental Entomology* 15, 1–6.

Cockbain, A.J. (1961) Fuel utilization and duration of tethered flight in *Aphis fabae* Scop. *Journal of Experimental Biology* 38, 163–174.

Colvin, J. (1995) The regulation of migration in *Helicoverpa armigera*. In: Drake, V.A. and Gatehouse, A.G. (eds) *Insect Migration: Tracking Resources through Space and Time*. Cambridge University Press, Cambridge, UK, pp. 265–277.

Colvin, J. and Gatehouse, A.G. (1993a) The reproduction-flight syndrome and the inheritance of tethered-flight potential in the cotton-bollworm moth, *Heliothis armigera*. *Physiological Entomology* 18, 16–22.

Colvin, J. and Gatehouse, A.G. (1993b) Migration and genetic regulation of the pre-reproductive period in the cotton-bollworm moth, *Helicoverpa armigera*. *Heredity* 70, 407–412.

Cooter, R.J. (1982) Studies on the flight of black-flies (Diptera: Simuliidae). I. Flight performance of *Simulium ornatum* Meigen. *Bulletin of Entomological Research* 72, 303–317.

Cooter, R.J. (1983) Studies on the flight of black-flies (Diptera: Simuliidae). II. Flight performance of three cytospecies in the complex of *Simulium damnosum* Theobald. *Bulletin of Entomological Research* 73, 275–288.

Cooter, R.J. (1993) The flight potential of insect pests and its estimation in the laboratory: techniques, limitations and insights. *The Gooding Memorial Lecture, given to The Central Association of Bee-Keepers, 13 September 1992*. The Central Association of Bee-Keepers, Hadleigh, Essex, UK.

Cooter, R.J. and Armes, N.J. (1993) Tethered flight technique for monitoring the flight performance of *Helicoverpa armigera* (Lepidoptera: Noctuidae). *Environmental Entomology* 22, 339–345.

Culin, J.D. and Alverson, D.R. (1986) A technique to mark adult *Heliothis zea* using rubidium chloride spiked artificial nectar sources. *Journal of Agricultural Entomology* 3, 56–60.

Dahmen, H.-J. and Zeil, J. (1984) Recording and reconstructing three-dimensional trajectories: a versatile method for the field biologist. *Proceedings of the Royal Society of London B* 222, 107–113.

Daly, J.C. and Gregg, P. (1985) Genetic variation in *Heliothis* in Australia: species identification and gene flow in the two pest species *H. armigera* (Hübner) and *H. punctigera* Wallengren (Lepidoptera: Noctuidae). *Bulletin of Entomological Research* 75, 169–184.

Danthanarayana, W. (ed.) (1986) *Insect Flight: Dispersal and Migration*. Springer-Verlag, Berlin/Heidelberg.

David, C. T. and Hardie, J. (1988) The visual responses of free-flying summer and autumn forms of the black bean aphid, *Aphis fabae*, in an automated flight chamber. *Physiological Entomology* 13, 277–284.

Dempster, J.P., Lakhani, K.H. and Coward, P.A. (1986) The use of chemical composition as a population marker in insects: a study of the Brimstone butterfly. *Ecological Entomology* 11, 51–65.

Dent, D. (1991) *Insect Pest Management*. CAB International, Wallingford, UK.

Dickson, R.C. (1959) Aphid dispersal over southern California deserts. *Annals of the Entomological Society of America* 52, 368–372.

Dingle, H. (1985) Migration. In: Kerkut, G.A. and Gilbert, L.I. (eds) *Comprehensive Insect Physiology, Biochemistry and Pharmacology, Volume 9, Behaviour*. Pergamon Press, Oxford, pp. 375–415.

Dingle, H. (1989) The evolution and significance of migratory flight. In:

Goldsworthy, G.J. and Wheeler, C.H. (eds) *Insect Flight*. CRC Press, Boca Raton, Florida, pp. 99–114.

Dingle, H. (1994) Genetic analyses of animal migration. In: Boake, C.R.G. (ed.) *Quantitative Genetic Studies of Behavioural Evolution*. University of Chicago Press, Chicago, pp. 145–164.

Drake, V.A. (1981a) Target density estimation in radar biology. *Journal of Theoretical Biology* 90, 545–571.

Drake, V.A. (1981b) Quantitative observation and analysis procedures for a manually operated entomological radar. *Commonwealth Scientific and Industrial Research Organisation, Australia, Division of Entomology Technical Paper* No. 19, 1–41.

Drake, V.A. (1990) Methods for studying adult movement in *Heliothis*. In: Zalucki, M.P. (ed.) *Heliothis: Research Methods and Prospects*. Springer-Verlag, New York, pp. 109–121.

Drake, V.A., (1993) Insect-monitoring radar: a new source of information for migration research and operational pest forecasting. In: Corey, S.A., Dall, D.J. and Milne, W.M. (eds) *Pest Control and Sustainable Agriculture*, CSIRO Publications, Melbourne, pp. 452–455.

Drake, V.A. and Farrow, R.A. (1983) The nocturnal migration of the Australian plague locust *Chortoicetes terminifera* (Walker) (Orthoptera: Acrididae): quantitative radar observations of a series of northward flights. *Bulletin of Entomological Research* 73, 567–585.

Drake, V.A. and Farrow, R.A. (1988) The influence of atmospheric structure and motions on insect migration. *Annual Review of Entomology* 33, 183–210.

Drake, V.A. and Gatehouse, A.G. (eds) (1995) *Insect Migration: Tracking Resources through Space and Time*. Cambridge University Press, Cambridge, UK.

Drake, V.A., Harman, I.T. Bourne, I.A. and Woods, G.J.C. (1994) Simultaneous entomological and atmospheric profiling: a novel technique for studying the biometeorology of insect migration. In: *Proceedings of 21st Conference on Agricultural and Forest Meteorology – 11th Conference on Biometeorology, 7–11 March, San Diego, California*. American Meteorological Society, Boston, pp. 444–447.

Eddlestone, F.K., Setter, J. and Schofield, P. (1984) Insect marking methods for dispersal and other ecological studies. *TDRI Information Service Annotated Bibliographies Series no. 4*, Tropical Development and Research Institute, London.

Edwards, J.S. (1987) Arthropods of alpine aeolian ecosystems. *Annual Review of Entomology* 32, 13–179.

Eitschberger, U., Reinhardt, R. and Steiniger, H. (1991) Wanderfalter in Europa. Zugleich Aufruf für eine internationale Zusammenarbit an der Erforschung des Wanderphänomens bei den Insekten. *Atalanta* 22 (1) 1–67 + figs. [with Eng., Fr. and Sp. translations].

Farrow, R.A. (1984) Detection of transoceanic migration of insects to a remote island in the Coral Sea, Willis Island. *Australian Journal of Ecology* 9, 253–272.

Farrow, R.A. (1990) Flight and migration in acridoids. In: Chapman, R.F. and Joern, A. (eds) *Biology of Grasshoppers*. John Wiley & Sons, New York, pp. 227–314.

Farrow, R.A. and Dowse, J.E. (1984) Method of using kites to carry tow nets in the upper air for sampling migratory insects and its application to radar entomology. *Bulletin of Entomological Research* 74, 87–95.

Ferguson, D.C. (1991) An essay on the long-range dispersal and biogeography of Lepidoptera, with special reference to the Lepidoptera of Bermuda. *Memoirs of the Entomological Society of Canada* 158, 67–84.

Fleischer, S.J., Gaylor, M.J., Hue, N.V. and Graham, L.C. (1986) Uptake and elimination of rubidium, a physiological marker, in adult *Lygus lineolaris* (Hemiptera: Miridae). *Annals of the Entomological Society of America* 79, 19–25.

Fleischer, S.J., Gaylor, M.J. and Hue, N.V. (1988) Dispersal of *Lygus lineolaris* (Hemiptera: Miridae) adults through cotton following nursery host destruction. *Environmental Entomology* 17, 533–541.

Forsse, E. (1987) Flight duration in *Ips typographus* L.: insensitivity to nematode infection. *Journal of Applied Entomology* 104, 326–328.

Gangwere, S.K., Chavin, W. and Evans, F.C. (1964) Methods of marking insects, with especial reference to Orthoptera (Sens. lat.). *Annals of the Entomological Society of America* 57, 662–669.

Garrison, R.W. (1978) A mark–recapture study of imaginal *Enallagma cyathigerum* (Charpentier) and *Argia vivida* Hagen (Zygoptera: Coenagrionidae). *Odonatologica* 7, 223–236.

Gary, N.E. (1971) Magnetic retrieval of ferrous labels in a capture–recapture system for honey bees and other insects. *Journal of Economic Entomology* 64, 961–965.

Gatehouse, A.G. (1987) Migration: a behavioural process with ecological consequences. *Antenna* 11, 10–12.

Gatehouse, A.G. and Hackett, D.S. (1980) A technique for studying flight behaviour of tethered *Spodoptera exempta* moths. *Physiological Entomology* 5, 215–222.

Gatehouse, A.G. and Woodrow, K.P. (1987) Simultaneous monitoring of flight and oviposition of individual velvetbean caterpillar moths (by Wales, Barfield and Leppla, 1985): a critique. *Physiological Entomology* 12, 117–121.

Gatehouse, A.G. and Zhang, X.X. (1995) Migratory potential in insects: variation in an uncertain environment. In: Drake, V.A. and Gatehouse, A.G. (eds) *Insect Migration: Tracking Resources through Space and Time*. Cambridge University Press, Cambridge, UK, pp. 193–242.

Gauthreaux, S.A. (ed.) (1980) *Animal Migration, Orientation and Navigation*. Academic Press, New York.

Gentry, C.R. and Blythe, J.L. (1978) Lesser peachtree borers and peachtree borers: a device for trapping, collecting and marking native moths. *Environmental Entomology* 7, 783–784.

Gibo, D.L. (1986) Flight strategies of migrating Monarch butterflies (*Danaus plexippus* L.). In: Danthanarayana, W. (ed.) *Insect Flight: Dispersal and Migration*. Springer-Verlag, Berlin/Heidelberg, pp. 172–184.

Goldsworthy, G.J. and Wheeler, C.H. (eds) (1989) *Insect Flight*, CRC Press, Boca Raton, Florida.

Gottwald, T.R. and Tedders, W.L. (1986) MADDSAP-1, a versatile remotely piloted vehicle for agricultural research. *Journal of Economic Entomology* 79, 857–863.

Greenstone, M.H., Eaton, R.R. and Morgan, C.E. (1991) Sampling aerially dispersing arthropods: a high-volume, inexpensive, automobile- and aircraft-borne system. *Journal of Economic Entomology* 84, 1717–1724.

Gregg, P.C. (1993) Pollen as a marker for migration of *Helicoverpa armigera* and *H. punctigera* (Lepidoptera: Noctuidae) from western Queensland. *Australian Journal of Ecology* 18, 209–219.

Gregg, P.C., Fitt, G.P., Coombs, M. and Henderson, G.S. (1993) Migrating moths (Lepidoptera) collected in tower-mounted light traps in northern New South Wales, Australia: species composition and seasonal abundance. *Bulletin of Entomological Research* 83, 563–578.

Gunning, R.V. and Easton, C.S. (1989) Pyrethroid resistance in *Heliothis armigera*

(Hübner) collected from unsprayed maize crops in New South Wales 1983–1987. *Journal of the Australian Entomological Society* 28, 57–61.

Hardie, J. (1993) Flight behaviour in migrating insects. *Journal of Agricultural Entomology* 10, 239–245.

Hardy, A.C. and Cheng, L. (1986) Studies in the distribution of insects by aerial currents. III. Insect drift over the sea. *Ecological Entomology* 11, 283–290.

Harlan, D.P and Roberts, R.H. (1976) Tabanidae: use of a self-marking device to determine populations in the Mississippi–Yazoo river delta. *Environmental Entomology* 5, 210–212.

Hartstack, A.W., Lopez, J.D., Muller, R.A. and Witz, J.A. (1986) Early season occurrence of *Heliothis* spp. in 1982: evidence of long range migration of *Heliothis zea*. In: Sparks, A.N. (ed.) *Long-Range Migration of Moths of Agronomic Importance to the United States and Canada: Specific Examples of the Occurrence and Synoptic Weather Patterns Conducive to Migration.* United States Department of Agriculture, Agricultural Research Service ARS-43, pp. 48–68.

Hayes, J.L. (1989) Detection of single and multiple trace element labels in individual eggs of diet-reared *Heliothis virescens* (Lepidoptera: Noctuidae). *Annals of the Entomological Society of America* 82, 340–345.

Hegmann, J.P. and Dingle, H. (1982) Phenotypic and genetic covariance structure in milkweed bug life-history traits. In: Dingle, H. and Hegmann, J.P. (eds) *Evolution and Genetics of Life Histories.* Springer-Verlag, New York. pp. 177–184.

Hendrix, W.H. and Showers, W.B. (1992) Tracing black cutworm and armyworm (Lepidoptera: Noctuidae) northward migration using *Pithecellobium* and *Calliandra* pollen. *Environmental Entomology* 21, 1092–1096.

Hendrix, W.H., Mueller, T., Phillips, J.R. and Davis, O.K. (1987) Pollen as an indicator of long-distance movement of *Heliothis zea* (Lepidoptera: Noctuidae). *Environmental Entomology* 16, 1148–1151.

Hill, J.K. (1992) Regulation of migration and migratory strategy of *Autographa gamma* (L.) (Lepidoptera: Noctuidae). PhD Thesis, University of Wales.

Hill, L. (1993) Colour in adult *Helicoverpa punctigera* (Wallengren) (Lepidoptera: Noctuidae) as an indicator of migratory origin. *Journal of the Australian Entomological Society* 32, 145–151.

Hirao, J. (1985) Recent trends in the occurrence and forecasting procedures in the brown planthopper. *Chinese Journal of Entomology* 4, 65–76.

Hobbs, S.E. and Wolf, W.W. (1989) An airborne radar technique for studying insect migration. *Bulletin of Entomological Research* 79, 693–704.

Hollinger, S.E., Sivier, K.R., Irwin, M.E. and Isard, S.A. (1991) A helicopter-mounted isokinetic aerial insect sampler. *Journal of Economic Entomology* 84, 476–483.

Hopper, K.R. (1991) Ecological applications of elemental labeling: analysis of dispersal, density, mortality and feeding. *Southwestern Entomologist* Suppl., 14, 71–83.

Hopper, K.R. and Woolson, E.A. (1991) Labeling a parasitic wasp, *Microplitis croceipes* (Hymenoptera: Braconidae), with trace elements for mark–recapture studies. *Annals of the Entomological Society of America* 84, 255–262.

Ives, P.M. (1981) Estimation of coccinellid numbers and movement in the field. *The Canadian Entomologist* 113, 981–997.

Johnson, C.G. (1969) *Migration and Dispersal of Insects by Flight.* Methuen, London.

Kennedy, J.S. (1966) The balance between antagonistic induction and depression of flight activity in *Aphis fabae* Scopoli. *Journal of Experimental Biology* 45, 215–228.

Kennedy, J.S. (1985) Migration, behavioural and ecological. In: Rankin, M.A. (ed.)

Migration: Mechanisms and Adaptive Significance. Contributions in Marine Science 27 (supplement), 5–26.

Kennedy, J.S. and Booth, C.O. (1963) Free flight of aphids in a laboratory. *Journal of Experimental Biology* 40, 67–85.

Kennedy, J.S. and Ludlow, A.R. (1974) Co-ordination of two kinds of flight activity in an aphid. *Journal of Experimental Biology* 61, 173–196.

King, A.B.S., Armes, N.J. and Pedgley, D.E. (1990) A mark–capture study of *Helicoverpa armigera* dispersal from pigeonpea in southern India. *Entomologia Experimentalis et Applicata* 55, 257–266.

Knight, A.L., Hull, L.A., Rajotte, E.G. and Fleischer, S.J. (1990) Labeling tufted apple bud moth (Lepidoptera: Tortricidae) with rubidium: effect on development, longevity and fecundity. *Annals of the Entomological Society of America* 82, 481–485.

Korman, A.K., Mallet, J., Goodenough, J.L., Graves, J.B., Hayes, J.L., Hendricks, D.E., Luttrell, R., Pair, S.D. and Wall, M. (1993) Population structure in *Heliothis virescens* (Lepidoptera: Noctuidae): an estimate of gene flow. *Annals of the Entomological Society of America* 86, 182–188.

Levy, R. and Cromroy, H.L. (1973) Concentration of some major and trace elements in forty-one species of adult and immature insects determined by atomic absorption spectroscopy. *Annals of the Entomological Society of America* 66, 523–526.

Lingren, P.D., Raulston, J.R., Henneberry, T.J. and Sparks, A.N. (1986) Night-vision equipment, reproductive biology, and nocturnal behaviour: importance to studies of insect flight, dispersal, and migration. In: Danthanarayana, W. (ed.) *Insect Flight: Dispersal and Migration.* Springer-Verlag, Berlin/Heidelberg, pp. 253–264.

Lingren, P.D., Westbrook, J.K., Bryant, V.M., Raulston, J.R., Esquivel, J.F. and Jones, G.D. (1994) Origin of corn earworm (Lepidoptera: Noctuidae) migrants as determined by *Citrus* pollen markers and synoptic weather systems. *Environmental Entomology* 23, 562–570.

López, J.D. (1979) Recovery in blacklight traps of marked bollworms released in a multiple cropped area. *Southwestern Entomologist* 4, 46–52.

Loxdale, H.D., Hardie J., Halbert, S., Footit, R., Kidd, N.A.C. and Carter, C.I. (1993) The relative importance of short- and long-range movement of flying aphids. *Biological Reviews* 68, 291–311.

MacKenzie, D.R., Barfield, C.S., Kennedy, G.C., Berger, R.D. and Taranto, D.J. (eds) (1985) *The Movement and Dispersal of Agriculturally Important Biotic Agents.* Claitor's Publishing Division, Baton Rouge, Louisiana.

Mascanzoni, D. and Wallin, H. (1986) The harmonic radar: a new method of tracing insects in the field. *Ecological Entomology* 11, 387–390.

McAnelly, M.L. and Rankin, M.A. (1986) Migration in the grasshopper *Melanoplus sanguinipes* (Fab). I. The capacity for flight in non-swarming populations. *Biological Bulletin (Woods Hole, Mass.)* 170, 368–377.

McCaffery, A.R., King, A.B.S., Walker, A.J. and El-Nayir, H. (1989) Resistance to synthetic pyrethroids in the bollworm, *Heliothis armigera* from Andhra Pradesh, India. *Pesticide Science* 27, 65–76.

McCorcle, M.D. and Fast, J.D. (1989) Prediction of pest distribution in the Corn Belt: a meteorological analysis. *Proceedings of the 9th Conference on Biometeorology & Aerobiology.* American Meteorological Society, Boston, pp. 298–302.

McKibben, G.H. (1985) Computer-monitored flight mill for the boll weevil (Coleoptera: Curculionidae). *Journal of Economic Entomology* 78, 1519–1520.

McManus, M.L. (1988) Weather, behaviour and insect dispersal. *Memoirs of the Entomological Society of Canada* 146, 71–94.

Meinzingen, W.F. (ed.) (1993) *A Guide to Migrant Pest Management in Africa*. FAO, Rome.

Mikkola, K. (1971) Pollen analysis as a means of studying the migrations of Lepidoptera. *Annales Entomologici Fennici* 37, 136–139.

Moriya, S. (1987) Automatic data acquisition system for study of the flight ability of brown-winged green bug, *Plautia stali* (Hemiptera: Pentatomidae). *Applied Entomology and Zoology* 22, 19–24.

Muirhead-Thomson, R.C. (1991) *Trap Responses of Flying Insects*. Academic Press, London.

Nakasuji, F. and Nakano, A. (1990) Flight activity and oviposition characteristics of the seasonal form of a migrant skipper, *Parnara guttata guttata* (Lepidoptera: Hesperiidae). *Researches in Population Ecology* 32, 227–233.

Neilson, W.T.A. (1971) Dispersal studies of a natural population of apple maggot adults. *Journal of Economic Entomology* 64, 648–653.

Noda, T. and Kamano, S. (1988) Flight capacity of *Spodoptera litura* (F.) (Lepidoptera: Noctuidae) determined with a computer-assisted flight mill: effect of age and sex of the moth. *Japanese Journal of Applied Entomology and Zoology* 32, 227–229.

Padgham, D. (1983) Flight fuels in the brown planthopper *Nilaparvata lugens*. *Journal of Insect Physiology* 29, 95–99.

Padgham, D.E., Cook, A.G. and Hutchison, D. (1984) Rubidium marking of the rice pests *Nilaparvata lugens* (Stål) and *Sogatella furcifera* (Horváth) (Hemiptera: Delphacidae) for field dispersal studies. *Bulletin of Entomological Research* 74, 379–385.

Peck, S.B. (1994) Aerial dispersal of insects between and to islands in the Galápagos Archipelago, Ecuador. *Annals of the Entomological Society* 87, 218–224.

Pedgley, D.E. (ed.) (1981) *Desert Locust Forecasting Manual, Volumes 1 and 2*. Centre for Overseas Pest Research, London.

Pedgley, D.E. (1982) *Windborne Pests and Diseases: Meteorology of Airborne Organisms*. Ellis Horwood, Chichester, UK.

Pedgley, D.E. (1990) Concentration of flying insects by the wind. *Philosophical Transactions of the Royal Society of London B* 328, 631–653.

Pedgley, D.E. (1993) Managing migratory insects pests – a review. *International Journal of Pest Management* 39, 3–12.

Pedgley, D.E. and Yathom, S. (1993) Windborne moth migration over the Middle East. *Ecological Entomology* 18, 67–72.

Pedgley, D.E., Reynolds, D.R., Riley J.R. and Tucker, M.R. (1982) Flying insects reveal small-scale wind systems. *Weather* 37, 295–306.

Pedgley, D.E., Reynolds, D.R. and Tatchell, G.M. (1995) Long-range insect migration in relation to climate and weather: Africa and Europe. In: Drake, V.A. and Gatehouse, A.G. (eds) *Insect Migration: Tracking Resources through Space and Time*. Cambridge University Press, Cambridge, pp. 3–29.

Pollard, E. and Yates, T.J. (1993) *Monitoring Butterflies for Ecology and Conservation: the British Butterfly Monitoring Scheme*. Conservation Biology Series no. 1. Chapman & Hall, London.

Preiss, R. and Kramer, E. (1984) Control of flight speed by minimization of the apparent ground pattern movement. In: Varju, D. and Schnitzler, H.U. (eds)

Localisation and Orientation in Biology and Engineering, Springer-Verlag, Berlin / Heidelberg, pp. 140–142.

Preiss, R. and Spork, P. (1993) Flight-phase and visual-field related optomotor yaw responses in gregarious desert locusts during tethered flight. *Journal of Comparative Physiology A* 172, 733–740.

Rainey, R.C. (1963) *Meteorology and the Migration of Desert Locusts*. WMO Techical Note No. 54. World Meteorological Organization, Geneva.

Rainey, R.C. (1974) Biometeorology and insect flight: some aspects of energy exchange. *Annual Review of Entomology* 19, 407–439.

Rainey, R.C. (1976) Flight behaviour and features of the atmospheric environment. In: Rainey, R.C. (ed.) *Insect Flight*. Symposia of the Royal Entomological Society, No. 7. Blackwell, Oxford. pp. 75–112.

Rainey, R.C., Browning, K.A., Cheke, R.A. and Haggis, M.J. (eds) (1990) *Migrant Pests: Progress, Problems and Potentialities*. The Royal Society, London. *Philosophical Transactions of the Royal Society of London B* 328, 515–755.

Rankin, M.A. (ed.) (1985) *Migration: Mechanisms and Adaptive Significance. Contributions in Marine Science* 27 (supplement).

Rankin, M.A. and Burchsted, J.C.A. (1992) The cost of migration in insects. *Annual Review of Entomology* 37, 533–559.

Rankin, M.A. and Singer, M.C. (1984) Insect movement: mechanisms and effects. In: Huffaker, C.B. and Rabb, R.L (eds) *Ecological Entomology*. Wiley, New York, pp. 185–216.

Rankin, M.A., McAnelly, M.L. and Bodenhamer, J.E. (1986) The oogenesis-flight syndrome revisited. In: Danthanarayana, W. (ed.) *Insect Flight: Dispersal and Migration*. Springer-Verlag, Berlin/Heidelberg, pp. 27–38.

Raulston, J.R. (1979) Tagging of natural populations of Lepidoptera for studies of dispersal, flight and migration. In: Rabb, R.L. and Kennedy, G.G. (eds) *Movement of Highly Mobile Insects: Concepts and Methodologies in Research*. University Graphics, North Carolina State University, Raleigh, pp. 354–358.

Reling, D. and Taylor, R.A.J. (1984) A collapsible tow net used for sampling arthropods by airplane. *Journal of Economic Entomology* 77, 1615–1617.

Resurreccion, A.N., Showers, W.B. and Rowley, W.A. (1988) Microcomputer-interfaced flight mill system for large moths such as black cutworm (Lepidoptera: Noctuidae). *Annals of the Entomological Society of America* 81, 286–291.

Riley, J.R. (1974) Radar observations of individual desert locusts (*Schistocerca gregaria*) (Forsk.). *Bulletin of Entomological Research* 64, 19–32.

Riley, J.R. (1979) Quantitative analysis of radar returns from insects. In: Vaughn, C.R., Wolf, W. and Klassen, W. (eds) *Radar, Insect Population Ecology, and Pest Management*, NASA Conference Publication 2070. Wallops Island, Virginia, pp. 131–158.

Riley, J.R. (1989) Remote sensing in entomology. *Annual Review of Entomology* 34, 247–271.

Riley, J.R. (1992) A millimetric radar to study the flight of small insects. *Electronics and Communication Engineering Journal* 4, 43–48.

Riley, J.R. (1993) Flying insects in the field. In: Wratten, S.D. (ed.) *Video Techniques in Animal Ecology and Behaviour*. Chapman & Hall, London, pp. 1–15.

Riley, J.R. and Reynolds, D.R. (1979) Radar-based studies of the migratory flight of grasshoppers in the middle Niger area of Mali. *Proceedings of the Royal Society of London B* 204, 67–82.

Riley, J.R. and Reynolds, D.R. (1983) A long-range migration of grasshoppers observed in the Sahelian zone of Mali by two radars. *Journal of Animal Ecology* 52, 167–183.

Riley, J.R. and Reynolds, D.R. (1986) Orientation at night by high-flying insects. In: Danthanarayana, W. (ed.) *Insect Flight: Dispersal and Migration*. Springer-Verlag, Berlin, pp. 71–87.

Riley, J.R. and Reynolds D.R. (1990) Nocturnal grasshopper migration in West Africa: transport and concentration by the wind, and the implications for air-to-air control. *Philosophical Transactions of the Royal Society of London B* 328, 655–672.

Riley, J.R. and Reynolds, D.R. (1993) Radar monitoring of locusts and other migratory insects. In: Cartwright, A. (ed.) *World Agriculture 1993*. Sterling Publications, London, pp. 51–53.

Riley, J.R. and Reynolds, D.R. (1995) Monitoring the movement of migratory insect pests. *Proceedings of the Third International Conference on Tropical Entomology, Nairobi, 30 October – 4 November 1994*. (in press).

Riley, J.R., Reynolds, D.R. and Farmery, M.J. (1983) Observations of the flight behaviour of the armyworm moth, *Spodoptera exempta*, at an emergence site using radar and infra-red optical techniques. *Ecological Entomology* 8, 395–418.

Riley, J.R., Cheng, X.-N., Zhang, X.-X., Reynolds, D.R., Xu, G.-M., Smith, A.D., Cheng, J.-Y., Bao, A.-D. and Zhai, B.-P. (1991) The long-distance migration of *Nilaparvata lugens* (Stål) (Delphacidae) in China: radar observations of mass return flight in the autumn. *Ecological Entomology* 16, 471–489.

Riley, J.R., Smith, A.D. and Bettany, B.W. (1990) The use of video equipment to record in three dimensions the flight trajectories of *Heliothis armigera* and other moths at night. *Physiological Entomology* 15, 73–80.

Roderick, G.K. and Caldwell, R.L. (1992) An entomological perspective on animal dispersal. In: Stenseth, N.C. and Lidicker, W.Z. (eds) *Animal Dispersal: Small Mammals as a Model*. Chapman & Hall, London, pp. 274–290.

Roffey, J. (1963) Observations of the night flight of the Desert Locust (*Schistocerca gregaria* Forskål). *Anti-Locust Bulletin* no. 39.

Rose, D.J.W. and Khasimuddin, S. (1979) Wide area monitoring of the African armyworm, *Spodoptera exempta* (Walker); (Lepidoptera: Noctuidae). In: Rabb, R.L. and Kennedy, G.G. (eds) *Movement of Highly Mobile Insects: Concepts and Methodology in Research*. University Graphics, North Carolina State University, Raleigh, pp. 212–219.

Rose, D.J.W., Page, W.W., Dewhurst, C.F., Riley, J.R., Reynolds, D.R., Pedgley, D.E. and Tucker, M.R. (1985) Downwind migration of the African armyworm moth, *Spodoptera exempta*, studied by mark-and-capture and by radar. *Ecological Entomology* 10, 299–313.

Rose, D.J.W., Dewhurst, C.F. and Page, W.W. (1997) *The African Armyworm Handbook: its Status, Biology, Ecology, Epidemiology and Management, with Particular Reference to Eastern Africa*. Desert Locust Control Organisation for Eastern Africa, Nairobi (in press).

Rosenberg, L.J. and Magor, J.I. (1983) Flight duration of the brown planthopper, *Nilaparvata lugens* (Homoptera: Delphacidae). *Ecological Entomology* 8, 341–350.

Rudd, W.G. and Gandour, R.W. (1985) Diffusion model for insect dispersal. *Journal of Economic Entomology* 78, 295–301.

Ruzicka, Z. (1984) Two simple recording flight mills for the behavioural study of insects. *Acta Entomologica Bohemoslovaca* 81, 429–433.

Schaefer, G.W. (1976) Radar observations of insect flight. In: Rainey, R.C. (ed.) *Insect*

Flight, Symposia of the Royal Entomological Society, No. 7. Blackwell, Oxford, pp. 157–197.

Schaefer, G.W. (1979) An airborne radar technique for the investigation and control of migrating insect pests. *Philosophical Transactions of the Royal Society of London B* 287, 459–465.

Schaefer, G.W. and Bent, G.A. (1984) An infra-red remote sensing system for the active detection and automatic determination of insect flight trajectories (IRA-DIT). *Bulletin of Entomological Research* 74, 261–278.

Schmidt-Koenig, K. (1993) Orientation of autumn migration in the Monarch butterfly. In: Malcolm, S.B. and Zalucki, M.P. (eds) *Biology and Conservation of the Monarch Butterfly*. Natural History Museum of Los Angeles County Science Series contrib. no. 38, 275–283.

Scott, R.W. and Achtemeier, G.L. (1987) Estimating pathways of migrating insects carried in atmospheric winds. *Environmental Entomology* 16, 1244–1254.

Seely, M.K., Mitchell, D. and Louw, G.N. (1985) A field technique using iridium-192 for measuring subsurface depths in free-ranging Namib Desert beetles. *South African Journal of Science* 81, 686–688.

Sellers, R.F. (1980) Weather, host and vector – their interplay in the spread of insect-borne animal virus diseases. *Journal of Hygiene (Cambridge)* 85, 65–102.

Service, M.W. (1993) *Mosquito Ecology: Field Sampling Methods*, 2nd edn. Elsevier Applied Science, London.

Showers, W.B., Smelser, R.B., Keaster, A.J., Whitford, F., Robinson, J.F., Lopez, J.D. and Taylor, S.E. (1989) Recapture of marked black cutworm (Lepidoptera: Noctuidae) males after long-range transport. *Environmental Entomology* 18, 447–458.

Smith, A.D., Riley, J.R. and Gregory, R.D. (1993) A method for routine monitoring of the aerial migration of insects by using a vertical-looking radar. *Philosophical Transactions of the Royal Society of London B* 340, 393–404.

Smith, M.A.H. and MacKay, P.A. (1989) Seasonal variation in the photoperiodic responses of a pea aphid population: evidence for long-distance movements between populations. *Oecologia* 81, 160–165.

Snow, J.W., Cantelo, W.W. and Bowman, M.C. (1969) Distribution of the corn earworm on St. Croix, US Virgin Islands, and its relation to suppression programs. *Journal of Economic Entomology* 62, 606–611.

Southwood, T.R.E. (1978) *Ecological Methods with Particular Reference to the Study of Insect Populations*, 2nd edn. Chapman & Hall, London.

Spillman, J.J. (1980) The design of an aircraft-mounted net for catching airborne insects. In: *Trends in Airborne Equipment for Agriculture and Other Areas. Proceedings of a Seminar organised by the United Nation Economic Commission for Europe, Warsaw, 18–22 September 1978*. Pergamon Press, Oxford, pp. 169–180.

Stadelbacher, E.A. (1991) Bollworm and tobacco budworm (Lepidoptera: Noctuidae): labeling of adults produced in wild geranium, *Geranium dissectum*, treated with rubidium or strontium chloride. *Journal of Economic Entomology* 84, 496–501.

Stimmann, M.W. (1974) Marking insects with rubidium: imported cabbageworm marked in the field. *Environmental Entomology* 3, 327–328.

Taylor, L.R. (1974) Insect migration, flight periodicity and the boundary layer. *Journal of Animal Ecology* 43, 225–238.

Taylor, L.R. (1985) An international standard for the synoptic monitoring and dynamic mapping of migrant pest aphid populations. In: MacKenzie, D.R.,

Barfield, C.S., Kennedy, G.C., Berger, R.D. and Taranto, D.J. (eds) *The Movement and Dispersal of Agriculturally Important Biotic Agents*. Claitor's Publishing Division, Baton Rouge, Louisiana, pp. 337–380.

Taylor, L.R. (1986) The distribution of virus disease and the migrant vector aphid. In: McLean, G.D., Garrett, R.G. and Ruesink, W.G. (eds) *Plant Virus Epidemics: Monitoring, Modelling and Predicting Outbreaks*. Academic Press, Sydney, pp. 35–57.

Taylor, L.R. and Taylor, R.A.J. (1983) Insect migration as paradigm for survival by movement. In: Swingland, I.R. and Greenwood, P.J. (eds) *The Ecology of Animal Movement*. Clarendon Press, Oxford. pp. 181–214.

Tedders, W.L. and Gottwald, T.R. (1986) Evaluation of an insect collecting system and an ultra-low-volume spray system on a remotely piloted vehicle. *Journal of Economic Entomology* 79, 709–713.

Treat, A.E. (1979) Moth-borne mites and their hosts. In: Rabb, R.L. and Kennedy, G.G. (eds) *Movement of Highly Mobile Insects: Concepts and Methodology in Research*. University Graphics, North Carolina State University, Raleigh, pp. 359–368.

Tucker, M.R. (1994) Inter- and intra-seasonal variation in outbreak distribution of the armyworm, *Spodoptera exempta* (Lepidoptera: Noctuidae), in eastern Africa. *Bulletin of Entomological Research* 84, 275–287.

Tucker, M.R., Mwandoto, S. and Pedgley, D.E. (1982) Further evidence for wind-borne movement of armyworm moths, *Spodoptera exempta*, in East Africa. *Ecological Entomology* 7, 463–473.

Turchin, P., Odendaal, F.J. and Rausher, M.D. (1991) Quantifying insect movement in the field. *Environmental Entomology* 20, 955–963.

Turner, R.J. and Bowden, J. (1983) An evaluation of the use of elytra and bodies in X-ray energy-dispersive spectroscopic studies of the red turnip beetle, *Entomoscelis americana* (Coleoptera: Chrysomelidae). *The Canadian Entomologist* 112, 609–614.

Turnock, W.J., Chong, J. and Luit, B. (1971) Scanning electron microscopy: A direct method of identifying pollen grains on moths. *Canadian Journal of Zoology* 56, 2050–2054.

Unruh, T.R. and Chauvin, R.L. (1993) Elytral punctures: a rapid, reliable method for marking Colorado potato beetle. *The Canadian Entomologist* 125, 55–63.

Urquhart, F.A. and Urquhart, N.R. (1979) Vernal migration of the monarch butterfly, *Danaus plexippus* (Lepidoptera: Danaidae) in North America from the overwintering site in the neo-volcanic plateau of Mexico. *The Canadian Entomologist* 111, 15–18.

Van Steenwyk, R.A. (1991) The use of elemental marking for insect dispersal and mating competitiveness studies: from the laboratory to the field. *Southwestern Entomologist* Suppl. 14, 15–23.

Van Steenwyk, R.A., Ballmer, G.R., Page, A.L., Ganje, T.J. and Reynolds, H.T. (1978) Dispersal of rubidium-marked pink bollworm. *Environmental Entomology* 7, 608–613.

Wada, T., Ogawa, Y. and Nakasuga, T. (1988) Geographical difference in mated status and autumn migration in the rice leaf roller moth, *Cnaphalocrocis medinalis*. *Entomologia Experimentalis et Applicata* 46, 141–148.

Wada, T., Ito, K. and Takahashi, A. (1994) Biotype comparisons of the brown plant-hopper, *Nilaparvata lugens* (Homoptera: Delphacidae) collected in Japan and the Indochina Peninsula. *Applied Entomology and Zoology* 29, 477–484.

Wainhouse, D. (1980) Dispersal of first instar larvae of the felted beech scale, *Cryptococcus fagisuga*. *Journal of Applied Ecology* 17, 523–532.

Wales, P.J., Barfield, C.S. and Leppla, N.C. (1985) Simultaneous monitoring of flight and oviposition of individual velvetbean caterpillar moths. *Physiological Entomology* 10, 467–472.

Walker, T.J. (1985) Permanent traps for monitoring butterfly migration: tests in Florida, 1979–84. *Journal of the Lepidopterists' Society* 39, 313–320.

Walker,T.J. (1991) Butterfly migration from and to peninsular Florida. *Ecological Entomology* 16, 241–252.

Wallin, H. and Ekbom, B.S. (1988) Movements of carabid beetles (Coleoptera: Carabidae) inhabiting cereal fields: a field tracing study. *Oecologia* 77, 39–43.

Wiens, J.A., Crist, T.O. and Milne, B.T. (1993) On quantifying insect movements. *Environmental Entomology* 22, 709–715.

Westbrook, J.K., Eyster, R.S., Wolf, W.W., Lingren, P.D. and Raulston, J.R. (1995) Migration pathways of corn earworm (Lepidoptera: Noctuidae) indicated by tetroon trajectories. *Agricultural and Forest Meteorology* 73, 67–87.

Williams, C.B. (1930) *The Migration of Butterflies*. Oliver & Boyd, Edinburgh.

Williams, C.B. (1965) *Insect Migration*, 2nd edn. Collins, London.

Wilson, K. and Gatehouse, A.G. (1993) Seasonal and geographical variation in the migratory potential of outbreak populations of the African armyworm moth, *Spodoptera exempta*. *Journal of Animal Ecology* 62, 169–181.

Woiwod, I.P. and Harrington, R. (1994) Flying in the face of change – the Rothamsted Insect Survey. In: Leigh, R.A. and Johnson, A.E. (eds) *Long-Term Experiments in Agricultural and Ecological Sciences*. CAB International, Wallingford, UK, pp. 321–342.

Wolf, W.W., Westbrook, J.K., Raulston, J., Pair, S.D. and Hobbs, S.E. (1990) Recent airborne observations of migrant pests in the United States. *Philosophical Transactions of the Royal Society of London B* 328, 619–630.

Wolfenbarger, D.A., Graham, H.M., Nosky, J.B. and Lindig, O.H. (1982) Boll weevil (Coleoptera: Curculionidae): marking with rubidium chloride sprays on cotton and dispersal from cotton. *Journal of Economic Entomology* 75, 1038–1041.

Wolfenbarger, D.A., Lukefahr, M.J. and Graham, H.M. (1973) LD_{50} values of methyl parathion and endrin to tobacco budworms and bollworms collected in the Americas and hypothesis on the spread of resistance in these lepidopterans to these insecticides. *Journal of Economic Entomology* 66, 211–216.

Woodrow, K.P., Gatehouse, A.G. and Davis, D.A. (1987) The effect of larval phase on flight performance of the African armyworm moth, *Spodoptera exempta* (Walker) (Lepidoptera: Noctuidae). *Bulletin of Entomological Research* 77, 113–122.

Young, J.R. (1979) Assessing the movement of the fall armyworm (*Spodoptera frugiperda*) using insecticide resistance and wind patterns. In: Rabb, R.L. and Kennedy, G.G. (eds) *Movement of Highly Mobile Insects: Concepts and Methodology in Research*. University Graphics, North Carolina State University, Raleigh, pp. 344–351.

Zalucki, M.P., Kitching, R.L., Abel, D. and Pearson, J. (1980) A novel device for tracking butterflies in the field. *Annals of the Entomological Society of America* 73, 262–265.

Evaluation of Factors Affecting Host Plant Selection, with an Emphasis on Studying Behaviour

6

S.D. Eigenbrode[1] and E.A. Bernays[2]
[1]*Department of Plant, Soil and Entomological Sciences, University of Idaho, Moscow, Idaho 83844-2339, USA;* [2]*Department of Entomology, University of Arizona, Tucson, AZ 85721, USA*

6.1. Introduction

Whether the motivation for study of insect–plant interactions is elucidation of ecological and evolutionary relationships, or development of resistant crop varieties, it is often necessary to identify the plant factors mediating host selection. This requires study of the relevant activities of the herbivores. In practice these are frequently inferred from outcomes such as the number of eggs laid, amounts of leaf discs consumed, or insect locations in choice arenas (many examples in Miller and Miller, 1986; Smith *et al.*, 1994). Such end point bioassays often are inferior to direct observations, which provide more and less-ambiguous information about the plant–insect interaction under study. Observational studies can be employed effectively at all stages of investigation – from the documentation of behaviours as they occur in the field, to the identification of specific plant factors that mediate these behaviours. In spite of the advantages, continuous observations of plant selection by insect herbivores is underutilized as a research tool because it is considered by many to be too difficult and laborious. In this chapter we provide examples of the effectiveness of direct observation for the study of plant–insect interactions, as well as a brief guide to methods. We hope thereby to encourage the use of observational methods to identify the factors mediating host plant selection by insect herbivores.

6.2. Observations of Host Selection Behaviour in the Field

Study of host selection ideally begins with observations of undisturbed insects in their natural habitat (Opp and Prokopy, 1986). The fundamental knowledge gained ensures the relevance of subsequent experimental work aimed at identifying mechanisms. Field observations can determine whether insects actively orientate to the host, and, after locating the plant, whether and how they examine the surface, make a test bite, or feed briefly and then reject the food or feed continuously. Such observations can indicate the use of specific cues; for example, visual or volatile cues during orientation, tactile or gustatory cues during examining. These can then be studied experimentally. Direct observation in the field can also determine the ecological importance of apparently inconsequential behaviour. For example, a reduced feeding rate that does not affect survival in the laboratory may interact with environmental stresses to greatly reduce survival in nature.

6.2.1. Factors unrelated to the host

Host plant selection may be influenced more by microclimate, vegetational associations, the presence of other insects, and other environmental factors than by the host plant. It is important to detect such influences before focusing a research effort on plant factors alone. For example, the speckled wood butterfly, *Pararge aegeria*, lays eggs on various species of grass, but the choice of species depends on the temperature of the leaves, which must be between 24 and 30°C. Thus in spring in Britain the host may be a grass species in a sunny environment, while in summer it may be a grass species that grows only in a shaded habitat (Shreeve, 1986). Similarly, extensive field observations of the tephritid *Anastrepha obliqua*, demonstrated that flies in a warm Mexican habitat avoided using preferred hosts even when there was plenty of fruit in which to oviposit if adjacent shade trees were absent (Aluja and Birke, 1993).

Host plant associations may also be influenced by interactions among arthropods on the plants. Direct observation revealed that mutualists influence host selection by a myrmecophilous lycaenid butterfly (Pierce and Elgar, 1985), and that the presence of predators influenced oviposition by *Heliconius* butterflies (Smiley, 1978). Papaj (1994) observed tephritid flies *Rhagoletis juglandis* and *R. boycei* near their host plants during oviposition and mating after the flies had alighted on walnut fruits, and found evidence for intraspecific competition for matings and for the host resource that could then be tested more explicitly in the laboratory.

6.2.2. Plant cues effective from a distance

Butterflies can be tracked during ovipositional flights (Root and Kareiva, 1984), and the flight paths used to determine the importance of distant cues using the 'flight path sampling technique' (Stanton, 1982). In this way Stanton showed that the butterfly, *Colias philodice*, had a significant pre-alighting preference for plants that were most acceptable after landing.

Smaller insects may also be tracked in flight in the field to observe their orientation to plant cues. Observations of tephritid flies demonstrated movements towards host trees and host fruit (Roitberg and Prokopy, 1982).

Aluja and Prokopy (1992) used a three-dimensional grid system to record the movements of individual *Rhagoletis pomonella* within blocks of trees and found that within a few metres, flies showed odour-mediated positive anemotactic responses to host odour emanating from vials.

Walking insects may also be observed responding to host plants at a distance. For example, although the grasshopper, *Brachystola magna*, does not orientate to host plants while walking in the field, continuous observations showed that favoured host plants were encountered more often than predicted by chance, indicating some distance effect (Bright *et al.*, 1994).

6.2.3. Short range orientation and examination of the plant surface

Field observations can indicate the relative importance of factors acting only very close to the plant, or after contact. This is feasible even with small insects, as demonstrated by Kennedy *et al.*'s (1959) pioneering investigations of winter host location by aphids. By following individual insects in the field, they found, for example, that *Aphis fabae* alighted with equal frequency on the winter host trees and non-host trees, eliminating a role of long distance attraction. Soon after contact, however, all of the aphids took off again from the non-host, while only a few took off again from the host plant. Antennation and movements of the insects on the plants were also recorded, and the study suggested the importance of contact for winter host assessment by this species in the field.

Recent work by Visser and Piron (1995) demonstrates the chemosensory sensitivity of an aphid to volatiles, and other laboratory studies (Nottingham *et al.*, 1991) show that aphids can respond to volatiles. How the sensory and behavioural capacity of aphids to respond to volatiles is used in host finding, however, must also be demonstrated using observations in the natural context.

A more recent example with a larger insect is Damman and Feeny's (1988) observations of Zebra swallowtail butterflies, *Eurytides marcellus*, under natural conditions. The host plant, *Asimina* spp., was approached by the butterflies twice as often as non-hosts. At closer range, the non-hosts were usually rejected prior to landing while hosts were rejected only about 25% of the time at this stage. Another 50% of these host plants were then rejected after contact. The pattern suggested a predominance of close range volatile and visual cues for detecting the host plant species and contact cues for discriminating among host plants. By contrast, Zalucki *et al.* (1990) demonstrated that wild, free-flying *Danaus plexippus* selected among potential host plants only after they had alighted.

Nymphs of *Locusta migratoria* observed over periods of hours in the field in Australia were found to reject non-host plants at palpation, and before contacting the internal constituents of a plant (Bernays *et al.*, 1976). Another acridid, *Taeniopoda eques*, observed continuously over long periods in Arizona rarely palpated many unacceptable plant species, while others were rejected after palpation or biting (Raubenheimer and Bernays, 1993).

6.2.4. Manipulations to facilitate field observations

MANIPULATING THE INSECTS

Commonly, wild insects are not available just when all is ready for a detailed study, or their physiological state is unknown and probably variable. A good example of the use of standardized insects is a study of foraging behaviour by female *Rhagoletis pomonella* (Roitberg *et al.*, 1982). Puparia collected from infested fruit in the field were subjected to standard conditions in the laboratory. Well-fed females of known age, maturity and experience were then used for field observations. Flies released on to trees were followed for up to 120 min. Movement to fruit, acceptance levels of fruit, search paths, and movement away from the tree were monitored. With only 14 pairs of flies, meaningful behaviour and significant differences in acceptability of different fruit types were found.

Garcia and Altieri (1992) followed individual flea beetles *Phyllotreta cruciferae* in the field. The beetles were trapped in the field immediately before the observations and marked with a fluorescent powder. Released individuals could then be followed for several minutes during short flights, and in contact with host plants. The goal of the studies was to determine the effects of mixed host and non-host vegetation on host plant acceptance by the beetles. The beetles moved more often on broccoli in polycultures indicating that olfactory or visual cues from non-hosts affected their behaviour. After landing on non-hosts, the insects almost always emigrated from the plot on their next flight, indicating plant surface cues were important.

Host plant variability often has greatest impact on neonate or early instar insects. Observations of these vulnerable early stages can be critical for understanding the relevant host plant factors. To facilitate this potentially difficult task, eggs or neonates may be placed directly on the plants in the field. Zalucki and Brower (1992) placed near-hatch eggs of *Danaus plexipus* directly on host plants and observed their behaviour continuously for several hours to determine the causes of the high mortality of these animals on their host plants. The neonates became mired in the latex exuding from feeding wounds. The observation that *Plutella xylostella* neonates moved more rapidly after being placed on the resistant than on the susceptible *Brassica* plants led to investigations elucidating the resistance factors (Eigenbrode *et al.*, 1991a, b). Bernays *et al.* (1983) placed eggs of *Chilo partellus* directly on resistant and susceptible *Sorghum* plants in the field. Neonates were followed continuously as they moved upwards, until final establishment in the whorl. Individuals were less successful in locating the whorl on resistant plants than on susceptible plants; plant morphological characters and surface waxes were both found to be important (Bernays *et al.*, 1985).

MANIPULATING THE PLANTS

Observations of insects on potted plants in the field can be effective. Åhman (1985) observed wild cecidomyiid midges (*Dasineura brassicae*) in this way.

She was able to record landings, locations, movements, ovipositor insertions, and stationary periods. Differences in landing rates, and time remaining on the plant to oviposit indicated that host discrimination depended on both long-distance and contact cues.

Plant density and arrangement can also be controlled to give better opportunities for replication and analysis. Field plots with standard arrays of plants in grids made it possible to refine the flight path analysis of ovipositing butterflies, and to determine the importance of visual cues (Jones, 1977). Similar techniques have been instrumental in establishing the importance of visual attraction for ovipositing butterflies (Rausher, 1978; Mackay and Jones, 1989; see Bernays and Chapman, 1994, for other references).

Circles of different plant varieties or species can be used to assess the role of plant volatiles in eliciting upwind movement and distance attraction of insects to hosts. For example, adult *Leptinotarsa decemlineata*, released in the centre of a 6 m diameter circle of plants, were observed to disperse generally upwind (Jermy *et al.*, 1988). Within a distance of 40 cm the beetles were able to discriminate between host and non-host plants in the circle, demonstrating that short range arrestment cues were more important than distance orientation for this insect.

FIELD ENCLOSURES

Large enclosures are useful for confining flying insects for observation in the field while maintaining near-normal environmental conditions. They may be placed over natural vegetation or plants may be manipulated within them (Papaj, 1986). There are problems with their use for some insects, however. It is common for moths in cages to lay a large proportion of their eggs on the walls, even when the cages are large (> 1 m^3) (Ramaswamy, 1988), throwing doubt on the value of data collected. Evidence is accumulating that many moths are extremely sensitive to host volatiles (Städler, 1995) and it seems likely that the build-up of volatiles within the cage can confound the results so that special efforts to ventilate the environment are needed. Cages are discussed further in Section 6.5.4.

6.2.5. *Observations with traps or model plants*

Finch (1986) describes some problems interpreting results of trapping experiments. A particular difficulty is determining whether insects are responding to trap cues with directed flight over longer distances or whether trapping is a result of short-range arrestment. Just how insects respond to traps, as determined by direct observations, can resolve these questions. For example, the aphid, *Cavariella aegopodii*, was observed to fly upwind to a host odour on colourless traps from approximately 1 m away (Chapman *et al.*, 1981). Cabbage root flies were observed to respond to models reflecting specific wavelengths of light from up to 50 cm in field cages (Prokopy *et al.*, 1983a, b).

Attraction into traps from greater distances has been observed less often. Cabbage root flies, *Delia radicum*, will fly upwind from 50 m towards host and non-host plants alike but observations with traps showed that directed odour-mediated attraction only occurs over much shorter distances (Finch and Skinner, 1982). In contrast, the closely related *Delia antiqua* was observed by Judd and Borden (1989) to take off in response to host odours in traps 50 m upwind.

6.3. Behavioural Observations in Experimental Bioassays

6.3.1. Advantages of behavioural vs. endpoint experiments

In experimental laboratory bioassays, behavioural observations are often more informative than end point tests. For example, two substrates may receive equal numbers of eggs in 24 h in an oviposition experiment, but direct observation may reveal that one of these is initially rejected several times. Such initial rejection may be significant ecologically. In addition, observations can reveal more about the specific plant cues affecting selection. Rejection of a substrate for oviposition, for example, may result from reduced contact, indicating a role for volatiles, or reduced oviposition after contact, indicating a role for surface factors. Similarly, a feeding insect may reject a substrate before contact, after palpation, or only after biting or ingesting. This information will speed identification of the important specific factors. Roessingh and Städler's (1990) study of physical factors affecting oviposition by *D. radicum* provides an excellent example of the value of behavioural studies. The flies laid more eggs on model plants with vertical folds than those with horizontal folds. Observations of behaviour revealed that the insects alighted and explored both model types with equal frequency, and that discrimination between the model types only occurred during the geotactic host runs that precede oviposition. The behavioural test provided much more information about the biology of the interaction.

6.3.2. Laboratory experiments evaluating factors affecting host finding behaviour

OBSERVATIONS IN CAGES

Harris *et al.* (1993) used observations to study the importance of distance cues for oviposition by the Hessian fly, *Mayetiola destructor*. The flies were standardized with respect to rearing conditions, age, feeding state and mating status and placed into arenas at a standard time of day; conditions of temperature, humidity and light intensity were controlled. Female flies were observed with grass models which varied in reflectance, volatiles released, or both. Frequency of approach to and landing on the models were monitored from 1100 to 1300 hours, which is the peak period of natural responsiveness. With relatively few flies unambiguous data were obtained demonstrating that visual cues influenced behaviour 25 cm away, while the presence of volatiles was only effective within a few centimetres. At close range, these effects were additive.

Todd *et al.* (1990) studied responses of the leafhopper, *Dalbulus maidis*, to plant volatiles and visual cues, using direct observations of the insects in a cage through which an air stream was directed. Individuals orientated more often to a green target in the presence of host odours in the air stream and less in the presence of non-host odours. If the target reflected wavelengths other than green, however, odours had no influence on landings.

WINDTUNNELS AND OLFACTOMETERS

Windtunnels deliver odour to test insects in moving air, and many different designs developed for different insects and experimental situations are given by Finch (1986), Baker and Cardé (1984), and Smith *et al.* (1994). These devices are almost invariably used with direct observations of insect orientation, walking, or flying responses. The odours can be delivered dispersed in a laminar flow, in discontinuous packets, or discrete continuous plumes, to emulate particular field conditions. They have been extremely powerful tools for studies of insect pheromones and are increasingly used to study host plant selection. With a windtunnel in combination with a locomotion compensator that automatically recorded orientation and speed of the insect, Visser and Avè (1978) showed the importance of odour mixtures for positive anemotaxis in walking Colorado potato beetles, *Leptinotarsa decemlineata*. When other plants were present and the integrity of the odour mixture lost, the attractiveness was neutralized (Visser and de Jong, 1987). Similar experiments showed the attractiveness of host odours for walking apterae of the aphid *Cryptomyzus korschelti* (Visser and Taanman, 1987).

Vertical windtunnels have been developed for use with flying aphids and whiteflies, and those with automatic compensation of windspeed to match flight speed (David and Hardie, 1988) facilitate data collection by automatically recording flight speed data. Using such a system, Nottingham and Hardie (1993) showed that summer-generation *Aphis fabae* alates exhibit stronger flight responses to a host than to a non-host plant.

Video recording of responses in windtunnels can facilitate data collection and analysis (see Wratten, 1994). Infrared cameras can produce images of nocturnal insects flying in red light. For example, Tingle *et al.* (1989) showed that the specialist noctuid *Heliothis subflexa*, showed upwind flight and landing responses in the presence of a surface extract made from the leaves of its host plant, but not that from other plants.

The simple Y-tube olfactometer, the more sophisticated four-directional olfactometer (Vet *et al.*, 1983) or similar devices are usually used to collect end point data. The positions of the insects in these devices indicate if odours are attractive or repellent, relative to one another or a neutral control. Direct observations of insects in olfactometers can be used to learn more about the behavioural mechanisms producing these distributions. Movement rates of insects can be monitored in such devices to measure relative attractiveness of volatiles. For example, Hibbard and Bjostad (1988) studied the responses of corn root worm larvae, *Diabrotica virgifera*, to host volatiles. Cryogenically collected volatiles were placed into tubes attached to the bottom of a petri dish. Larvae were placed in a ring around the periphery of the dish and the rate of movement into the sample tubes monitored. The experiments demonstrated the attractiveness to the larvae of corn seedling volatiles.

6.3.3. Experiments evaluating factors effective after contact with the plant

FACTORS AT OR NEAR THE PLANT SURFACE

Observational experimental studies have demonstrated the importance of sensory contact with the leaf surface during host plant selection by insects. For example, rejection of potential host plants by different grasshopper species often occurs after palpation, and before actual biting. Chapman and Sword (1993) confined individual *Schistocerca americana* in ventilated plastic cylinders containing wire perches and observed them (ten at a time) in a controlled environment. The grasshoppers were standardized for state by allowing them to have a meal on a known acceptable plant before being presented with a variety of alternative plants. Such observations have helped focus attention on the plant surface in grasshopper feeding and the role of the palps in detecting surface deterrents (Blaney and Chapman, 1970; Woodhead, 1983; Woodhead and Chapman, 1986). In addition, studies of this kind have shown that palpation rejection of non-hosts by *S. americana* (Chapman and Sword, 1993) and *Locusta migratoria* (Blaney and Simmonds, 1985) increases in frequency with experience, implicating associative learning or sensitization.

Studies on the brown planthopper, *Nilaparvata lugens*, by Cook *et al.* (1987) and Woodhead and Padgham (1988) used videoanalysis. After exploration of the leaf surface with the tip of the labium, the insects moved more and probed into the plant less on certain resistant rice varieties than on a susceptible one. Since all these host selection activities observably precede contact with the internal leaf tissues, the leaf surface is implicated as the site of the relevant resistance factors. Application of surface waxes from the resistant varieties on to the susceptible one increased planthopper activity and caused far more insects to move off the plants in the first 30 min, and it was subsequently shown that hydrocarbon and carbonyl compound fractions were important (Woodhead and Padgham, 1988).

Host discrimination often begins during the very first seconds of contact with the plant. Harrison (1987) compared host preferences among four geographic populations of *L. decemlineata* by directly observing their feeding behaviour. Newly emerged adults from these populations that had fed as larvae on their respective regionally-dominant hosts, differed strongly in the amount of time spent feeding during a first meal on each of the four host plants. Harrison showed that increased palpation and test biting, within the first few seconds of contact, were associated with eventual reduced consumption of potential hosts. Eigenbrode *et al.* (1991b) obtained detailed data on the behaviour of neonate *P. xylostella* on resistant and susceptible cabbages and found that surface waxes of resistant plants elicited reduced acceptance in the caterpillars during the first 5 min of contact. Subsequent work, using the same behavioural bioassay (unpublished data), has identified components of the surface waxes that elicit reduced acceptance.

Discrimination during oviposition also can begin during the first contacts. Harris and Rose (1989) quantified the behaviour of individual ovipositing female Hessian fly, *Mayetiola destructor*, on their preferred host, wheat, and on a non-host, oats. During a 4-h period, the insects laid 10 times more eggs on wheat than on oats in choice tests, but in similar no-choice tests, all the insects eventually oviposited, even on the less preferred oats. It was found that in the first 5 min of contact, abdominal arching (enhancing ovipositor contact with the surface) and antennation occurred about three times more frequently on wheat. These differences waned with continued contact. Analysis of transition probabilities, obtainable from the continuous behavioural data, also showed that arching was more likely to be followed by antennation on wheat, resulting in the ovipositor contacting the surface of wheat plants more frequently. Chemical or physical features of the leaf, detected by the ovipositor, were in this way shown to be critical for host selection by *M. destructor*. Subsequent work using end point bioassay revealed the role of surface lipids in this discrimination (Foster and Harris, 1992).

Experimental observations can also reveal the ecological complexity affecting host selection. Kostal and Finch (1994), using careful observational experiments, showed how *D. radicum* oviposition on host plants on bare ground and with vegetational background was influenced by distance visual cues in combination with contact cues from host and non-host plants.

FACTORS WITHIN THE PLANT

Behavioural observations can reveal if plant internal constituents rather than surface factors influence host plant selection by insects. For example, rejection of cassava by the grasshopper, *Zonocerus variegatus*, was found to occur only after the first incision when hydrogen cyanide was released (Bernays *et al.*, 1977). The time course of the response can indicate if reduced acceptability is due to postingestive toxicity or even nutritional factors. Glendinning and Slansky (1994) plotted time spent feeding by caterpillars of *Anticarsia gemmatalis* and found that feeding rate declined on initially acceptable food after minutes due to noxious postingestive effects. Champagne and Bernays (1991) showed that on a nutritionally unsuitable plant, the grasshopper, *Schistocerca americana*, took shorter meal lengths over a matter of hours, and the food was eventually rejected.

Insects that feed on vascular tissues often show discrimination only after contact with xylem, phloem, or other internal tissues (e.g. Ogecha *et al.*, 1992). These activities can be monitored electronically using the equipment discussed in Section 6.5.2.

6.3.4. *Development of surrogate plants*

Potential chemical cues can be tested by applying these to plants and then observing insects (e.g. Nault and Styer, 1972), but plant models allow more controlled experimental manipulation of specific plant factors. Coupled with behavioural observations plant models can provide subtle and powerful tools to elucidate the important factors and their potential interactions. For example, J. Miller and his co-workers (e.g. Harris and Miller, 1982) showed the importance of plant shape in combination with volatiles for the onion fly using simple surrogate onion plants; Prokopy and his co-workers (e.g. Prokopy, 1968) used model fruit to demonstrate colour and shape parameters influencing apple maggot fly.

Plant models show that certain physical features such as vertical folds that mimic veins may be required to elicit natural behaviour (Gupta and Thorsteinson, 1960; Roessingh and Städler, 1990). The importance of smaller scale features such as trichomes in host acceptance behaviour is well documented (Bernays and Chapman, 1994), and surrogate plants may require such features to be effective in behavioural bioassays. Bernays *et al.* (1985) made precise plant models using moulds of actual leaves or stems to study the behaviour of newly hatched larvae of *Chilo partellus*. The small spines along the edges of sorghum leaf blades were found to be part of the suite of cues used in finding the feeding site.

It has been found that the presence of paraffin waxes can be important in the responsiveness of insects to sign stimuli in their host plants (e.g. Städler and Schöni, 1990; Spencer, 1996), so that the addition of test chemicals to such waxes on leaf or plant models is an improvement in testing procedures.

6.4. Integrating Behavioural Bioassay with Other Methods

Behavioural assays are powerful in themselves, but their potential may be greatest in combination with other techniques. There are excellent examples of efforts spanning years and involving numerous studies that have been extremely successful in combining behavioural observations and other types of experimentation. Our understanding of foraging by tephritid flies has been built through such efforts by R.J. Prokopy and his students and colleagues (Prokopy and Roitberg, 1989; Prokopy, 1993). The work of J.R. Miller and others with *Delia antiqua* oviposition (Harris and Miller, 1982, 1988, 1991), and Städler and others working with other anthomyiids has been similarly productive (Städler and Schöni, 1990; Roessingh *et al.*, 1992) and relied on creative integration of behavioural observation and end point bioassays. All the fly work owes much to the original behavioural descriptions provided by Zohren (1968) working with *D. brassica*.

Observations can validate rapid end point bioassays in a complete study. Foster and Harris (1992) used end point bioassays to study chemical factors stimulating Hessian fly oviposition, but the validity of these end point assays was first confirmed with continuous (20 min) observations to determine that flies did most of their discrimination after contact with test substrates. Trials evaluating resistance in cereal crops to lepidopteran stem borers often employ dispensers that drop neonates into the whorl of the plant, but this procedure is not valid for *Chilo partellus*, as was shown by behavioural observations in the field. It was found that this insect must climb the plant and then enter the whorl before feeding and that this establishment phase was critical for detection of some kinds of plant resistance (Bernays *et al.*, 1983). Thus, screening by placing insects directly in the whorl is not the appropriate end point test for resistance to *C. partellus*.

End point assays will sometimes provide the initial data upon which subsequent observational studies are based. Observed egg distributions in the field, for example, can suggest studies of oviposition behaviour. In host plant resistance work, differences in damage to different accessions are frequently the first data obtained, but these can then lead to behavioural studies to understand mechanisms. Eigenbrode *et al.* (1993, 1994) determined that damage to fruit of *Lycopersicon* accessions by *Spodoptera exigua* was not correlated with caterpillar densities. Subsequent assays then revealed the animals fed less on the fruit of a resistant accession.

Observations can remove the ambiguity of both choice and no-choice endpoint assays. In end point no-choice tests, the effects of deprivation can reduce or eliminate differences (Dethier, 1982), but continuous observations can discover, for example, differences in the latency to feed, which could have a large impact in the field. If necessary, end point tests can be designed to capture these important events, e.g. number of eggs laid after a short period of time. In choice tests, feeding or oviposition on one test substrate can influence the response to a subsequently encountered substrate, thus the order of presentation is critical. While the order of presentation can be controlled with some insect species (Singer, 1986), continuous observations of insects in the choice arena to determine the sequences of events leading to discrimination are superior. In one study, grasshoppers showed no discrimination between sucrose-impregnated discs and discs impregnated with sucrose and caffeine in a 1-h choice test, but observation showed that initially the caffeine discs were rejected (Bernays, unpublished).

Reliance on end point assays has produced difficulties in assessing the modalities of host plant resistance. 'Non-preference' (Painter, 1951) or 'antixenosis' (Kogan and Ortman, 1978) includes all resistance characteristics influencing pest insect behaviour. Unfortunately, end point bioassays of feeding cannot generally distinguish between antixenosis and antibiosis (resistance that depends on postingestive effects), and this has led to criticism of the usefulness of these terms. Direct behavioural observations, by contrast, potentially distinguish between antibiosis and antixenosis unambiguously, so the difficulty is largely methodological. The distinction between antixenosis and antibiosis is important biologically and has implications for integration of resistant crops into pest management (Eigenbrode and Trumble, 1994) and is also critical for the eventual elucidation of the actual resistance mechanism.

Numerous other experimental techniques used to study host selection can be effectively combined with observational studies. These include gut or faecal analyses to identify foods that have been actually eaten (e.g. Bernays and Chapman, 1970), ablation experiments to identify the role of olfaction or contact chemoreception (e.g. de Boer *et al.*, 1992), electrophysiology to determine sensitivity to specific chemicals (e.g. Roessingh *et al.*, 1992), and automated behavioural recordings (see Sections 6.5.1 and 6.5.2).

6.5. Methodology Considerations for Behavioural Bioassay of Host Plant Selection by Insects

Methods for recording and analysing animal behaviours are reviewed in detail by Martin and Bateson (1993). Opp and Prokopy (1986) and Wyatt (this volume, Chapter 3) provide excellent reviews with specific reference to insect behaviour. Some considerations specific to insect–plant interactions can be mentioned here.

6.5.1. Recording tracks and movement

Foraging butterflies can sometimes be tracked in the field (Stanton, 1982; Zalucki and Kitching, 1982; Root and Kareiva, 1984) and their movements preserved by attaching flags to plants contacted. Search paths of tephritid flies can be reconstructed using similar methods (Roitberg *et al.*, 1982). Aluja *et al.* (1989) and Zalucki and Kitching (1982) have developed methods for recording and analysing insect movement in three dimensions.

At closer range or on the plant, movements may be recorded by simple mechanical tracing to a variety of sophisticated video or computer assisted methods. In the laboratory, a clear plastic window above a leaf or other substrate permits manual tracings of insect movements (e.g. Sakurai, 1988; Strnad and Dunn, 1990). Insects can also be recorded using video equipment and the movement tracks traced later on to a digitizing tablet for analysis (Eigenbrode *et al.*, 1991a). Completely automated video tracking systems have become increasingly available and affordable in recent years (Hoy, 1994; 'Ethovision' by Noldus Information Technologies). These systems convert digitized video recordings to linear tracks, and calculate rates of movement, turning angles, and other summary statistics. These are suitable for walking or flying insects, in field conditions, cages or wind tunnels. To use these systems for studying insect movements on plants, subject and background must contrast sufficiently for automated recording of tracks from video images. For cryptic insects, this can pose a difficulty. One approach on leaves is to backlight the leaf and record the silhouettes of the insects.

For recording movements on a neutral substrate in the presence of volatile or visual cues, 'servo-sphere' locomotion compensating devices (Kramer, 1976; Thiery and Visser, 1986) are ideal, but expensive to construct and maintain. They have the advantage that there are few limits to the movement range of the insect being observed, whereas all other methods require that the insects be confined by some kind of barrier. A disadvantage is that they cannot be used to study the movement of insects on plant surfaces. A system developed by Eigenbrode *et al.* (1989) permits this, and also facilitates recording other behaviour simultaneously.

Insect movements can even be tracked within the soil using the approach developed by Villani (Villani and Gould, 1986). Soil sections containing test insects and host plant material are exposed to X-rays in the laboratory during the observation period. Lead chips attached to the insects produce a record of their movements on film.

6.5.2. *Recording host selection behaviour*

WHAT TO RECORD AND ANALYSE

The basic principles and techniques for behavioural research with insects in the field and laboratory are outlined by Wyatt (this volume, Chapter 3). For the study of host selection by herbivores, some behavioural parameters have proven especially useful. These include:

1. Time spent walking in relation to time spent feeding or ovipositing – assays as short as 5 min or less per insect have proved useful for some species (Eigenbrode *et al.*, 1991a).
2. Length of time spent feeding after first encounter – most insects feed in discrete bouts, with rests or movements in the gaps between the bouts. Thus, one possible measure is the length of the first feeding bout on a substrate. If bouts are irregular or occur in bursts with longer gaps between the bursts, then the feeding bouts can be grouped together as meals and the first meal lengths on different foods compared. Methods for distinguishing between bouts and meals have been recently discussed by Simpson (1990) and Berdoy (1993).
3. Structure of feeding over a longer period – lengths of feeding bouts and gaps may vary on different hosts (Glendinning and Slansky, 1994), or feeding bouts or meal lengths may decrease more over time on less acceptable or nutritionally suitable hosts (Champagne and Bernays, 1991; Bernays and Chapman, 1994).
4. Locomotion away from the plant or model after antennation, palpation, test biting or probing, or ovipositor extension in ovipositing insects (Chapman *et al.*, 1988).
5. Conditional probabilities of transitions between behaviours – these can be a sensitive measure of host plant acceptability (Harris and Rose, 1989).

Depending on the information needed, each individual insect may be observed continuously for from less than a minute (Harrison, 1987) to several hours (Chapman and Sword, 1993; Bernays *et al.*, 1994). Continuous recording will not always be necessary or may be impractical; in such cases behaviours may be recorded only at intervals (Martin and Bateson, 1993). Interval sampling sacrifices information about frequencies, durations, and transitions, but provides a reliable estimate of the proportion of time spent in each behaviour during an observational period (Blaney *et al.*, 1988). Relevance of interval sampling can be confirmed with preliminary continuous observations (e.g. Simmonds *et al.*, 1994).

RECORDING THE DATA: USE OF AUTOMATION

Manual recording methods will often be sufficient or preferable but automation can sometimes improve efficiency and ease of monitoring some behaviour continuously.

Electronic recording of mandibular movements (Jones *et al.*, 1981), or electrical resistance between a chewing insect and plant (Bowdan, 1984; Blust and Hopkins, 1990; Saxena and Kahn, 1991) are two such techniques. For phloem- or xylem-feeding insects, the important activities of the stylets during penetration and feeding must be recorded indirectly. Electronic feeding monitors (McLean and Kinsey, 1964; Tjallingii, 1978) measure changes in electrical impedance in a circuit that includes the insect and plant on which it is feeding. Plotted, these changes produce an electrical penetration graph (EPG), the patterns of which can be correlated with probing, phloem ingestion, and salivation (Tjallingii and Esch, 1993; Walker and Perring, 1994). Interpretation of EPGs is still being refined, but even simple correlations with probing or phloem ingestion are useful in understanding host plant resistance. For example, the technique has been used to compare greenbug biotype feeding patterns on resistant and susceptible barleys (Ogecha *et al.*, 1992), and to demonstrate that phloem characteristics produce resistance to the aphid, *Nasonovia ribisnigri*, in lettuce (van Helden *et al.*, 1993, 1995). Because recordings are made from insects harnessed to a fine gold or platinum wire, and a small current is used to produce the EPG, an obvious concern is the disruption of the insect's normal behaviour. Hardie *et al.* (1992) documented changes in *Aphis fabae* behaviour as a result of tethering.

6.5.3. *Experimental design and analysis*

In addition to the general principles of experimental design and statistical analysis for ecological research (Scheiner and Gurevitch, 1993), there are some issues particular to behavioural experiments (Martin and Bateson, 1993). Perhaps most importantly, the extreme variability of behavioural data reduces the statistical power of experiments. Variability can be reduced by standardizing the physiological and motivational state of the insects, as in several examples cited earlier in this chapter. There can be subtle problems with this approach. Eigenbrode *et al.* (1991b) used only neonate *P. xylostella* that had descended on silks soon after hatching in their bioassays. This was considered reasonable, as most insects would eventually perform this behaviour, but it remains possible that insects with delayed spinning behave differently during establishment on host plants.

Another approach is to standardize individuals genetically, but this obviously potentially erodes the relevance of experimental results. The once widely accepted practice of using genetically uniform test insects in host plant resistance work (Gallun and Khush, 1980) ignores the genetic variability in insect populations that is a requirement for the development of biotypes (Gould, 1983). Where feasible, it is better to include genetic variability as an effect in the experimental design so it can be included in the statistical treatment.

In spite of these precautions individual variation in behaviours is likely to remain large, necessitating where feasible large sample sizes to increase statistical power. In behavioural work, this can be difficult if observations are long. The temptation to increase apparent power by treating several observations from a single individual as independent must be avoided as this constitutes pseudoreplication (Hurlbert, 1984).

Martin and Kraemer (1987) provide several methodological approaches to extremely variable behavioural data. All involve obtaining additional information from each subject, either by establishing a behavioural baseline before applying a treatment, averaging the responses of each animal, or some combination of these. Long observations potentially reduce variability among individuals by obtaining a more representative sample of each insect's behaviour, averaged through fluctuations in motivational state.

Behavioural data are often highly skewed, requiring the use of appropriate non-parametric statistical tests (Siegel and Castellan, 1988), and presentation of medians, box plots, or actual data points rather than means and standard errors.

6.5.4. Avoiding artefacts in controlled experiments

Any manipulation of insects or plants to facilitate observations or to reduce environmental variability can produce artefacts. This problem can be especially critical in behavoural bioassays. Although the best check for artefacts is sufficient knowledge of the natural behaviours of unmanipulated insects on plants in the field, some sources of artefacts that should be considered include:

- Cages
  - Size (too small may trigger escape responses in insects).
  - Ventilation (volatiles may be artificially concentrated).
  - Shading (reduced light intensity influences insects and plants).
- Artificial lighting
  - Intensity (ambient greenhouse light intensities low; UV absent or low).
  - Spectral quality (artificial usually different from natural light).

- Frequency (AC – induced pulsing in intensity can influence insect behaviour).
- Directionality (effects on behaviour must be considered).
- Temperature and humidity.
- Diurnal cycles (can affect insects and plants).
- State of insects
    - Experiential effects (sensitization, habituation, associative learning).
    - Physiological effects (egg loads, hunger, developmental stage).
- State of plants or plant tissues
    - Developmental stage and plant part.
    - Nutrients, water, diurnal cycles (all affect plant chemistry).
    - Effects of cutting (can be severe, and different in different species).
    - Damage (induced defences).
- Observer effects (movements can be detected by insects during observations).

6.6. Conclusions

Evaluation of factors affecting host plant selection behaviour will be most realistic and most rewarding if direct observations are used as a major part of the work. With the knowledge gained from early observations, the most realistic and meaningful assays can be set up. Later the laboratory assay can be considerably more informative, and less ambiguous, if observations are incorporated into the experimental design, than if they are not. With care, and the use of new tools for recording, the task is not as formidable as many believe, and the reward greatly exceeds the effort expended.

6.7. References

Åhman, I. (1985) Oviposition behaviour of *Dasineura brassicae* on a high versus a low-quality *Brassica* host. *Entomologia Experimentalis et Applicata* 39, 247–253.

Aluja, M. and Birke, A. (1993) Habitat use by adults of *Anastrepha obliqua* (Diptera: Tephritidae) in a mixed mango and tropical plum orchard. *Annals of the Entomological Society of America* 86, 799–812.

Aluja, M. and Prokopy, R.J. (1992) Host search behaviour by *Rhagoletis pomonella* flies: inter-tree movement patterns in response to wind-borne fruit volatiles under field conditions. *Physiological Entomology* 17, 1–8.

Aluja, M., Prokopy, R.J., Elkinton, J.S. and Laurence, F. (1989) A novel approach for tracking and quantifying the movement patterns of insects in three dimensions under semi-natural conditions. *Environmental Entomology* 18, 1–7.

Baker, T.C. and Cardé, C.E. (1984) Windtunnels in pheromone research. In: Hummel, H.E. and Miller, T.A. (eds) *Techniques in Pheromone Research*. Springer-Verlag, New York, pp. 75–110.

Berdoy, M. (1993) Defining bouts of behaviour: a three-process model. *Animal Behaviour* 46, 387–396.

Bernays, E.A. and Chapman, R.F. (1970) Food selection by *Chorthippus parallelus* (Zetterstedt) in the field. *Journal of Animal Ecology* 39, 383–394.

Bernays, E.A. and Chapman, R.F. (1994) *Host Plant Selection in Phytophagous Insects.* Chapman & Hall, New York 312pp.

Bernays, E.A., Chapman, R.F., McDonald, J and Salter, J.E.R. (1976) The degree of oligophagy in *Locusta migratoria* (L.). *Ecological Entomology* 1, 223–230.

Bernays, E.A., Chapman, R.F, Leather, E.M., McCaffery, A.R. and Modder, W.W.D. (1977) The relationship of *Zonocerus variegatus* (L.) (Acridoidea: Pyrgomorphidae) with cassava (*Manihot esculenta*). *Bulletin of Entomological Research* 67, 391–404.

Bernays, E.A., Chapman, R.F. and Woodhead, S. (1983) Behaviour of newly hatched larvae of *Chilo partellus* (Swinhoe) (Lepidoptera: Pyralidae) associated with their establishment in the host-plant, sorghum. *Bulletin of Entomological Research* 73, 75–83.

Bernays, E.A., Woodhead, S. and Haines, L. (1985) Climbing by newly hatched larvae of the spotted stalk borer, *Chilo partellus,* to the top of sorghum plants. *Entomologia Experimentalis et Applicata* 39, 73–79.

Bernays, E.A., Bright, K.L., Gonzalez, N. and Angel, J. (1994) Dietary mixing in a generalist herbivore: tests of two hypotheses. *Ecology* 75, 1997–2006.

Blaney, W. M. and Chapman, R.F. (1970) The functions of the maxillary palps of Acrididae (Orthoptera). *Entomologia Experimentalis et Applicata* 13, 363–376.

Blaney, W.M. and Simmonds, M.S.J. (1985) Food selection by locusts: the role of learning rejection behaviour. *Entomologia Experimentalis et Applicata* 39, 273–278.

Blaney, W.M., Simmonds, M.S.J., Ley, S.V. and Jones, P.S. (1988) Insect antifeedants: a behavioural and electrophysiological investigation of natural and synthetically derived clerodane diterpenoids. *Entomologia Experimentalis et Applicata* 46, 267–274.

Blust, M. H. and Hopkins, T. L. (1990) Feeding patterns of a specialist and a generalist grasshopper: electronic monitoring on their host plants. *Physiological Entomology* 15, 261–267.

Bowdan, E. (1984) An apparatus for the continuous monitoring of feeding by caterpillars in choice, or non-choice tests (automated cageteria test). *Entomologia Experimentalis et Applicata* 36, 13–17.

Bright, K.L., Bernays, E.A. and Moran, V.C. (1994) Foraging patterns and dietary mixing in the field by the generalist grasshopper *Brachystola magna.* (Orthoptera: Acrididae). *Journal of Insect Behavior* 7, 779–794.

Champagne, D. and Bernays, E.A. (1991) Phytosterol unsuitability as a factor mediating food aversion learning in the grasshopper, *Schistocerca americana. Physiological Entomology* 16, 391–400.

Chapman, R. F. and Sword, G. (1993) The importance of palpation in food selection by a polyphagous grasshopper (Orthoptera: Acrididae). *Journal of Insect Behavior* 6, 79–91.

Chapman, R.F., Bernays, E.A. and Simpson, S.J. (1981) Attraction and repulsion of the aphid, *Cavariella aegopodii,* by plant odors. *Journal of Chemical Ecology* 7, 881–888.

Chapman, R.F., Bernays, E.A. and Wyatt, T. (1988) Chemical aspects of host-plant specificity in three *Larrea*-feeding grasshoppers. *Journal of Chemical Ecology* 14, 561–580.

Cook, A. G., Woodhead, S., Magalit , V.F. and Heinrichs, E.A. (1987) Variation in feeding behaviour of *Nilaparvata lugens* on resistant and susceptible rice varieties. *Entomologia Experimentalis et Applicata* 32, 227–235.

Damman, H. and Feeny, P. (1988) Mechanisms and consequences of selective oviposition by the zebra swallowtail butterfly. *Animal Behaviour* 36, 563–573.

David, C.T. and Hardie, J. (1988) The visual response of free-flying summer and autumn forms of the black bean aphid *Aphis fabae* in an automated flight chamber. *Physiological Entomology* 13, 277–284.

de Boer, G., Schmitt, A., Zavod, R. and Mitscher, L.A. (1992) Feeding stimulatory and inhibitory chemicals from an acceptable nonhost plant for *Manduca sexta*: improved detection by larvae deprived of selected chemosensory organs. *Journal of Chemical Ecology* 18, 885–895.

Dethier, V.G. (1982) Mechanisms of host-plant recognition. *Entomologia Experimentalis et Applicata* 31, 49–56.

Eigenbrode, S. D. and Shelton, A. M. (1990) Behavior of neonate diamondback moth larvae (Lepidoptera: Plutellidae) on glossy-leafed resistant genotypes of *Brassica oleracea*. *Environmental Entomology* 19, 1566–1571.

Eigenbrode, S. D. and Trumble, J. T. (1994) Plant resistance to insects in integrated pest management in vegetables. *Journal of Agricultural Entomology* 11, 201–224.

Eigenbrode, S. D., Barnard, J. and Shelton, A.M. (1989) A system for quantifying behavior of neonate caterpillars and other small, slow-moving animals. *Canadian Entomologist* 121, 1125–1126.

Eigenbrode, S.D, Stoner, K.A., Shelton, A.M. and Cain, W.C. (1991a) Characteristics of leaf waxes of *Brassica oleracea* associated with resistance to diamondback moth. *Journal of Economic Entomology* 83, 1609–1618.

Eigenbrode, S. D., Espelie, K.E., and Shelton, A.M. (1991b) Behavior of neonate diamondback moth larvae on leaves and leaf waxes of resistant and susceptible cabbages. *Journal of Chemical Ecology* 17, 1691–1704.

Eigenbrode, S.D., Shelton, A.M., Kain, W.C., Leichtweis, H. and Spittler, T.D. (1993) Controlling lepidopteran pests in cabbage by inducing leaf glossiness with S-ethyldipropylthiocarbamate herbicide. *Entomologia Experimentalis et Applicata,* 69, 41–50.

Eigenbrode, S.D., Trumble, J.T. and White, K.K. (1994) Fruit-based tolerance to beet armyworm in *Lycopersicon* accessions. *Environmental Entomology* 23, 937–942.

Eigenbrode, S.D., Moodie, S. and Castagnola, T. (1995) Generalist predators interact with insect resistance in glossy cabbage. *Entomologia Experimentalis et Applicata* (in press).

Finch, S. (1986) Assessing host-plant finding by insects. In: Miller, J.R. and Miller, T.A. (eds) *Insect–Plant Interactions*, Springer-Verlag, New York, pp. 23–64.

Finch, S. and Skinner, G. (1982) Upwind flight by the cabbage root fly, *Delia radicum*. *Physiological Entomology* 7, 387–399.

Foster, S.P. and Harris, M.O. (1992) Foliar chemicals of wheat and related grasses influencing oviposition by Hessian fly, *Mayetiola destructor* (Say) (Diptera: Cecidomyiidae). *Journal of Chemical Ecology* 18, 1965–1980.

Gallun, R.L. and Khush, G.S. (1980) Genetic factors affecting expression and stability of resistance. In: Maxwell, F.G. and Jennings, P.R. (eds) *Breeding Plants Resistance to Insects*. Springer-Verlag, New York, pp. 63–86.

Garcia, M.A. and Altieri, M.A. (1992) Explaining differences in flea beetle *Phyllotreta cruciferae* Goeze densities in simple and mixed broccoli cropping systems as a function of individual behavior. *Entomologia Experimentalis et Applicata* 62, 201–209.

Glendinning, J. and Slansky, F. (1994) Interactions of allelochemicals with dietary constituents: effects on deterrency. *Physiological Entomology* 19, 173–186.

Gould, F. (1983) Host variability and herbivore pest management. In: Denno, R.F. and McClure, M.S. (eds) *Variable Plants and Herbivores in Natural and Managed Ecosystems*. Academic Press, New York, pp. 597–654.

Gupta, P.D. and Thorsteinson, A.J. (1960) Food plant relationships of the diamond-back moth *Plutella maculipennis* (Curt.). III. Sensory regulation of oviposition of the adult female. *Entomologia Experimentalis et Applicata* 3, 305–314.

Hardie, J., Holyoak, M., Taylor, N.J. and Griffiths, D.C. (1992) The combination of electronic monitoring and video-assisted observations of plant penetration by aphids and behavior effects of polygodial. *Entomologia Experimentalis et Applicata* 62, 233–239.

Harris, M.O. and Miller, J.R. (1982) Synergism of visual and chemical stimuli in the oviposition behavior of *Delia antiqua* (Meigen) (Diptera: Anthomyiidae). In: Visser, H. and Minks, A. (eds) *Proceedings of the 5th International Symposium on Insect Plant Relationships*. Pudoc, Wageningen, pp. 117–122.

Harris, M.O. and Miller, J.R. (1988) Host-acceptance behaviour in an herbivorous fly, *Delia antiqua*. *Journal of Insect Physiology* 34, 179–190.

Harris, M.O. and Miller, J.R. (1991) Quantitative analysis of ovipositional behavior: effects of a host-plant chemical on the onion fly (Diptera: Anthomyidae). *Journal of Insect Behavior* 4, 773–792.

Harris , M.O. and Rose, S. (1989) Temporal changes in the egglaying behaviour of the Hessian fly. *Entomologia Experimentalis et Applicata* 53, 17–29.

Harris, M.O., Rose, S. and Malsch, P. (1993) The role of vision in the host plant-finding behaviour of the Hessian fly. *Physiological Entomology* 18, 31–42.

Harrison, G.D. (1987) Host-plant discrimination and evolution of feeding preference in the Colorado potato beetle *Leptinotarsa decemlineata*. *Physiological Entomology* 12, 407–415.

Hibbard, B.E. and Bjostad, L.B. (1988) Behavioral responses of Western corn rootworm larvae to volatile semiochemicals from corn seedlings. *Journal of Chemical Ecology* 14, 1522–1540.

Hoy, J. B. (1994) Follow that roach: exploiting data from desktop video clips. *Advanced Imaging* January.

Hurlbert, S.H. (1984) Pseudoreplication and the design of ecological field experiments. *Ecological Monographs* 54, 187–211.

Jermy, T., Szentesi, A. and Horvath, J. 1988. Host plant finding in phytophagous insects: the case of the *L. decemlineata*. *Entomologia Experimentalis et Applicata* 49, 83–98.

Jones, C.G., Hoggard, M.P. and Blum, M.S. (1981) Pattern and process in insect feeding behavior: a quantitative analysis of Mexican bean beetle, *Epilachna varivestis*. *Entomologia Experimentalis et Applicata* 30, 254–264.

Jones, R.E. (1977) Movement patterns and egg distribution in cabbage butterflies. *Journal of Animal Ecology* 46, 195–212.

Judd, G.J.R. and Borden, J.H. (1989) Distant olfactory response of the onion fly, *Delia antiqua*, to host-plant odour in the field. *Physiological Entomology* 14, 429–441.

Kennedy, J.S., Booth, C.O. and Kershaw, W.J.S. (1959) Host finding by aphids in the field II. *Aphis fabae* Scop. (gynoparae) and *Brevicoryne brassicae* L. with a re-appraisal of the role of host-finding behavior in virus spread. *Annals of Applied Biology* 47, 424–444.

Kogan, M. and Ortman, E.E. (1978) Antixenosis – a new term proposed to replace

Painter's 'non-preference' modality of resistance. *Bulletin of the Entomological Society of America* 24, 175–176.

Kostal, V. and Finch, S. (1994) Influence of background on host-plant selection and subsequent oviposition by the cabbage root fly (*Delia radicum*). *Entomologia Experimentalis et Applicata* 70, 153–163.

Kramer, E. (1976) The orientation of walking honeybees in odour fields with small concentration gradients. *Physiological Entomology* 1, 27–37.

Mackay, D.A. and Jones, R.E. (1989) Leaf shape and host-finding behaviour of two ovipositing monophagous butterfly species. *Ecological Entomology* 14, 423–431.

Martin P. and Bateson, P. (1993) *Measuring Behaviour: An Introductory Guide*, Cambridge University Press, 238pp.

Martin, P. and Kraemer, H.C. (1987) Individual differences in behaviour and their statistical consequences. *Animal Behaviour* 35, 1366–1375.

McLean, D.L. and Kinsey, M.G. (1964) A technique for electronically recording aphid feeding and salivation. *Nature* 202, 1358–1359.

Miller, J.R. and Miller, T.A. (1986) *Insect–Plant Interactions*. Springer-Verlag, New York 342pp.

Nault, L.R. and Styer, W.E. (1972) Effects of sinigrin on host selection by aphids. *Entomologia Experimentalis et Applicata* 15, 423–437.

Noldus, L. P. J. J., van de Loo, E. L. H. M. and Timmers, P. H. A. (1989) Computers in behavioural research. *Nature* 341, 767–768.

Nottingham, S.F. and Hardie, J. (1993) Flight behaviour of the black bean aphid, *Aphis fabae*, and the cabbage aphid, *Brevicoryne brassicae*, in host and non-host plant odour. *Physiological Entomology* 18, 389–394.

Nottingham, S.F., Hardie, J., Dawson, G.W., Hicke, A.J., Pickett, J.A., Wadhams, C.M. and Woodcock, C.M. (1991) Behavioural and physiological responses of aphids to host and non-host plant volatiles. *Journal of Chemical Ecology* 12, 1231–1242.

Ogecha, J., Webster, J.A. and Peters, D.C. (1992) Feeding-behavior and development of biotype-E, biotype-G, and biotype-H of *Schizaphis graminum* (Homoptera: Aphididae) on Wintermalt and Post barley. *Journal of Economic Entomology* 85, 1522–1526.

Opp, S.B. and Prokopy, R.J. (1986) Approaches and methods for direct behavioral observation and analysis of plant–insect interactions. In: Miller, J.R. and Miller, T.A. (eds) *Insect–Plant Interactions*. Springer-Verlag, New York, pp. 1–22.

Painter, R. (1951) *Insect Resistance in Crop Plants*. Macmillan, New York, 520pp.

Papaj, D.R. (1986) Conditioning leaf shape preference discrimination by chemical cues in the butterfly *Battus philenor*. *Animal Behavior* 34, 1281–1288.

Papaj, D. R. (1994) Use and avoidance of occupied hosts as a dynamic process in tephritid flies. In: Bernays, E.A. (ed.) *Insect–Plant Interactions, Vol V*. CRC Press, Boca Raton, Florida, pp. 25–46.

Pierce, N.E. and Elgar, M.A. (1985) The influence of ants on host-plant selection by *Jalmenus evagoras*, a myrmecophilous lycaenid butterfly. *Behavioral Ecology and Sociobiology* 16, 209–222.

Prokopy, R.J. (1968) Visual responses of apple maggot flies, *Rhagoletis pomonella* (Diptera: Tephritidae): orchard studies. *Entomologia Experimentalis et Applicata* 11, 403–422.

Prokopy, R. J. (1993) Levels of quantitative investigation of tephritid fly foraging behavior. In: Aluja, M. and Liedo. P. (eds) *Fruit Flies: Biology and Management*. Springer, New York. pp. 165–171.

Prokopy, R.J. and Roitberg, B.D. (1989) Fruit fly foraging behavior. In: Robinson, A.S.

and Hooper, G. (eds) *Fruit Flies: Their Biology, Natural Enemies and Control.* Elsevier, Amsterdam, pp. 293–306.

Prokopy, R.J., Collier, R.H. and Finch, S. (1983a) Leaf color used by cabbage root flies to distinguish among host plants. *Science* 221, 190–192.

Prokopy, R.J., Collier, R.H. and Finch,S. (1983b) Visual detection of host plants by cabbage root flies. *Entomologia Experimentalis et Applicata* 34, 85–89.

Ramaswamy, S.B. (1988) Host finding by moths: sensory modalities and behaviours. *Journal of Insect Physiology* 34, 235–249.

Raubenheimer, D. and Bernays, E.A. (1993) Patterns of feeding in the polyphagous grasshopper, *Taeniopoda eques. Animal Behavior* 45, 153–167.

Rausher, M.D. (1978) Search image for leaf shape in a butterfly. *Science* 200, 1071–1073.

Roessingh, P. and Städler, E. (1990) Foliar form, colour and surface characteristics influence oviposition behaviour in the cabbage root fly *Delia radicum. Entomologia Experimentalis et Applicata* 57, 93–100.

Roessingh, P., Städler, E., Fenwick, G.R., Lewis, J.A., Nielsen, J.K., Hurter, J. and Ramp, T. (1992) Oviposition and tarsal chemoreceptors of the cabbage root fly are stimulated by glucosinolates and host plant extracts. *Entomologia Experimentalis et Applicata* 65, 267–282.

Roitberg, B. and Prokopy, R. (1982) Influence of intertree distance on foraging behaviour of *Rhagoletis pomonella* in the field. *Ecological Entomology* 7, 437–442.

Roitberg, B.D., van Lenteren, J.C., van Alphen, J.J.N., Galis, F. and Prokopy, R.J. (1982) Foraging behavior of *Rhagoletis pomonella*, a parasite of hawthorn (*Crataegus viridis*), in nature. *Journal of Animal Ecology* 51, 307–325.

Root, R.B. and Kareiva, P.M. (1984) The search for resources by cabbage butterflies (*Pieris rapae*): ecological consequences and adaptive significance of Markovian movements in a patchy environment. *Ecology* 65, 147–165.

Sakurai, K. (1988) Leaf size recognition and evaluation of some attelabid weevils. *Behaviour* 106, 279–300.

Saxena, R.C. and Khan, Z.R. (1991) Electronic recording of feeding behavior of *Cnaphalocrocis medinalis* (Lepidoptera: Pyralidae) on resistant and susceptible rice cultivars. *Annals of the Entomological Society of America* 84, 316–318.

Scheiner, S.M. and Gurevitch, J. (1993). *The Design and Analysis of Ecological Experiments.* Chapman & Hall, New York, 445pp.

Shreeve, T.G. (1986) Egg-laying by the speckled wood butterfly (*Pararge aegeria*): the role of female behaviour, host abundance and temperature. *Ecological Entomology* 11, 229–236.

Siegel, S and Castellan, N.J. (1988) *Nonparametric Statistics for the Behavioral Sciences,* 2nd edn. McGraw-Hill, New York, 399pp.

Simmonds, M.S.J., Blaney, W.M., Mithen, R., Birck, A.N.E. and Lewis, J. (1994) Behavioural and chemosensory responses of the turnip root fly (*Delia floralis*) to glucosinolates. *Entomologia Experimentalis et Applicata* 71, 41–57.

Simpson, S.J. (1990) The pattern of feeding. In: Chapman, R.F. and Joern, A. (eds) *Biology of Grasshoppers.* Chapman & Hall, New York, pp. 73–104.

Singer, M.C. (1986) The definition and measurement of oviposition preference in plant-feeding insects. In: Miller, J.R. and Miller, T.A. (eds) *Insect–Plant Interactions.* Springer-Verlag, New York, pp. 65–94.

Smiley, J. (1978) Plant chemistry and the evolution of host specificity: new evidence from *Heliconius* and *Passiflora. Science* 201, 745–747.

Smith, C.M., Khan, Z.R. and Pathak, M. (1994) *Techniques for Evaluating Insect*

Resistance in Crop Plants. CRC Press, Boca Raton, 320pp.

Städler, E. (1995) Oviposition behavior of insects influenced by chemoreception. In: Kurihare, K., Suzuki, N. and Ogawa, H. (eds) *Olfaction and Taste XI*. Springer Verlag, Tokyo, pp. 821–826.

Städler, E. and Buser, H.-R. (1984) Defense chemicals in leaf surface wax synergistically stimulate oviposition by a phytophagous insect. *Experientia* 40, 1157–1159.

Städler, E. and Schöni, R. (1990) Oviposition behavior of the cabbage root fly, *Delia radicum* (L.). *Journal of Insect Behavior* 3, 195–210.

Stanton, M.L. (1982) Searching in a patchy environment: Food plant selection by *Colias p. eriphyle* butterflies. *Ecology* 63, 839–853.

Strnad, S.P. and Dunn, P.E. (1990) Host search behavior of neonate western corn rootworm (*Diabrotica virgifera virgifera*). *Journal of Insect Physiology* 36, 201–205.

Thiery, D. and Visser, J.H. (1986) Making of host plant odour in the olfactory orientation of the Colorado potato beetle. *Entomologia Experimentalis et Applicata* 41, 165–172.

Tingle, F.C., Heath, R.R. and Mitchell, E.R. (1989) Flight response of *Heliothis subflexa* females to an extract from groundcherry. *Journal of Chemical Ecology* 15, 221–232.

Tjallingii, W.F. (1978) Electronic recording of penetration behavior by aphids. *Entomologia Experimentalis et Applicata* 24, 521–530.

Tjallingii, W.F. and Esch, T.H. (1993) Fine-structure of aphid stylet routes in plant-tissues in correlation with EPG signals. *Physiological Entomology* 18, 317–328.

Todd, J.L., Phelan, P.L. and Nault, L.R. (1990) Interaction between visual and olfactory stimuli during host-finding by leafhopper, *Dalbulus maidis*. *Journal of Chemical Ecology* 16, 2121–2134.

Unwin, D.M.and Martin, P. (1987) Recording behaviour using a portable microcomputer. *Behaviour* 101, 87–100.

van Helden, M., van Heest, H.P.N.F., van Beek, T.A. and Tjallingii, W.F. (1995) Development of a bioassay to test phloem samples from lettuce for resistance to *Nasonovia ribisnigri* (Homoptera: Aphididae). *Journal of Chemical Ecology* 21, 761–774.

van Helden, M., Tjallingii, W.F. and van Beek, T.A. (1993) Tissue localisation of lettuce resistance to the aphid *Nasonovia ribisnigri* using electrical penetration graphs. *Entomologia Experimentalis et Applicata* 68, 269–278.

Vet, L.E.M., van Lenteren, J.C., Heymans, M. and Meelis, E. (1983) An airflow olfactometer for measuring olfactory responses of hymenopterous parasitoids and other small insects. *Physiological Entomology* 8, 97–106.

Villani, M.G. and Gould, F. (1986) Use of radiographs for movement analysis of the corn wireworm, *Melanotus communis* (Coleoptera: Elateridae). *Environmental Entomology* 15, 462–464.

Visser, J.H. and Avè, D.A. (1978) General green leaf volatiles in the olfactory orientation of the Colorado beetle, *Leptinotarsa decemlineata*. *Entomologia Experimentalis et Applicata* 24, 738–749.

Visser, J.H. and de Jong, R. (1987) Plant odour perception in the *L. decemlineata*: chemoattraction towards host plants. In: Labeyrie, V., Fabres, G. and Lachaise, D. (eds) *Insects–Plants*. Junk, Dordrecht, pp. 129–134.

Visser, J.H. and Piron, P.G.M. (1995) Olfactory antennal responses to plant volatiles in apterous virginoparae of the vetch aphid *Megoura viciae*. *Entomologia Experimentalis et Applicata* 77, 37–46.

Visser, J.H. and Taanman, J.W. (1987) Odour-conditioned anemotaxis of apterous aphids in response to host plants. *Physiological Entomology* 12, 473–479.

Walker, G.P. and Perring, T.M. (1994) Feeding and oviposition behavior of whiteflies (Homoptera: Aleyrodidae) interpreted from AC electronic feeding monitor wave-forms. *Annals of the Entomological Society of America* 87, 363–374.

Wayadande, A.C. and Nault, L.R. (1996) Leafhoppers on leaves: an analysis of feeding behavior using conditional probabilities. *Journal of Insect Behavior* 9, 3–22.

Woodhead, S. (1983) Surface chemistry of sorghum bicolor and its importance in feeding by *Locusta migratoria*. *Physiological Entomology* 8, 345–352.

Woodhead, S. and Chapman, R.F. (1986) Insect behaviour and the chemistry of plant surface waxes. In: Juniper, B. and Southwood, T.R.E. (eds) *Insects and the Plant Surface*. Edward Arnold, London, pp. 123–136.

Woodhead, S. and Padgham, D.E. 1988. The effect of plant surface characteristics on resistance of rice to the brown planthopper, *Nilaparvata lugens*. *Entomologia Experimentalis et Applicata* 47, 15–22.

Wratten, S.D. (ed.) (1994) *Video Techniques in Animal Ecology and Behaviour*. Chapman & Hall, London, 211pp.

Zalucki, M.R. and Brower, L.P. (1992) Survival of first instar larvae of *Danaus plexippus* (Lepidoptera) in relation to cardiac glycoside and latex content of *Asclepias humistrata* (Asclepiadaceae). *Chemoecology* 3, 81–93.

Zalucki, M.P. and Kitching, R.L. (1982) The analysis and description of movement in adult *Danaus plexippus* L. (Lepidoptera: Danainae). *Behaviour* 80, 174–198.

Zalucki, M.P., Brower, L.P. and Malcolm, S.B. (1990) Oviposition by *Danaus plexippus* in relation to cardenolide content of three *Asclepias* species in the southeastern USA. *Ecological Entomology* 15, 231–240.

Zohren, E. (1968) Laboruntersuchungen zur Massenzucht, Lebensweise, Eiablage und Eiablageverhalten der Kohlfliege *Chortophila brassicae* Bouche (Diptera: Anthomyidae). *Zeitschrift fur Angewelte Entomologie* 62, 139–188.

7 Statistical Aspects of Field Experiments

J.N. Perry

Department of Entomology and Nematology, Rothamsted Experimental Station, Harpenden, Herts AL5 2JQ, UK

7.1. Introduction

Here, the quantitative techniques used in the planning, design and analysis of field experiments in agricultural and ecological entomology are addressed. Because of space limitations this cannot be a comprehensive account of the statistical aspects of design and analysis. It is indeed a rather personal view from my own work, and so the chapter focuses on important techniques and problems that I believe have received too little coverage elsewhere. For a fuller treatment of the fundamentals of design and analysis, three older books and one relatively modern text are recommended, respectively: Fisher (1925) *Statistical Methods for Research Workers*; Fisher (1935) *The Design of Experiments*; Cochran and Cox (1950) *Experimental Designs*; and Mead and Curnow (1983) *Statistical Methods in Agriculture and Experimental Biology*. These, better than most modern texts, outline the principles behind the techniques of modern applied statistics, and it is these principles that I believe biologists find difficult, rather than the calculations themselves. The means to do the calculations are, in any case, available to almost all workers through the PC and modern packages. Armed with a good understanding of such principles all biologists can make their experiments more precise, their sampling schemes more efficient and the interpretation of their data more incisive.

The emphasis on the design of experiments and on more empirical statistical modelling unfortunately precludes discussion of the importance of mathematical modelling; nor is there space to discuss the many classes of statistical regression models now commonly used. Some elaboration of several of the themes in this chapter may be found in Perry (1989, 1994a). For clarity, each statistical term is italicized the first time it is defined.

© CAB INTERNATIONAL 1997. *Methods in Ecological and Agricultural Entomology* (D.R. Dent and M.P. Walton eds)

7.2. The Concept of Population: Entomological and Statistical

To understand the area where agricultural and ecological entomology meets the quantitative science of statistics, or as it should more properly be termed biometry, we must first appreciate the language each discipline uses, especially for those concepts which are common to both. Nowhere is this more important than in the concept of a population, for this is crucial in both sciences. Entomologists have struggled with the term 'population' over the years; Johnson (1969) described it as an imprecise concept depending on the variable mobility of an insect species. Dempster's (1975) discussion of populations rightly emphasized the added complication that all species, and also the habitats in which they live, are patchy in their distribution. Taylor (1986) noted that a population could include both the single colonizing individual and the entire population. Lest this seems too negative, we immediately note that these ideas encompass a spatial and temporal hierarchy of scales, and immigration and emigration between patches, all of which are addressed through recent interest in the metapopulation dynamics of animal (Hanski and Gilpin, 1991) and plant (Czaran and Bartha, 1992) populations. Indeed, metapopulation dynamics resolves the difficulties in the biologist's definition of 'population' and builds a bridge towards quantitative treatment of population data because it recognizes the stochastic nature of the presence of individuals in an area, the probability of the incidence of a sub-population in a patch unit, and the spatial and temporal heterogeneity of population distribution.

Pragmatically, Ruesink (1980) defines a 'sample' to consist of a small (observed) collection drawn from a larger population about which information is desired. Indeed, the practical importance of a population is manifest only through sampling. Sampling takes time and is expensive because pests and diseases are highly variable in space and time. Therefore, it is essential to develop the most efficient sampling schemes possible, and applied statistics has a long tradition in the development of methods to allow for spatio-temporal variation. In particular, in North America, most of the important pests are introduced species (Woods, 1974) that can multiply rapidly, especially on crops such as cotton, maize and soya, often grown in subtropical monocultures which cover wide geographic regions. So, for many years, agriculture in North America has had to develop and has benefited from intensive pre-treatment sampling programmes.

Most of the early developments in the modern science of statistics came from such agricultural applications (Fisher, 1925, 1935). However, there is a considerable difference between the biologist's concept of population, which relates wholly to a real and observable phenomenon, and the statistician's, which relates to a real phenomenon but in an abstract and literally unobservable way, through the concept of infinitely repeatable sampling. Let us take the definitions of *population, frequency distribution* and *sample* directly from Fisher's (1925) classic *Statistical Methods for Research Workers*. The book begins with a description of statistics as 'the study of *populations,* or aggregates of individuals, rather than individuals' and provides a good working definition: 'if an observation, such as a simple measurement, be repeated indefinitely, the aggregate of the results is a population of measurements'. Fisher introduced biological variability by noting that the study of variation leads immediately to the concept of a *frequency distribution*, an idea which is 'applicable ... to infinite populations. ... Only an infinite population can exhibit accurately, and in their true proportion, the whole of the possibilities arising from the causes actually at work, and which we wish to study. The actual observations can only be a *sample* of such possibilities.' Hence the population *parameters*, those quantities which technically describe the population, are constant and unvarying, but unknowable and unobservable. So, for the statistician, a statement about the population mean, say μ, such as: 'the probability that μ > 20 is 0.1', is strictly meaningless. By contrast, consider a *sample statistic*, that is a value derived from the data. For example, the value of the *sample mean*, say m, is the arithmetic average of the values over each sample unit; m is always just an estimate, is variable, and may be unrepeatable in the short term. A statement such as: 'the probability that m is within 5 units of the true, unknown mean, μ, is 0.95' is then perfectly sensible. The link between population and sample comes because m is an *estimate* of μ.

Many authors stress the requirements of independence of the sample units and note that it is the repeatability of sampling under identical circumstances which allows the concept of probability to be adequately defined, and a valid statistical theory developed. In practice, the strict accuracy of such statistical assumptions is never met, but for most crop protection applications an approximate adherence is sufficient.

7.3. Sampling Distributions

Associated with each population frequency distribution, for example that of the wing length of some aphid species, may be several sample statistics, such as the sample mean length, or the proportion of samples expected to exceed some threshold length. For each summary statistic, there is also an exact small-sample *sampling distribution*. For example, 'Student' (1908) discovered the exact sampling distribution of the sample mean if the underlying population distribution is a Normal, and denoted it as the t-distribution. From this, and other distributions like the chi-squared and the binomial, came tables of significance (Fisher and Yates, 1938) and *significance testing* for relatively small sets of experimental data. Especially important was Fisher's discovery of the F-distribution, and its use in the *analysis of variance*.

A biologist's attention naturally focuses on m, the mean of a sample of, say, n units, and on s^2, the *sample variance*. The statistics m and s^2 are by no means the only estimates possible, but for a Normal distribution they best estimate the underlying true, but unknown, population mean, μ, and variance, σ, respectively. Fisher (1925) showed that the variability of the estimate, *m*, was inversely proportional to sample size, since its variance is σ^2/n. This implies that the precision of an estimate may be improved by replication, which is of crucial importance in the design of experiments and sample surveys. Of course, since the statistic *m* is itself variable, it must, like all statistics, have a frequency distribution.

Much modern statistics appeals to the 'Central-limit' theorem, which states that for *any* population distribution, the distribution of the sample mean, *m*, becomes ever closer to a Normal with mean μ, as the sample size, *n*, is increased. The lack of restriction on the underlying population distribution is particularly important for applications in crop protection where many distributions represent counts of insect or weed pests, or proportions of plants diseased. These are almost invariably *skew*, asymmetric distributions, in which the mean count may be considerably larger than the median. While early work recognized the importance of the Poisson distribution for counts in biological research, the distribution to be expected under the assumption of a random spatial distribution of individuals, studies soon found evidence of non-randomness. Over fifty years ago, Bliss (1941) found such evidence even in small, apparently uniform plots, and concluded that the Poisson distribution would be found only under the simplest and most uniform of conditions. There are other, yet more skewed distributions, in which the sample mean may be very sensitively affected by a few large counts, such as the negative binomial (Fisher, 1941) or log-normal (Finney, 1940), and the logarithmic series (Fisher *et al.*, 1943). These are commonly used to describe insect populations, at the species or community level, respectively. However, whatever the degree of skewness, the central-limit theorem provides reassurance that the distribution of the sample mean will tend towards the symmetric, well-behaved and ubiquitously tabulated Normal distribution.

7.3.1. Significance testing and multiple comparisons

One of the worst barriers to the quality of data interpretation, and to the understanding of what data analysis should aspire towards, lies with the current obsession with significance testing.

Perry (1986) noted that significance tests have a limited role in biology because: significance relates to plausibility not to biological importance; the outcome of a test depends as much on the amount of replication as the size of the effect studied; a null hypothesis may be *known* to be false before an experiment is done – testing this hypothesis is redundant and the significance level meaningless; in agricultural or ecological entomology, the really critical single experiment is rare; a theory may be strictly false but useful practically; quotation of a result that a hypothesis was rejected gives no information concerning the magnitude of effects; and, finally, all tests are based on assumptions that rarely hold in practice. By contrast, the primary interest in field experimentation is to estimate the magnitude of treatment effects. Here, a typical relevant question is: 'by how much does fecundity increase for each unit increase in food intake, and at what level does this relationship cease to be linear?'

A significance test is mainly useful to provide confidence that unjustified conclusions are not being made from the data. This is particularly useful at the stage of exploring competing statistical models, to inform a choice of which is 'best'. Normally the criterion for this will be: 'what is the simplest model we can find that still provides an adequate description of the data, consistent with our biological understanding of the process?' Once this best model has been found, attention should focus on estimating the parameters of this model, and hence the magnitude of the biological effects.

Many leading applied statisticians (Pearce, 1983; Mead, 1990) point out that for well-planned research, the comparisons of interest will have been decided well before the experiment; often the treatments will have a factorial structure and possibly the levels of these will form a quantitative series (Cox, 1982). For such data, most statisticians now agree that there is no place whatsoever for multiple-comparison procedures, such as Duncan's multiple-range test. Some (Nelder, 1971; Perry, 1986) see no role for these procedures in the interpretation of data, even for unstructured data.

Much more use could be made of the consistency of results. For example, experimental results might demonstrate that of 11 out of 13 species studied, treatment A gave increased numbers, relative to treatment B, but it might be that for only 4 of these 13 analyses could statistical significance be demonstrated. The judgement of the biologist, that the results are consistent over the species and signify a real effect, should surely outweigh purely abstract concerns that for 9 of the species no significance could be demonstrated. Sometimes, it is possible to formalize this reasoning using a further significance test! Here, a null hypothesis that for each species there was an equal chance of increased numbers under either treatment would be rejected, given independence of the results for each species, at approximately the 1% level.

It is difficult to say who or what is to blame for the preponderance of significance testing. Some might say it is journal editors, abetted by advisory boards and referees, whom authors believe, rightly or wrongly, demand positive results from significance tests as a prerequisite for publication. Others blame statistical packages, especially those that appear to equate providing a good product with the saturation of output with as many summary statistics as possible, and their uncritical use by biologists who, through no fault of their own, may have little idea of the precise methods used by the package. Yet others point to the cook-book approach of many service courses in statistics, that teach by providing statistical recipes for analysis at the expense of good understanding of the principles underlying them. Whatever the case, in my view entomologists could improve their analyses greatly by:

1. Plotting their data more often, especially in preliminary analyses.
2. Avoiding, where possible, the automatic and routine running of

packages on their data; doing fewer analyses, but spending more of their own time in doing or checking calculations.

3. Placing more reliance on consistency of results over time, sites, etc., than is currently the case.

4. Placing more confidence in their own biological judgement as to whether a putative effect is real.

7.4. Components of Variance and Sampling Plans

7.4.1. Basic definitions

Early experimentation recognized that yield variability or pest abundance was likely to vary more between samples the further the sample units were placed apart. Rather than give full details here, the reader is referred to Perry (1989) for a full description of what Yates and Zacopany (1935) called the concept of *variance components*. Briefly, the idea is that the mean per sampling unit is considered to be the sum of two parts: one, with variance component V_s, which varies independently from unit to unit within a plot, and another, with variance component V_p, which varies between plots but affects all units within the plot equally. The variance of the sample mean per sample unit, $(V_p + V_s/k)$, depends on the number, k, of units per plot. The corresponding analysis of variance is hierarchical, with effects being estimated in different *strata* as appropriate.

Equally, while blocking allows for changes in variability in agricultural experimentation where this variability is not the prime interest, identical concepts of hierarchies of variability may motivate the ecological study of spatial pattern at various scales for its own sake. Thus, Bliss (1941) predated Greig-Smith's (1952) analysis and Taylor's (1961) power-law, both later independently rediscovered, with his analysis of variance components.

7.4.2. An example to compare sampling strategies

Bliss (1941) used Fleming and Baker's (1936) *Popillia japonica* data to illustrate how large areas could be subdivided by successive agglomerations of sample units into a hierarchy. He considered 2304 of the 1 foot square sample units of Fleming and Baker's original data, arranged in a square of side 48 feet (see Perry, 1989, Fig. 1), and grouped into blocks of size 4×4 feet, 8×8 feet and 16×16 feet, i.e. four different strata comprising units and three increasing sizes of blocks. Using a square root transformation followed by an analysis of variance, Bliss (1941) was able to isolate the component of variance attributable to each stratum (s_1^2 to s_4^2, respectively). Recently, powerful residual maximum-likelihood (REML) methodology and software allows such calculations to be done easily (e.g. Robinson, 1987).

The importance of this scheme regarding sampling is that it allows a comparison of the precision by which the sample mean is estimated, given various competing sampling strategies. Sampling schemes often involve a natural hierarchy of units, e.g. leaves within plants, plants within rows, rows within plots; the use of variance components can inform a choice regarding where to put most sampling effort within such a hierarchy. The different strategies may be compared with respect to their expected variance of the mean per sample unit, to which the various strata contribute different amounts, dependent on the replication within a stratum and the randomization scheme employed. Consider, for example, the six sampling strategies, (i)–(vi) in Fig. 7.1. Comparing (i) and (ii), we see that the smaller the area the more efficient the estimate, as we would expect; and comparing (i) and (iii), that improved replication increases efficiency. There may be a trade-off

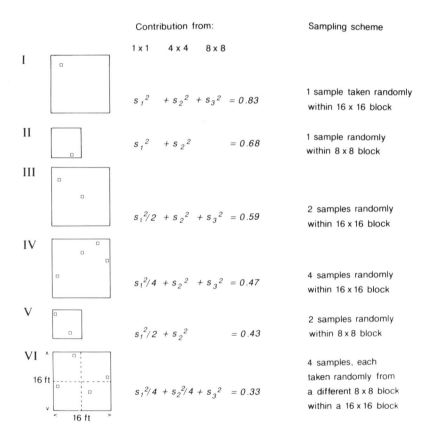

	Contribution from:			Sampling scheme
	1 x 1	4 x 4	8 x 8	
I	$s_1^2 + s_2^2 + s_3^2 = 0.83$			1 sample taken randomly within 16 x 16 block
II	$s_1^2 + s_2^2 = 0.68$			1 sample randomly within 8 x 8 block
III	$s_1^2/2 + s_2^2 + s_3^2 = 0.59$			2 samples randomly within 16 x 16 block
IV	$s_1^2/4 + s_2^2 + s_3^2 = 0.47$			4 samples randomly within 16 x 16 block
V	$s_1^2/2 + s_2^2 = 0.43$			2 samples randomly within 8 x 8 block
VI	$s_1^2/4 + s_2^2/4 + s_3^2 = 0.33$			4 samples, each taken randomly from a different 8 x 8 block within a 16 x 16 block

Fig. 7.1. Six sampling strategies for *Popillia japonica* data of Fleming and Baker (1936), with components of variance estimated by Bliss (1941), labelled (i) to (vi) in the text, from top down. In each case: the scheme is described in words, in terms of the randomization and block sizes; the expected variance of the mean per sample unit is calculated, by adding contributions from each of three block sizes; a visual example is given of one realization of the scheme.

between size of area and replication, as in the comparison of (iv) and (v). The rules for randomization may be subtle, as in (vi), but are crucial to the calculation of efficiency, as shown in the comparison of (iv) and (vi). The form of blocking in (vi) used to be known as *local control*; Cochran (1938) explains lucidly how this can only improve precision, sometimes (as here) substantially.

7.4.3. Sampling plans

The above discussion is designed to shed light on the question: how many replicates should I take? This is a question I am frequently asked but never have the remotest idea how to answer. This is not because the calculations are difficult technically, but because the information on which they rely is so seldom available. In a previous era workers like Marshall (1936), Ladell (1938) and Fleming and Baker (1936) might intensively sample an entire field before doing an experiment, to seek the most efficient design possible. Only with such information could the question be answered with any certainty. But the unpredictability of populations ensures that the information may be only of ephemeral relevance. Hence, despite the current obsession with efficiency savings, such a labour-intensive approach is seldom acceptable. But it *is* often worth while considering the per unit costs of time and material to take one sample and to identify it, and the costs of time and material to travel to the sample unit. In addition, the cost of inaccurate results should be allowed for if this is quantifiable. In terms of variance reduction, more samples are undoubtedly better, but with a law of diminishing returns.

See Perry (1989) for a fuller discussion of sampling efficiency and optimal sampling, and Yates and Zacopany (1935) for a discussion of the benefits of *random samples*. Operator error (Beall, 1939; Bliss, 1941; Anscombe, 1950; Harrington, 1987) should be guarded against, possibly by judicious blocking. Subsampling (Anscombe, 1950) may require the application of a specific formula due to Cochran (1953), and given in Perry (1989).

7.5. The Design of Field Experiments

7.5.1. Block and treatment structure

So far, we have considered surveys, and not yet experiments. Both surveys and experiments may involve sampling, which for both may be done within some hierarchical structure comprising several strata, often termed a *blocking structure*. However, in experimentation there is in addition a *treatment structure*, whose effects we are interested to assess on some *response variable* of interest. The treatments must be assigned, using some procedure, to the sample units; usually this will be done in relation to the blocking structure. Usually, the treatments are *balanced* in relation to one another, so that one does not occur too often at the expense of the others, especially in areas of the field which influence strongly the response to treatments, either favourably or unfavourably. There are often many ways in which this may be done. Fisher (1925, 1935) set out the principles of *randomization*, now well known to biologists, to create balanced experimental designs in which each treatment had a fair chance of expressing its potential effects on the response variable.

7.5.2. Examples of different randomizations

Once the design is set and the results gathered, we must consider precisely where, in the ANOVA table, the treatment terms will appear, relative to the various strata defined by the blocking structure. It is often not appreciated that the rules for the construction of the ANOVA table and, indeed, the mathematical justification for the *F*-tests so commonly used, depend completely on the way in which the treatments are randomized to the sample units. The basic rule is that differences between treatments should be compared with respect to the estimate of variance that is appropriate to the units in the stratum over which they were randomized. This is a wordy and difficult concept that is easier to appreciate with an example.

Suppose there are four benches in a glasshouse, on each of which there are two cages. Within each cage there are four compartments, in each of which there are ten larvae of some insect species. The blocking structure forms a hierarchy, in which compartments are *nested* within cages (the lower, compartment stratum), and cages are nested within benches (the higher, cage stratum), reflected in the ANOVA table, from which we could extract the components of variation, were there no treatments involved (Table 7.1(a)). Note that ascribing of the sources of variation is easy, once all the variability has been categorized into between- and within-components. Suppose we wish to compare the effect of two diets, A and B, on these ten larvae; the response variable used could be the combined weight gain of all ten larvae (using, instead, the average weight gain would give identical mean squares in the ANOVA table). Firstly, consider the result of assigning the diets randomly to the cages within benches; an example of such a randomization is given in Fig. 7.2(a). Following the above rule, the single degree of freedom for the effect of diets must be tested against the estimate of variance made between cages, in the cage stratum. That degree of freedom, for treatments, is gained at the expense of residual variation, and is lost from the particular residual mean square against which it is tested, i.e. that between cages (Table 7.1(b)), which reduces it from four to three. We cannot expect any great precision in the estimated effect of diets, with so few degrees of freedom available for the estimate of its variability. However, now consider an alternative randomization, to compartments within cages, so that each diet occurs twice in each cage (Fig. 7.2(b)). Now, using the above rule, the effect of diets is tested against the estimate of variance made between compartments, in the compartment stratum. The single degree of

Table 7.1. ANOVAs for two randomizations of two treatments in an imaginary diet experiment.

Source of variation	Degrees of freedom	
(a) Variability due to blocking structure alone – no treatments imposed		
Between benches	3	⎫
Between cages, within benches	4	⎬ cage stratum
Total for cage stratum (between cages)	7	⎭
Between compartments, within cages	24	⎫ compartment stratum
Total for compartment stratum	31	⎭
(b) Diets assigned randomly to cages		
Between benches	3	⎫
Between diets	1	⎬ cage stratum
Between cages, within benches	3	⎬
Total for cage stratum (between cages)	7	⎭
Between compartments, within cages	24	⎫ compartment stratum
Total for compartment stratum	31	⎭
(c) Diets assigned randomly to compartments		
Between benches	3	⎫
Between cages, within benches	4	⎬ cage stratum
Total for cage stratum (between cages)	7	⎭
Between diets	1	⎫
Between compartments, within cages	23	⎬ compartment stratum
Total for compartment stratum	31	⎭

freedom lost from that residual mean square (Table 7.1(c)), reduces it from 24 to 23. By comparison with the original randomization, such a reduction is easier to accept, and the precision of the experiment will almost certainly be improved greatly, and nothing has been lost. This example demonstrates the importance of giving careful thought to the design of experiments. Indeed, I would recommend that a skeleton ANOVA be done at the design stage of every experiment. I find this invaluable as a pointer to judge the likely precision.

The variability between compartments in the first randomization (Table 7.1(b)) cannot be utilized to test between diets, and, so far as this experiment goes, is a wasted resource, although the information could be used to help design future experiments. The records from individual compartments are

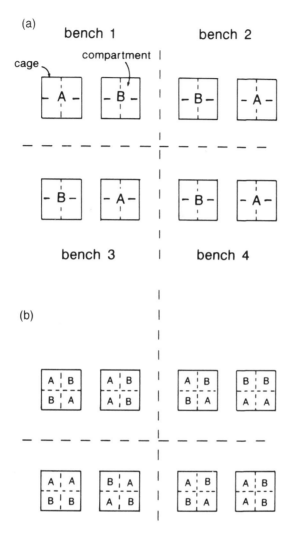

Fig. 7.2. Examples illustrating two randomization schemes for the imaginary experiment comparing diets A and B. (a) Diets randomized to cages within benches; (b) diets randomized to compartments within cages, with each diet occurring twice in each cage.

termed *pseudoreplicates*; their misuse as true replicates can be extremely mis-leading, as pointed out forcefully and amusingly by Hurlbert (1984). In general terms, the more samples that are taken in a stratum lower than that in which the treatment of interest is assessed, the more accurately the mean over the units in that stratum may be estimated. However, the variance of a treatment mean assessed in a higher stratum is affected not only by the vari-ation between units in that lower stratum but also by the variation between units in the higher one; increasing the number of samples in the lower stratum is therefore tantamount to invoking the law of diminishing returns.

Of course, there may not always be this flexibility in the decision as to which units the treatments should be randomized. For example, it might have been decided for experimental reasons that odours from the food constituting diet A could not be allowed to cross between compartments, because of possible interference with the larval response to diet B. Then, diets might have *had* to be randomized to cages. When more than one treatment factor is involved, restrictions on the randomization of one set of factors that do not apply to others can lead to split-plot, or other more complex designs, as is outlined below.

7.5.3. Factorial experiments

As Fisher (1935, Chapter VI) showed, *factorial* experiments use resources more efficiently than would be the case if one question were asked at a time. As an example, consider the experiment of Kennedy *et al.* (1994) in which the response variable of interest was the number of a particular species of cereal aphid natural enemy sampled by pitfall traps. There were two treatment *factors*: first, a pesticide application; and second, the use of barriers to exclude the aphid's natural enemies. Each of these treatments occurred at several *levels*. For the treatment factor representing pesticide application, the levels studied were: the effects of Pirimicarb, Dimethoate and an untreated control. For the other factor the levels studied were: large unbarriered plots and two types of small barriered plots, one with three pitfall traps and one with one. The resulting experiment with two factors, each at three levels, is often known as a 3×3, or sometimes a 3^2 experiment. In the simplest such factorial experiment, all the combinations of the different levels of each factor occur in combination with each other on at least one plot (and usually on only one) in every *replicate* or block. In this case there are three blocks, in each of which there are $3^2 = 9$ combinations, so that, after randomization, on the first plot in the first block we might have the combination of Dimethoate occurring on a large unbarriered plot; on the second plot, the combination of Pirimicarb on a small barriered plot with one trap, and so on.

7.5.4. Interactions

There is not only a gain in efficiency in conducting one experiment to study both factors, instead of one experiment for each, but it is also the only way in which the *interaction* between the two factors may be quantified. An interaction is a difference in the effect of one particular factor according to which of the levels of another factor it occurs with. For the above example, it might well be the case that the effect of the toxic chemical Dimethoate, compared to the control, might be different for the small barriered plots, where natural enemies are excluded from recolonization following pesticide application, than for the large unbarriered plots, where there are no such restrictions.

This interaction may be formally studied in the usual analysis of variance (Table 7.2(a)). However, a very useful informal impression of the strength of an interaction may be gained graphically (Fig. 7.3). Each line is drawn to connect the value of the response variable (ordinate) for a particular level of one factor, over the different levels of the other factor. The latter factor forms the abscissa, and the levels, if qualitative, are spaced an arbitrary distance apart (usually equally-spaced) on the x-axis. The result is several lines, one for each level of the first factor, between each pair of levels of the second factor. An interaction is indicated by any gross divergence, or non-parallelism, of these lines, between any of these pairs. In particular, the interaction is strong if the lines between a particular pair have opposite slope. In the hypothetical example shown (Fig. 7.3) there is no interaction between the effect of barriers and the effect of the pesticide factor corresponding to the difference between Pirimicarb and control, but there is an interaction between barriers and the difference between Dimethoate and control, shown by the two bold lines. From this example, it may be seen that an interaction is essentially measuring the difference of a difference. The effects of either of the two single factors from which a two-factor interaction is formed are termed *main effects*. It is often not sufficiently well understood

Table 7.2. ANOVAs for natural enemy experiment.

Source of variation	Degrees of freedom
(a) Actual experiment – full randomization of both treatment factors	
Blocks	2
Pesticides	2
Barriers	2
Barriers × Pesticides (interaction)	4
Residual	16
Total	26
(b) Split-plot design – pesticides randomized to main plots, barriers to sub-plots	
Blocks	2
Pesticides	2
Residual (main plots)	4
Total (main plots)	8
Barriers	2
Barriers × Pesticides (interaction)	4
Residual (sub-plots)	12
Total (sub-plots)	26

In part (b), Blocks, Pesticides, and Residual (main plots) form the main-plot stratum; Total (main plots), Barriers, Barriers × Pesticides (interaction), Residual (sub-plots), and Total (sub-plots) form the sub-plot stratum.

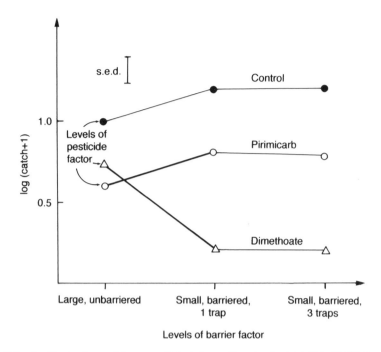

Fig. 7.3. An illustration of the possible interaction between the factors 'pesticide' and 'barrier' in the natural enemy experiment. The main source of the interaction between these factors is evident from the non-parallelism of the two bold lines at the lower left of the figure.

that once an interaction has been demonstrated to have a substantial, significant value, the main effect of each of the single factors is virtually meaningless, because they are each formed from averages over levels of the other factor. It is like saying 'the average risk of death from cancer due to smoking is 0.2' when it is known that there is an interaction between risk and gender; the only way to meaningfully interpret the risk data would be if we specified whether we were speaking of a man or a woman – since these are mutually exclusive, the average value of 0.2 is meaningless. Therefore, the more large interactions that are indicated by the results of an experiment, the more specific we must be in referring to the effects of combinations of treatments, and the more complex must be our interpretation. Since, if there are three or more factors in an experiment, there may be three-factor or higher-order interactions present, interpretation may become even more difficult, but thankfully the strength of interactions usually declines with their order. Hence, large three-factor interactions are relatively rare compared with those involving two factors.

7.5.5. *Randomization for several treatment factors*

The experimental design for the example above (Fig. 7.4) is relatively straightforward, except for a complication with regard to the different-sized plots that we will return to later. Suppose, however, that it had been impractical to apply the pesticide treatments individually to the small barriered plots, perhaps because of restrictions on the tractor boom width. Then, to achieve balance, each group of three plots (one large and the two small to the right or left of it; Fig. 7.4) would need to be treated with the same level of insecticide. Such a group, which receives the same level of a particular factor, is termed a *main plot*. The plots within this group, over which the other factor is randomized, are termed *sub-plots*. The experiment as a whole is termed a *split-plot* experiment, because each main plot is 'split' into sub-plots. For a split-plot experiment, the randomization rules for the two treatment factors therefore differ. The three levels of the pesticide treatment would be randomized to the three groups of three (main) plots within each block, while the three levels of the barrier treatment would be randomized to the three sub-plots within each main plot. Because the effect of each treatment factor is

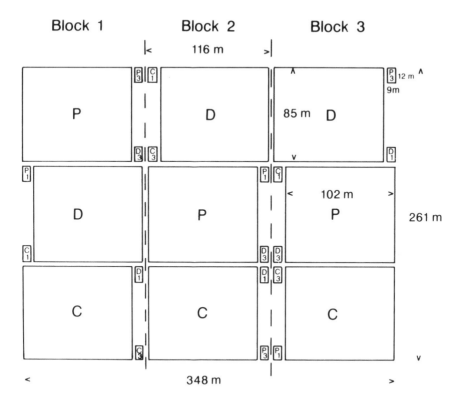

Fig. 7.4. Experimental plan for the natural enemy experiment.

estimated in the stratum over which its levels are randomized, the pesticide and barrier factors are assessed in different strata (Table 7.2(b)). As in most split-plot experiments, the factor which could only be randomized to main plots (pesticides) has fewer degrees of freedom associated with its appropriate residual mean square, and would probably be estimated with less precision than the factor (barriers) which could be randomized over the smaller sub-plot units. Note that the interaction between the two factors is estimated in the lower of the strata of the factors concerned. A fuller description of the principles underlying factorial experiments is in Yates (1937).

7.5.6. Spatial considerations

More and more applications, particularly in ecological entomology, include the phenomenon of spatial dynamics, dispersal, etc. Often, the requirement to fully randomize treatments of a spatial nature cannot be completely met, particularly when the area available for the experiment is strictly limited. Consider the design of the natural enemy experiment described above (Fig. 7.4). We would ideally have wished to randomize the position of *each* of the nine plots within each block. However, given the overall width and length of a block, 116 m and 261 m, respectively; the width and length of a large unbarriered plot, 102 m and 85 m, respectively; and of a small barriered plot, 9 m and 12 m, respectively, three things are clear. Firstly, that the plots must be placed in three groups of three, each comprising one large and two small, otherwise they could not be fitted into the block. Secondly, that there is no space to allow the smaller plots to be placed in any other way than one above the other. (It was desirable to space the small plots as far from each other as possible, but this could have been done better than in the arrangement shown.) Thirdly, that with these restrictions on randomizations, the only arrangements possible are: (i) placing the large plot to the left of the two smaller ones; and (ii) placing the large plot to the right. There is no need to restrict the levels of the two small barriered plots (one and three traps) to be randomized within these groups of three, as may be seen from the plan for the block on the left-hand side (Fig. 7.4), and we should never impose restrictions on the randomization that are not essential.

The crucial message is that designs with spatial components may be more complex than those without, and usually require much thought; they may also be awkward to randomize. There are more sophisticated statistical analyses available that account for such restricted randomizations; whether they are worth the effort of using must depend on the degree of the restrictions imposed, and the loss of balance that results. This topic is discussed more fully later in the chapter.

7.5.7. Time as a factor in experiments

Many experiments must be replicated through time because the effects of treatments may alter with, for example, stages in insect phenology, seasonal temperature, photoperiodic effects, etc. Consider, for instance, a very simple experiment, with one treatment factor, representing, say, the effect of the presence or absence of a fungal pathogen on the number of aphids of some particular pest species, measured 24 h after application of the treatment. Suppose the treatments are randomized to two plots in each of three blocks. The ANOVA is straightforward (Table 7.3(a)), but how should the data be analysed if we then decide to replicate over four weekly occasions?

Temporally, sample units may be autocorrelated if placed too close together in time. Then, the information in successive samples is less than that in two separate samples. One example might be automatically collected meteorological data sampled too often relative to the natural rate of change of the variable concerned. Less striking, but still requiring allowance in the analysis, are autocorrelated spatio-temporal processes like the spread of virus disease by an aphid vector (Tatchell and Plumb, 1992) or the persistence of weed patches in a field (Wilson and Brain, 1991). Techniques now exist to analyse autocorrelated *repeated measures*, often on the same individual, within a classical analysis of variance framework (Crowder and Hand, 1990). For insects, sampling frequency will obviously differ according to whether within- or between-generational processes are studied.

If we find or can assume that a repeated measures analysis is unnecessary because the degree of autocorrelation is small, then a simpler analysis may be provided by a split-plot type of ANOVA, where the time units form

Table 7.3. ANOVAs for imaginary fungal pathogen experiment.

Source of variation	Degrees of freedom	
(a) Simple unreplicated experiment, three blocks, two plots per block		
Blocks	2	
Treatment	1	
Residual	2	
Total	5	
(b) Same experiment, replicated on four occasions		
Blocks	2	plot stratum
Treatment	1	
Residual (plots)	2	
Total (plots)	5	
Occasion	3	
Occasion × Treatment (interaction)	3	occasion stratum
Residual (occasions)	12	
Total (occasions)	23	

the lower stratum, and those units within occasions the higher stratum. Hence, for our example (Table 7.3(b)), the overall comparison between treatments is still done on a plot basis (formed essentially from the average of the differences between the six plots over time) and so the degrees of freedom for treatments, in the plot stratum, remain as above for Table 7.3(a). This is as it should be, because, as noted above, there is no guarantee that much extra information will be gained through the replication in time. By contrast, treating the plots as main plots and the occasions as sub-plots enables the estimation of the treatment × occasion interaction, i.e. the difference in the effect of treatment over time, and estimates it in the correct, lower, stratum. Note that this design is particularly useful for the case where it is already *known* that a treatment has an effect and what is its approximate magnitude but where the main aim of the experiment is to determine how that effect *changes* through time. In this design the time × treatment interaction is estimated with more degrees of freedom than the treatment main effect.

A formal statistical objection to this analysis may be made, because of the fact that it is impossible to randomize occasions, one of the treatment factors. Clearly, we cannot randomly allocate an occasion, such as 25 June – it always occurs on that day! Furthermore, we have stated that it is valid randomization that provides the justification for the ANOVA *F*-tests. While formally true, the argument is pedantic, and the justification may be made empirically, so long as the data have what statisticians term a well-behaved correlation structure.

7.5.8. A short illustrative case study

In a recent study, the main aim was to compare organic and conventional farming systems on 13 matched pairs of fields. Samples were taken from five transects in each field, each transect consisting of five sample units taken at set distances away from a hedge that formed a boundary. Of subsidiary interest was the effect of the distance from the hedge. After over two months of the ten month study, it was reported that owing to the labour-intensive nature of the identification, samples from only three fields had been sorted. Constructing the ANOVA for the study was instructive (Table 7.4(a)). There were two strata, with the comparison of main interest being estimated in the higher stratum, with relatively few degrees of freedom for error. By contrast, there was a huge number of degrees of freedom for the residual in the other stratum, where the effect of distance from the hedge was to be tested. Clearly, as time was limited, samples from all the pairs of fields could not be sorted, yet that would reduce still further the precision of the study; it was essential to find a way to obtain this information. One solution was to bulk the samples along the five transects for each field, for each particular distance from the hedge, giving five bulked samples per field, and then to subsample a known proportion of this bulked sample. This would reduce the amount of work required in sorting by four-fifths and

Table 7.4. ANOVAs for farming systems case study.

Source of variation	Degrees of freedom

(a) Original design: 13 pairs of fields, 5 transects, 5 units per transect

Field pairs	12	
Farming systems	1	field stratum
Residual (fields)	12	
Total	25	
Distance from hedge	4	
Farming system × Distance from hedge (interaction)	4	unit stratum
Residual (units)	616	
Total (units)	649	

(b) As above, but with samples bulked from 5 transects for each distance from hedge

Field pairs	12	
Farming systems	1	field stratum
Residual (fields)	12	
Total	25	
Distance from hedge	4	
Farming system × Distance from hedge (interaction)	4	unit stratum
Residual (units)	96	
Total (units)	129	

(c) As above, but with all 25 samples for a field bulked

Field pairs	12
Farming systems	1
Residual (fields)	12
Total	25

lead to the ANOVA in Table 7.4(b), with plenty of precision left for the effect of subsidiary interest. Of course, since the systems must be compared on the basis of variation between fields, this change does not affect the analysis of the field stratum. If that procedure was *still* found to be too time-consuming, then information on distance from the hedge would have to be sacrificed. This could be done by bulking all 25 samples from each field and, again, subsampling, leading to a single-stratum analysis (Table 7.4(c)). Interestingly, one of the foremost exponents of modern experimental design, Mead (1990), has argued powerfully that too many degrees of freedom in the residual implies a waste of resources, and that these could often be utilized by extending the treatment structure of the design to include other factors of interest, and their interactions.

7.6. Discrete Data

One of the most notable differences between statistical techniques encountered typically in entomology, rather than, say, in agronomy or environmental sciences, is that for the former the data are often discrete counts of individuals, while in other disciplines variables usually have a continuous scale. Incidence data, in which purely presence or absence are recorded, are also often found in entomological applications, whether describing the insects themselves or the presence of some disease they have vectored, and are another example of discrete data. Counts are bounded below by zero, and, as noted above, are usually described by highly skewed distributions. Of course, this implies that the sample variance will not be constant, the data will display *heteroscedasticity*, there will be some relationship between sample variance and sample mean, and this must be accounted for in the analysis of data. Hence, Bartlett (1936) suggested a *logarithmic* transformation to overcome heteroscedasticity. Many specific two-parameter statistical distributions were proposed around this time as underlying models for count data. Whereas, arguably, they have been proved useful as theoretical ecological models, they have been used relatively infrequently in applied biology (but see Hughes and Madden, 1992; Blackshaw and Perry, 1994). Perhaps this is because, for the estimation of effects, the precise transformation used to reduce heteroscedasticity or equalize the variance–mean relationship matters little. The transformation improves the standard errors of the estimates, but often different transformations differ only marginally in these estimates.

Of far more importance to the precision of estimates is the *scale* on which effects are *additive*. Here, the contribution of Williams (1937) can hardly be overestimated. He too proposed a logarithmic transformation for biological counts, because he had noted that effects were *multiplicative* on the natural scale, and so became additive on the newly transformed scale. Similarly, the greatest contribution to the analysis of incidence data came with the development of *probits* (Bliss, 1934; Finney, 1942, 1947) and *logits* (Berkson, 1944), because these provided, for the first time, scales on which effects were found to be additive. Contrast this with the arcsine transformation (Fisher and Yates, 1938), a means of reducing variance heterogeneity for the binomial distribution, that lacks such a scale and is now, rightly, used seldom.

7.7. Variance–Mean Relationships

For many years (Bliss, 1941; Yates and Finney, 1942) it has been known that to improve the efficiency of monitoring insect pests and of sample surveys in applied entomology, a knowledge of the relationship between sample variance, s^2, and sample mean, m, is essential (Perry, 1994a). Briefly, the standard error of a sample mean, s^2/n, expressed usually as some proportion, say q, of that mean, varies through time with mean population density. To keep this proportion constant, and thus ensure a similar precision for monitoring throughout a season, requires the sample size, n, itself to vary with m. Knowledge of the relationship between s^2 and m enables the value of n to be found for given values of q and m.

In addition, many attempts have also been made to relate such empirical relationships to studies of fundamental population dynamics in insect ecology (Taylor, 1986). The difficulty with these approaches is that they require a vast amount of good data to discriminate between competing models, and a wide variety of models could give rise to the same variance–mean relationship (Perry, 1988, 1994b). Clark and Perry (1994) give a brief historical overview.

Taylor (1961) found the s^2–m relation to be a power law, $s^2 = am^b$, with wide application to taxa in ecology and crop protection. The law is usually expressed as a straight line on logarithmic axes:

$$\log(s^2) = \log(a) + b\log(m)$$

with *parameters*, $\log(a)$, the intercept, and b, the slope, estimated by *simple linear regression*. The ubiquity of the relationship was shown by Taylor *et al.* (1978). Finch *et al.* (1975, 1978) gave the expression for n, namely:

$$\log(n) = (b-2)\log(m) + \log(a) - 2\log(q)$$

This method, termed sampling for a *fixed precision level*, has been widely used for many years, particularly in the USA (Taylor, 1984). As noted above, the logarithmic scale for mean density is to be preferred; fortunately, the method of Finch *et al.* (1975, 1978) is approximately equivalent to ensuring that the standard error of $\log(m)$ remains constant. Variants of this method (Southwood, 1966) have used a variety of variance–mean relationships. Karandinos (1976) invoked the central limit theorem to fix the width of the *confidence interval* for the unknown mean, again as a proportion of the mean, but his method depends on the accuracy of the Normal approximation and should not be used for small sample sizes.

Taylor's power-law has engendered much controversy (Downing, 1986; Taylor *et al.*, 1988). To be a credible basis for sampling schemes, $\log(a)$ and b should be invariant for as many sources of variation as possible, such as year, sampling method, location, life stage, spatial scale, etc. Initial evidence that b appeared species-specific prompted Taylor to give the wrong impression concerning this invariance. In fact, as Taylor *et al.* (1988) emphasized, neither $\log(a)$ nor b can be assumed to be invariant unless all these sources of variation are controlled, but there is good evidence that the relationship remains reasonably consistent as long as the sampling method is unaltered and there has been no environmental change to the life stage concerned. Again, despite earlier comments, b is no more a species-specific parameter and no more an index of aggregation than is $\log(a)$. Knowledge of both b and $\log(a)$ is essential.

Perry (1981) investigated some statistical problems in fitting the relationship generally. Problems of fitting at small densities were addressed by Perry and Woiwod (1992). Clark and Perry (1994) have recently studied artefacts caused by small sample size, proposed correction for bias in parameter estimates and recommended that the number of samples making up each variance–mean pair should be no less than 15. Families of Adès frequency distributions, which are consistent with Taylor's power-law at all population densities, were developed by Perry and Taylor (1988) and have been used successfully to predict leatherjacket population frequencies (Blackshaw and Perry, 1994).

7.8. Spatial Aspects of Large-scale Experiments

Two simultaneous changes are occurring, one in agricultural applications, the other in ecological entomology, that increasingly force experimenters to study and account for spatial aspects explicitly. In both areas new techniques of design and analysis are required.

7.8.1. Spatial aspects in agricultural entomology

In agriculture, there are more and more studies in which the paramount aim is to assess environmental effects or to compare whole-farm systems; examples include the LIFE experiment at Long Ashton Research Station, the MAFF LINK Integrated Farming Systems (IFS) Study, the Nummela experiment in Jokioinen, Finland, and the University of Kansas NESA project (see references in Glen *et al.*, 1995). These present a great challenge in experimental design because they require large plots, of the order of whole fields, otherwise the effects studied are not representative. This is especially true when studying highly mobile natural enemies such as parasitoids or coccinellids. However, adequate replication within such restrictions requires considerable land resources, especially as between-field heterogeneity is likely to be far greater than that usually encountered between plots in conventional field experiments. This is costly, and may be inconvenient, causing problems in management that are not encountered in traditional small-plot experiments. Thus far, some of the designs proposed for such experiments within the UK have been poor compromises which may well lack the power or the realism to yield convincing results. The statistical community must take the blame for this; we must be more vocal in arguing for adequate land resources – if the experiments are important, they merit proper designs. Further problems occur because of the non-standard response variables involved. For instance, for the IFS study, the response variable of major interest is profitability, which few have any experience in analysing. Environmental assessment studies may involve many variables, that may be difficult to combine. Practical difficulties of space are compounded when factors need to be incorporated to measure effects such as 'relative isolation', perhaps measured in terms of distance of a plot from a hedge, or the nearest source of recolonizing, overwintering insects. Randomization of such factors requires care. In the natural enemy example given previously (Fig. 7.4) the field was no larger than sufficient to accommodate all the plots, the hedges bordered close to the sides of the plots, and no allowance could have been made to study such a factor.

7.8.2. *Spatial aspects in experimental design for ecology*

In ecological entomology, spatial scaling (Schneider, 1994) and the study of movement and dispersal, particularly with regard to the persistence of populations and the colonization of patchy habitats, is of increasing importance. Wiens (1989) presented a powerful case that the dependence of ecological effects on spatial scale presents the most challenging problem in ecology today. Ideally, work should be done at several scales simultaneously, and there is a choice between whether these scales should be decided on the basis of natural units (e.g. aphids on leaves; leaves within plants; plants within plots; plots within fields), or purely by measurement to achieve equal unit sizes for each stratum (e.g. Greig-Smith, 1952). Woiwod (1991) contrasts monitoring over a very large, synoptic scale with field-scale studies, for long-term data. Loxdale *et al.* (1993) and Wynne *et al.* (1994) discuss the crucial importance of scale for highly mobile aphid populations.

Field tests of ecological hypotheses concerned with metapopulation dynamics often require treatment factors that are spatial in nature. For example, we may wish to estimate probabilities of colonization in habitats at various distances from a known source of overwintering insects. To ensure adequate replication in designs which compare different degrees of isolation of patches requires a very considerable area; there are other serious problems of design. As an example, consider an experiment where the patches are a sown grass–herb mixture, representing semi-natural habitats, within a cereal field. A small amount of thought confirms that designs with the most efficient use of space place the most isolated patches at the outside of the available area, with the more crowded towards the centre (Street and Street, 1987). Statistical purists might argue that it is not then possible to randomize the treatment factor for isolation, but this would be as pedantic an argument as that voiced originally against the valuable fan designs of Nelder (1962). The requirement to compare patch sizes that are sufficiently different as to be likely to yield real differences in persistence may result in a range where the largest patch covers an area over one hundred times that of the smallest; this makes it difficult to replicate the larger patch sizes adequately. If we are also required to ensure that there is scope for movement to nearest neighbours between all patch sizes, the area required is still larger. Suppose our experiment compares square patches of three sizes (sides of 1, 6 and 15 units) and with three degrees of isolation (inter-patch distances of 3, 10 and 32 units). A fully randomized design could just be accommodated in a square of side 600 units comprising four different quarters (the top-right quarter is illustrated in Fig. 7.5). If, for example, for a persistent, viable population a minimum of 4 m² was required, then the total area required would be a square field of 144 ha, which would not be easy to obtain. The design in Fig. 7.5 was proposed while planning the

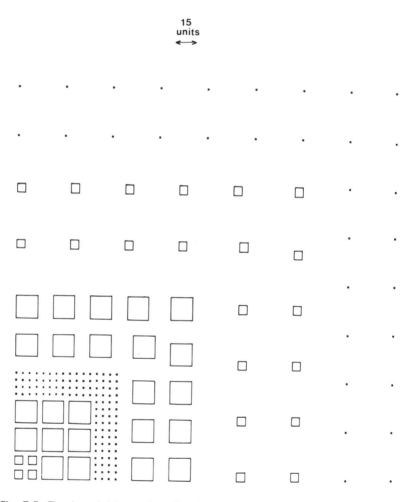

Fig. 7.5. The top-right quarter of a design proposed to compare the dynamics of populations on three sizes of square patches of semi-natural habitat, at three levels of isolation, within a cereal field.

Farmland Ecology experiment at Rothamsted and Long Ashton (Perry, 1995a), but was rejected as too impractical. Dropping the factor for isolation would reduce the area required; another candidate design for the same experiment (Fig. 7.6) was rejected because patches were never nearest neighbours to others of their own size. An ingenious design to study relative isolation, in which the number of neighbours of a patch within a given distance varied gradually from one at the outside to eight at the centre (Fig. 7.7) was rejected because it used only one patch size. The design eventually chosen (Perry, 1995a and Fig. 13.2 therein) was necessarily a compromise of the ideals described above, and was selected partly for ease of management. Perhaps such experiments would be more properly described as arenas, upon which other manipulative experiments may be imposed.

Fig. 7.6. A design proposed to compare the dynamics of populations on three sizes of square patches of semi-natural habitat, within a cereal field.

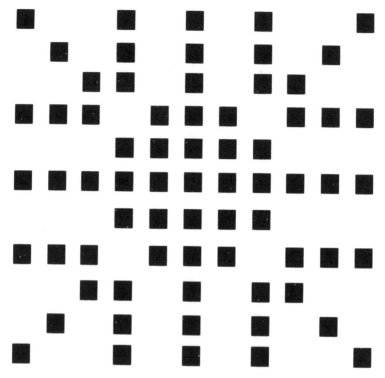

Fig. 7.7. A design proposed to compare the dynamics of populations on square patches of semi-natural habitat, at eight levels of isolation, within a cereal field.

7.8.3. Spatial pattern in ecology

The measurement of spatial pattern of vegetation and the detection of non-random arrangements was an early stimulus for initiating the discipline of plant ecology (Greig-Smith, 1979), while ecological modelling increasingly incorporates spatial heterogeneity (Hassell and May, 1973; Perry and Gonzalez-Andujar, 1993) as an essential framework to reflect the complex aggregation and regularity of populations in nature. Mapped data, where the location of each individual is known, provide the greatest spatial information, but in entomology this is usually only available at second hand, for instance in vectored disease maps. Because of the mobility of insects, abundance data usually occur in the form of counts. A series of papers dealing with counts (Perry 1995a, 1997) and maps (Perry 1995b) introduced a new method to detect and measure spatial pattern, termed Spatial Analysis by Distance IndicEs (SADIE). Two advantages of SADIE for counts are its improved intuitive basis compared with traditional, more abstract, mathematical approaches, and its use of all the spatial information in the sample.

Briefly, for any given arrangement, the minimum distance is calculated that sampled individuals would have to move to achieve an extreme pattern, such as complete regularity. The value of this distance, termed D, may be found using the 'transportation algorithm' from the operational research literature. This is compared with E_a, the value expected from random permutations of the observed arrangement, and an index of aggregation, I_a, is formed from D/E_a. Intuitively, we would expect relatively large values of I_a from aggregated, clumped or clustered arrangements, values around unity for spatially random patterns, and smaller values for more regular, uniform arrangements. Randomization tests, based on comparison of D with the values of the distance to regularity obtained from the random permutations, enable formal statistical tests of the null hypothesis of randomness. Perry (1995a) emphasized the difference between departures from the Poisson distribution of counts in a frequency distribution, and spatial non-randomness in the arrangement of those counts. Most previous approaches (Taylor, 1984, 1986; Blackshaw and Perry, 1994) considered only the properties of the frequency distribution, such as the relationship between its variance and mean. As a brief example, although the set of counts of moths in six light traps (0, 1, 4, 56, 484, 4095) may be highly skewed and obviously non-Poisson, their spatial arrangement may be completely random. Conversely, a set of counts of carabid beetles in pitfall traps (0, 0, 1, 1, 2, 2, 2, 2, 3, 3, 5) may conform closely to a Poisson distribution, but if sampled in that order along a line transect they show an obvious linear trend departing strongly from randomness. Recently (Perry, 1996) has described an algorithm that simulates, for specified locations and for a given set of counts, arrangements of those counts with any given levels of aggregation or regularity. Such simulations have wide applicability, both in ecological modelling and to compare the efficiencies of different sampling plans for pest monitoring. Perry (1996) compared two proposed sampling plans for the detection of cyst-nematodes, and showed that their performance was crucially dependent on the degree of spatial aggregation in the data.

7.8.4. Geostatistical methods and insect spatial patterns

Sound modern statistical techniques exist to analyse two-dimensional spatially *autocorrelated* yield data from field experiments (Cullis and Gleeson, 1991). By contrast, the current fashion to apply geostatistical techniques such as kriging or variograms (Matheron, 1976; Webster and Oliver, 1989) to the analysis of spatial pattern in ecology requires great caution. Such approaches were developed originally for physical variables studied commonly in soil science, such as fertility and chemical content, that are measured on continuous scales and display a stationary, stable covariance structure over a wide area. Counts of individual insects are not continuous, but discrete, are often distributed exceedingly patchily, and frequently comprise a majority of zero values. By contrast with physical variables, such population counts are highly dynamic, usually shifting ceaselessly in space and time for evolutionary reasons (Taylor, 1986). Such variables might not be expected to possess the stable spatial covariance structure assumed by geostatistical methods. They are often characterized instead by isolated clusters, which may be acting as metapopulations with varying degrees of inter-cluster dispersal (Perry and Gonzalez-Andujar, 1993). While it is true that the main aim of geostatistical analysis, that of local estimation, has some overlap with that of the analysis of spatial pattern, I believe that in general the contribution of geostatistical methods to pattern analysis for population count data may be limited.

7.9. Acknowledgements

I thank Wilf Powell for help during many discussions of possible experimental designs for the Farmland Ecology experiment, and John Gower likewise for components of variance.

7.10. References

Anscombe, F.J. (1950) Soil sampling for potato root eelworm cysts. *Annals of Applied Biology* 37, 286–295.

Bartlett, M.S. (1936) Some notes on insecticide tests in the laboratory and in the field. *Journal of the Royal Statistical Society Supplement* 3, 185–194.

Beall, G. (1939) Methods of estimating the population of insects in a field. *Biometrika* 30, 422–439.

Berkson, J. (1944) Application of the logistic function to bioassay. *Journal of the American Statistical Association* 39, 357–365.

Blackshaw, R.P. and Perry, J.N. (1994). Predicting leatherjacket population frequencies in Northern Ireland. *Annals of Applied Biology* 124, 213–219.

Bliss, C.I. (1934) The method of probits – a correction. *Science* 79, 409–410.

Bliss, C.I. (1941) Statistical problems in estimating populations of Japanese beetle larvae. *Journal of Economic Entomology* 34, 221–232.

Clark, S.J. and Perry, J.N. (1994). Small sample estimation for Taylor's power law. *Environmental and Ecological Statistics* 1, 287–302.

Cochran, W.G. (1938) The information supplied by the sampling results. Appendix to W.R.S. Ladell's 'Field experiments on the control of wireworms'. *Annals of Applied Biology* 25, 341–389.

Cochran, W.G. (1953) *Sampling Techniques*. Wiley, New York.

Cochran, W.G. and Cox, G.M. (1950) *Experimental Designs*. Wiley, New York.

Cox, N.R. (1982) Some aspects of the analysis and presentation of data from comparative trials. *Proceedings of the 35th New Zealand Weed and Pest Control Conference*, 289–293.

Crowder, M.J. and Hand, D.J. (1990) *Analysis of Repeated Measures*. Chapman & Hall, London.

Cullis, B.R. and Gleeson, A.C. (1991) Spatial analysis of field experiments – an extension to two dimensions. *Biometrics* 47, 1449–1460.

Czaran, T. and Bartha, S. (1992) Spatiotemporal dynamic models of plant populations and communities. *Trends in Ecology and Evolution* 7(2), 38–42.

Dempster, J.P. (1975) *Animal Population Ecology.* Academic Press, London.

Downing, J.A. (1986) Spatial heterogeneity: evolved behaviour or mathematical artefact? *Nature* 323, 255–257.

Finch, S., Skinner, G. and Freeman, G.H. (1975) The distribution and analysis of cabbage root fly egg populations. *Annals of Applied Biology* 79, 1–18.

Finch, S., Skinner, G. and Freeman, G.H. (1978) Distribution and analysis of cabbage root fly pupal populations. *Annals of Applied Biology* 88, 351–356.

Finney, D.J. (1940) On the distribution of a variate whose logarithm is normally distributed. *Journal of the Royal Statistical Society Supplement* 7, 155–161.

Finney, D.J. (1942) The analysis of toxicity tests on mixtures of poisons. *Annals of Applied Biology* 29, 82–94.

Finney, D.J. (1947) *Probit Analysis.* Cambridge University Press, Cambridge.

Fisher, R.A. (1925) *Statistical Methods for Research Workers.* Oliver and Boyd, Edinburgh.

Fisher, R.A. (1935) *The Design of Experiments.* Oliver and Boyd, Edinburgh.

Fisher, R.A. (1941) The negative binomial distribution. *Annals of Eugenics* 11, 182–187.

Fisher, R.A. and Yates, F. (1938) *Statistical Tables for Biological, Agricultural and Medical Research.* Oliver and Boyd, Edinburgh.

Fisher, R.A., Corbet, A.S. and Williams, C.B. (1943) The relation between the number of species and the number of individuals in a random sample of an animal population. *Journal of Animal Ecology* 12, 42–58.

Fleming, W.E. and Baker, F.E. (1936) A method for estimating populations of larvae of the Japanese beetle in the field. *Journal of Agricultural Research* 53, 319–331.

Glen, D.M., Greaves, M.P. and Anderson, H.M. (eds) (1995) *Ecology and Integrated Farming Systems.* Wiley, Chichester.

Greig-Smith, P. (1952) The use of random and contiguous quadrats in the study of the structure of plant communities. *Annals of Botany* 16, 293–316.

Greig-Smith, P. (1979) Pattern in vegetation. *Journal of Ecology* 67, 755–779.

Hanski, I. and Gilpin, M. (1991) Metapopulation dynamics: brief history and conceptual domain. *Biological Journal of the Linnean Society* 42, 3–16.

Harrington, R. (1987) Varying efficiency in a group of people sampling cabbage plants for aphids (Hemiptera: Aphididae). *Bulletin of Entomological Research* 77, 497–501.

Hassell, M.P. and May, R.M. (1973) Stability in insect host–parasite models. *Journal of Animal Ecology* 42, 693–726.

Hughes, G. and Madden, L.V. (1992) Aggregation and incidence of disease. *Plant Pathology* 41, 657–660.

Hurlbert, S.H. (1984) Pseudoreplication and the design of ecological field experiments. *Ecological Monographs* 54, 187–211.

Johnson, C.G. (1969) *Migration and Dispersal of Insects by Flight.* Methuen, London.

Karandinos, M.G. (1976) Optimum sample size and comments on some published formulae. *Bulletin of the Entomological Society of America* 22, 417–421.

Kennedy, P.J., Carter, N., Walters, K.F.A., Perry, J.N., Powell, D. and Powell, W. (1994) Design of field trials and its influence on the detection of pesticide side effects. *Proceedings of the 5th European Congress of Entomology,* Poster 276.

Ladell, W.R.S. (1938) Field experiments on the control of wireworms. With Appendix, 'The information supplied by the sampling results', by W.G. Cochran. *Annals of Applied Biology* 25, 341–389.

Loxdale, H.D., Hardie, J., Halbert, S., Footit, R., Kidd, N.A.C. and Carter, C.I. (1993) The relative importance of short- and long-range movement of flying aphids. *Biological Reviews* 66, 291–311.

Marshall, J. (1936) The distribution and sampling of insect populations in the field with special reference to the American bollworm, *Heliothis obsoleta. Annals of Applied Biology* 23, 133–152.

Matheron, G. (1976) A simple substitute for the conditional expectation: the disjunctive kriging. In: Guarascio, M., David, M. and Huijbregts, C. (eds) *Advanced Geostatistics for the Mining Industry.* Reidel, Dordrecht, Germany, pp. 221–236.

Mead, R. (1990) *Journal of the Royal Statistical Society, series A* 153, 151-201.

Mead, R. and Curnow, R.N. (1983) *Statistical Methods in Agriculture and Experimental Biology.* Chapman & Hall, London.

Nelder, J.A. (1962) New kinds of systematic designs for spacing experiments. *Biometrics* 18, 283–307.

Nelder, J.A. (1971) Contribution to the discussion of the paper by O'Neill and Wetherill. *Journal of the Royal Statistical Society series B* 36, 218–250.

Pearce, S.C. (1983) *The Agricultural Field Experiment.* Wiley, New York.

Perry, J.N. (1981) Taylor's power law for dependence of variance on mean in animal populations. *Applied Statistics* 30, 254–263.

Perry, J.N. (1986) Multiple-comparison procedures: a dissenting view. *Journal of Economic Entomology* 79, 1149–1155.

Perry, J.N. (1988) Some models for spatial variability of animal species. *Oikos* 51, 124–130.

Perry, J.N. (1989) Population variation in entomology: 1935–1950. I. Sampling. *The Entomologist* 108, 184–198.

Perry, J.N. (1994a) Sampling and applied statistics for pests and diseases. *Aspects of Applied Biology* 37, 1–14.

Perry, J.N. (1994b) Chaotic dynamics can generate Taylor's power law. *Proceedings of the Royal Society of London, Series B* 257, 221–226.

Perry, J.N. (1995a) Spatial aspects of animal and plant distribution in patchy farmland habitats. In: Glen, D., Greaves, M. and Anderson, H.M. (eds) *Ecology and Integrated Arable Farming Systems.* Wiley, Chichester, pp. 221–242.

Perry, J.N. (1995b) Spatial analysis by distance indices. *Journal of Animal Ecology* 64, 303–314.

Perry, J.N. (1996) Simulating spatial patterns of counts in agriculture and ecology. *Computers and Electronics in Agriculture* 15, 93–109.

Perry, J.N. (1997) Measuring the spatial pattern of animal counts with indices of crowding and regularity. *Ecology* (submitted).

Perry, J.N. and Gonzalez-Andujar, J.L. (1993) Dispersal in a metapopulation neighbourhood model of an annual plant with a seedbank. *Journal of Ecology* 81, 453–463.

Perry, J.N. and Taylor, L.R. (1988) Families of distributions for repeated samples of animal counts. *Biometrics* 44, 881–890.

Perry, J.N. and Woiwod, I.P. (1992) Fitting Taylor's power law. *Oikos* 65, 538–542.

Robinson, D.L. (1987) Estimation and use of variance components. *The Statistician* 36, 3–14.

Ruesink, W.G. (1980) Introduction to sampling theory. In: Kogan, M. and Herzog, D.C. (eds) *Sampling Methods in Soybean Entomology.* Springer-Verlag, New York, pp. 61–78.

Schneider, D.C. (1994) *Quantitative Ecology – Spatial and Temporal Scaling*. Academic Press, New York.

Southwood, T.R.E. (1966) *Ecological Methods*. Methuen, London.

Street, A.P. and Street, D.J. (1987) *Combinatorics of Experimental Design*. Oxford University Press, Oxford.

'Student' (1908) The probable error of a mean. *Biometrika* 6, 1–25.

Tatchell, G.M. and Plumb, R.T. (1992) Spread and infectivity of aphids as carriers of barley yellow dwarf virus in southern England in 1988–1990. *Pflanzenschutz Nachrichten Bayer* 45, 443–454.

Taylor, L.R. (1961) Aggregation, variance and the mean. *Nature* 189, 732–735.

Taylor, L.R. (1984) Assessing and interpreting the spatial distributions of insect populations. *Annual Review of Entomology* 29, 321–357.

Taylor, L.R. (1986) Synoptic dynamics, migration and the Rothamsted Insect Survey. *Journal of Animal Ecology* 55, 1–38.

Taylor, L.R., Woiwod, I.P. and Perry, J.N. (1978) The density dependence of spatial behaviour and the rarity of randomness. *Journal of Animal Ecology* 47, 383–406.

Taylor, L.R., Perry, J.N., Woiwod, I.P. and Taylor, R.A.J. (1988) Specificity of the spatial power-law exponent in ecology and agriculture. *Nature* 332, 721–722.

Webster, R. and Oliver, M.A. (1989) Optimal interpolation and isarithmic mapping of soil properties. VI. Disjunctive kriging and mapping the conditional probability. *Journal of Soil Science* 40, 497–512.

Williams, C.B. (1937) The use of logarithms in the interpretation of certain entomological problems. *Annals of Applied Biology* 24, 404–414.

Wiens, J.A. (1989) Spatial scaling in ecology. *Functional Ecology* 3, 385–397.

Wilson, B.J. and Brain, P. (1991) Long-term stability of distribution of *Alopecurus myosuroides* Huds. within cereal fields. *Weed Research* 31, 367–373.

Woiwod, I.P. (1991) The ecological importance of long-term synoptic monitoring. In: Firbank, L.G., Carter, N., Darbyshire, J.F. and Potts, G.R. (eds) *The Ecology of Temperate Cereal Fields*. Blackwell Scientific Publications, Oxford, pp. 275–301.

Woods, A. (1974) *Pest Control: A Survey*. McGraw-Hill, London.

Wynne, I.R., Howard, J.J., Loxdale, H.D. and Brookes, C.P. (1994) Population genetic structure during aestivation in the sycamore aphid *Drepanosiphum platanoidis* (Schrank) (Hemiptera: Drepanosiphidae). *European Journal of Entomology* 91, 375–383.

Yates, F. (1937) *The Design and Analysis of Factorial Experiments* – Technical Communication No. 35 of the Commonwealth Bureau of Soils, Harpenden, England. CAB, Farnham Royal, Slough, Berks.

Yates, F. and Finney, D.J. (1942) Statistical problems in field sampling for wireworms. *Annals of Applied Biology* 29, 156–167.

Yates, F. and Zacopany, B.A. (1935) The estimation of the efficiency of sampling, with special reference to sampling for yields in cereal experiments. *Journal of Agricultural Science* 25, 545–577.

8 Injury, Damage and Threshold Concepts

J.D. Mumford and J.D. Knight
Centre for Environmental Technology, Imperial College of Science, Technology and Medicine, Silwood Park, Ascot, Berkshire SL5 7PY, UK

8.1. The Decision Problem

Since the late 1950s the integrated pest management (IPM) concept has relied on the economic threshold credited to Stern *et al.* (1959) as one of the basic paradigms of agricultural entomology. It is a combination of ecological and economic principles that predicts insect abundance and damage and control performance, and describes the response of farmers or other decision makers. It attempts to model the decision problem and process of pest managers.

The overriding principle of IPM is that pest management inputs, especially pesticides, are only used when needed. The economic threshold is used to determine when this need arises. At its simplest, if the benefit of the control exceeds the cost, then it should be applied. The threshold is based on a series of functions (Southwood and Norton, 1973), illustrated in Figs 8.1–8.6.

The future pest population must be estimated, based either on current field sampling or on historical patterns, or some combination of these data. A figure such as Fig. 8.1 could be based entirely on past historical records, or it could be an extrapolation from field samples up to the present time, modifying past population records. The extent of the future prediction of the population depends on the amount of damage that may be done. If damage in the next few days would exceed the cost of control, then the prediction period can be quite short. In other cases, in which damage is relatively slight but cumulative, then the prediction period may need to be longer.

Injury is the physical impact of an insect on the plant, for example the area of leaf consumed per leaf feeding insect (Fig. 8.2). It is not directly an indication of the amount of loss because individual plants, and certainly a crop as a whole, may tolerate some injury, in which case damage does not necessarily follow to the same extent (Stern *et al.*, 1959). This compensation

Number of pests

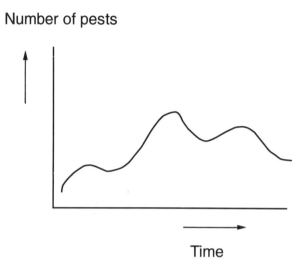

Time

Fig. 8.1. Pest population development over time from now.

Injury

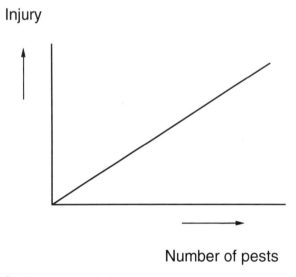

Number of pests

Fig. 8.2. Injury per pest at a particular time.

may occur because of excess productive capacity in the plants at certain times, or because another plant can take up the resources not used by an injured plant.

Individual crops have varying degrees of susceptibility to damage from insect injury at different growth stages. They tend to be most susceptible at the seedling stage and at flowering and seed setting. Figure 8.3 illustrates the different levels of susceptibility a crop may exhibit at different stages of development.

Control performance must be added to the population prediction, based on experimental evidence of kill rates under various conditions. The predicted pest population with and without a control applied is shown in Fig. 8.4. The relative damage suffered by the crop is the product of each of these

Damage

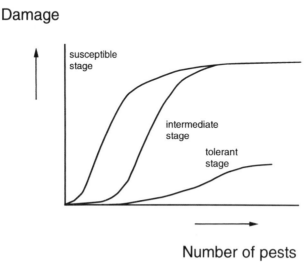

Fig. 8.3. Damage per pest at various crop stages.

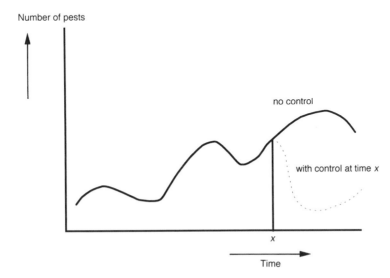

Fig. 8.4. Pest population development over time with and without control at a specific time.

two population curves and the time–growth stage dependent damage function, illustrated in Fig. 8.3, for the crop–pest combination.

The central comparison in the economic threshold concept is the value of the net revenue with and without control at a particular time. The damage function without control shows the returns expected from the present population of pests, up to the present and for the predicted future development of the population through various stages of crop susceptibility. These curves help to answer the critical question facing farmers and entomologists, which is when to start applying a pest control input such as a pesticide? Figure 8.6 illustrates the point at which applying a control becomes profitable – the economic threshold. The same principle could be used for other tactical inputs, such as inundative release biocontrol or mechanical control, but it is not directly relevant for strategic decisions, such as the use of resistant varieties, or avoidance of pests by manipulating planting dates, etc.

The relationship between with and without control curves can change over time, due mainly to the changing susceptibility of the crop at different growth stages. They may also change from season to season because pest development and crop development may not always be in phase (Fig. 8.7), due, for example, to early or late planting of the crop in particular years. The economic threshold would be very different in these two cases (Fig. 8.8).

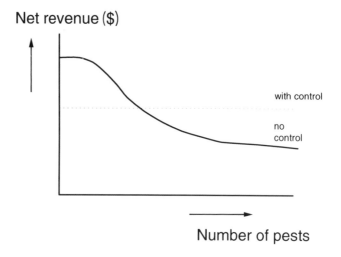

Fig. 8.5. Net revenue with and without a specific control input for a range of pest populations now.

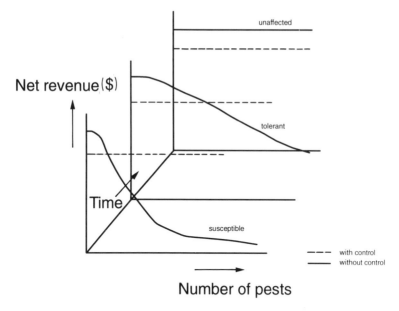

Fig. 8.6 Net revenue with and without a specific control input for a range of pest populations at different stages of crop susceptibility.

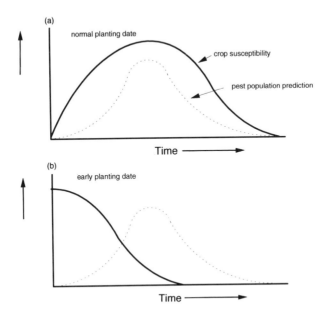

Fig. 8.7. (a) Pest development and crop susceptibility are in phase. (b) Pest development is out of phase with crop susceptibility.

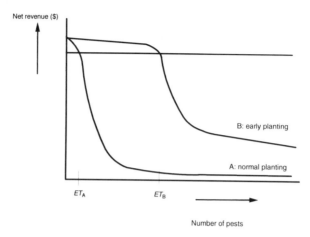

Fig. 8.8. The economic threshold (ET) for hypothetical perfect control (horizontal line) would be very different for the two cases shown in Fig. 8.7.

8.2 Examples and Methods for Determining Damage Relationships and Thresholds

The damage relationships between pests and crops are fundamental to the calculation of economic thresholds and therefore great care has to be taken in their measurement if the resulting threshold is to be of any practical use.

In most cases, the damage relationship between a pest and the crop it attacks is determined using field experimentation. The purpose of the experiments is to investigate, in a controlled manner, the effect of different numbers of pests or the timing or site of pest attack. The experiments are designed to eliminate, as far as practicably possible, the effects of any other factors that could affect the yield and or quality of the crop. The reduction in yield (or quality) can then be related to the pest numbers, timing, site of attack or any combination of these three factors.

Depending on the way in which the insect attacks the plant, different damage relationships will result. Insects that cause losses by consuming the part of the plant that is harvested such as grain borers or fruit feeders like codling moth larvae will generally have a linear damage relationship (Fig. 8.9(a)), i.e. two codling moth larvae may damage two apples, 10 larvae may damage 10 apples. Insects which feed by eating leaves or sucking sap will reduce the amount of reserves available for producing the harvestable component either directly removing sugars or by reducing the photosynthetic area and therefore sugar production; these pests will generally have a non-linear damage function. The first few insects will have no noticeable effect on yield and only large numbers will start to cause losses (Fig. 8.9(b)). Examples of this type of pest would be lepidopteran larvae causing defoliation and aphids removing plant sap by feeding. Extremely high numbers

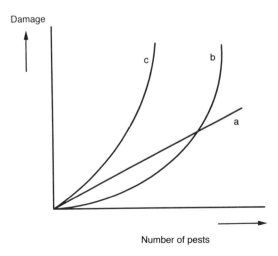

Fig. 8.9. Different types of damage functions for pests causing (a) damage to fruits, (b) leaf feeding pests and (c) stem borers.

may cause the death of the plant and therefore complete yield loss. The third type of insect damage is where the insect feeds on a crucial part of the plant such as the stem which if damaged can cause the death of the entire plant, for example stem borers. In this case one insect can reduce the yield very severely and will show a non-linear damage relationship (Fig. 8.9(c)); often after feeding on one stem the insect will move on to another and kill or damage that plant too. Insects may cause significant damage indirectly by transmitting plant diseases, for example aphids transmitting virus diseases. In this case the damage relationship will be a function of the number of viruliferous insects, the efficiency with which the insects transmit the virus and the time at which the plant is infected. It is therefore no simple matter to design suitable experiments to determine the damage relationships and great care should be taken to ensure that the design of experiment chosen will give statistically significant results from which to work.

Damage may also not be immediately obvious, for example the effects of spider mites in orchard crops may cause a reduction in the lifetime yield of the tree by slowly weakening it. The yield difference may not be measurable in a single year but only over the lifetime of the tree. This is very difficult and expensive to measure accurately and therefore the setting of economic thresholds is also extremely hard. The measurement of thresholds is also extremely difficult in crops such as cocoa where the crop is being produced throughout the year and the pest populations have to be monitored throughout as there is always a susceptible stage present.

The calculation of the damage relationship is the first step in determining the economic threshold since without this it is not possible to calculate the cost of the insect damage. The next step is the calculation of the effectiveness of any control measure that may be taken, such as the application of an insecticide. Field experiments similar to those above for the determination of the damage relationship have to be done to see what reduction in pest numbers, and consequently saving of yield, is achieved by each of the control measures. This may involve experimenting with the timing and application rate or method to see which gives the greatest level of control. The same care has to be taken with these experiments as with the first set. Careful experimental design can often make it possible to combine both sets of experiments into one trial.

The determination of a reliable economic threshold will involve many trials to ensure that sufficient information on the effects of differing site characters and climatic factors can be assessed and incorporated in the recommendation.

The final step of determining when the cost of control is offset by the savings made by applying the control is then a relatively simple process of comparing the costs and benefits of the treatment.

The cost of control measures has historically been limited to the cost of the chemical, its application and the cost of any scouting that has to be undertaken. This has ignored the other costs that could be associated with the use of certain control measures such as environmental damage. Proposals have been made to include the environmental risk of using a particular pesticide in the calculation of the economic injury level (EIL) (Higley and Wintersteen, 1992). Once the cost of the environmental risk is calculated, this is added to the cost of the pesticide and application and the economic injury level is adjusted accordingly. The way in which the cost of the environmental risk is calculated is very important and the factors that are considered can influence the result markedly and consequently make some pesticides much more attractive than others because they are judged to be environmentally less damaging.

In some situations the calculation of the economic threshold becomes redundant because of particular constraints within the farming system. If the farmer cannot tolerate any pests, for instance, the presence of quarantine pests, then there is little need to calculate thresholds and a calendar spray programme is the most appropriate approach. In some situations there may be insufficient resources or trained personnel to operate a threshold system all of the year and in this case a combination of a threshold and calendar spraying may be used.

A number of examples of 'threshold' systems will be described in order to illustrate how different situations have led to the adoption of different strategies.

8.2.1. Calendar spraying

In the first example, that of insect control in New Zealand orchards, there are many important factors within the system that influence the decision making of the grower. The market that exists for the crop is the export market and as such the crop has to satisfy the import regulations of the country into which the apples are going. There are, therefore, a number of quarantine pests that cannot be tolerated at any level or the crop will be rejected for export. Secondly, many apple pests do not produce any reduction in yield but cause cosmetic damage resulting in the downgrading or rejection of the fruit. Thirdly, calendar spraying for the control of fungal pathogens is normal and occurs throughout the time that insects need to be controlled. Many of the apple growers are relatively new to the industry and rely heavily on advice from horticultural merchant field representatives and want a simple recommendation that is relatively risk free and does not require a great deal of knowledge to operate (Stewart and Mumford, 1995).

The pressure to produce a 'clean crop' is very high and the stakes at risk are very large since the discovery of a quarantine pest will result in the entire crop being deemed unfit to export and it can then only be sold on the internal market probably for juicing. The price differential between the two is so great that the grower is unlikely to tolerate any risk of quarantine pests being detected in his crop. The fact that fungicides are being applied on a calendar basis every 7–10 days means that the cost of applying an insecticide is greatly reduced as it is effectively applied free, only the cost of the chemical has to be considered. Since the risk of damage from insects in any year is extremely high and the cost of application is nil, the adoption of a calendar spraying programme is the only acceptable strategy for the grower.

In this particular case the 'economic injury level' is in effect one insect in the grower's entire crop since the presence of one quarantine pest will result in the rejection of the entire crop. Since the cost of detecting the one insect in the crop would be very high, the economic threshold is effectively reduced to a level of zero insects.

8.2.2. Standard operating procedures

The standard operating procedure can be viewed as a hybrid between a threshold and calendar spray. In general some sort of threshold is used to trigger the start of a spray programme.

An example of the standard operating procedure approach is for the control of a range of pests that attack cotton in Côte d'Ivoire in Africa. In the past, chemicals for pest control in cotton have been supplied by the government but this subsidy is now being withdrawn. Consequently, farmers are having to buy their own chemicals and are realizing the cost of control. In order to reduce the control costs the farmers wish to time their applications correctly, which they can do by using economic thresholds to make their decisions. However, the level of farmer education is relatively low and the concept of using a threshold for each of the many pests that are found in the cotton crop could be a deterrent and make them less likely to change from their traditional calendar spraying. The solution to the problem was to identify what the key pests were early in the season and use a conservative economic threshold for these pests as an indicator of when to start the spray programme, which would consist of a minimum of four sprays which control all the insect pests. The end of the programme is set as the last of the four sprays. A conservative threshold was chosen because of the imperfect knowledge of the system and the unwillingness of the farmers to increase the risk of damage to their crops. Different schemes were tried which relied on using the numbers of bollworms, or the numbers of leaf pests, as the trigger for the start of the control programme. Each gave similar returns to the traditional calendar method but with reduced numbers of sprays in each case. The different strategies were developed for use in different regions of the country where the balance of pest types varies from north to south (Ochou, 1994).

A similar situation exists in Egypt where chemical control of cotton pests is carried out by the government. The start of the spraying season for the control of bollworms is determined by a threshold of eight or more *Pectinophora* in a pheromone trap, the first spray is applied 5 days later. The subsequent sprays are determined by a combination of timing and thresholds but the thresholds are relatively low so that the spray programme is effectively calendar based from this point on. The first threshold, determined from experimental trials, is useful and serves to prevent unnecessary early application of sprays. The later thresholds have little value since they are just about always exceeded and in practice the spray application is determined by the number of days since the last spray.

8.2.3. Thresholds

The final examples show how thresholds can be developed and implemented. The first example is the use of a threshold for the control of cutworms, *Agrotis* spp., in vegetable crops in the UK. In this particular case the threshold is used to indicate when is the best time to treat. The economic thresholds were calculated from the results of field experiments in the usual way but the timing of the treatment is determined by the use of a mathematical model which simulates the development of the pest using a day degree or phenological model (Bowden *et al.*, 1983).

The traditional method of controlling the pest was with an insecticide spray targeted at the third instar larvae, the most vulnerable stage. The correct time for spraying was judged by scouting the field for the presence of larvae in the crop which was difficult to do and resulted in poor timing of insecticide application. Spraying too early does not kill the eggs and results in larvae emerging after the spray, and spraying too late is also ineffective since the spray will not kill the late instars that drop from the plant and feed on the roots. The objective was to provide a more reliable method of timing the insecticide application, in this case a model of the development of the pest.

The start of the model is triggered when catches of adult moths in pheromone traps indicate the peak of egg laying. The model then uses daily temperature data to calculate the development of the eggs and larvae by determining the number of day-degrees that have been accumulated at any one time. When the model indicates that the threshold number of day-degrees has been reached, indicating that most of the larvae have reached the third instar, a warning is issued to growers to spray.

During the course of the development of this model it was realized that there was a strong correlation between rainfall and mortality of early stage larvae – the higher the rainfall the higher the mortality. The forecast was then developed further to include the effects of rainfall on the survival of larvae and, depending on the number of larvae estimated to have survived, advice was given on whether or not to spray. The threshold for applying insecticide is dependent on the crop that is being grown and the history of the area. For very susceptible crops, such as leeks, the threshold for spraying or overhead irrigation is set at 10–15 larvae per square metre, for potato crops, which are less susceptible, the threshold is set at 15 or more larvae per square metre. In this particular case the economic threshold was determined by field experimentation to determine the damage relationship but the action threshold is calculated from a model of the development of the pest and mortality due to rainfall and does not rely on scouting.

The final example is drawn from a lepidopterous pest of cotton – the tobacco budworm (*Heliothis virescens* (Fabr.)). It has been recognized for some time that economic injury levels change through time with the developmental stage of the crop. For this reason a range of damage functions using regression equations have been developed to provide economic injury levels for the appropriate stage of development of the crop (Stone and Pedigo, 1972; Pedigo *et al.*, 1986). However, this is a more complicated approach to the control procedure and still results in EILs being used over a range of plant ages other than the one for which they were developed with the resulting potential for error. One method that can be used to overcome this approach is to develop a continuous function to describe the relationship between the physiological age of the plant and the numbers of insects, which allows the EIL to calculated easily for each combination of conditions. Such an approach has been adopted for use with *H.virescens* on cotton in the USA (Ring *et al.*, 1993).

A randomized block design for the main plots with split-plot design for the other factors was used to measure the effect of the treatments on the yield of the cotton. Factors examined in the experiment were: level of infestation; cultivar; two insect species; and stage of development. A regression equation that provided the best fit to the data that described the damage relationship for each insect was calculated and used in the calculation of the economic injury level. The results showed that the EIL for each combination of the two pests (*H.virescens* and *Helicoverpa zea*) and the two different varieties was significantly different which required a different model to be used for each. The response surface revealed an overcompensation at low levels of injury with increasing levels of yield loss with increasing injury. The response curve illustrated in Fig. 8.10 shows how the damage relationship varies with time, with overcompensation occurring at low injury levels,

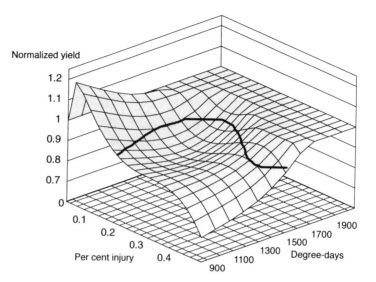

Fig. 8.10. Response surface depicting cotton yield as a function of injury, by *Heliothis virescens,* and plant age. The black line shows the economic injury level for a crop value of $250 ha^{-1} and control costs of $20 ha^{-1}. (After Ring *et al.*, 1993.)

shifting the EIL upward, and rapid changes occurring in the EIL as the plant approached 30 days after initiation of anthesis.

The EILs are heavily influenced by economics and could therefore occur at very nearly any point on the response surface. Using values of $20 ha^{-1} for control costs and a market value of $250 ha^{-1} for the crop (typical values for rainfed cotton) the dynamic EIL can be calculated and superimposed on the figure. This clearly shows how the EIL varies over the course of the season and with variable levels of damage.

The work of Ring *et al.* (1993) shows the potential of using a response surface rather than a single curve or series of curves. However, there are a number of points that should be borne in mind. Firstly, not all the factors associated with crop production were included in this experiment, natural enemies were removed and other pests also. The variable effects of weather are also not included in the model. More complex experiments could be conducted to quantify all these components or a mechanistic model could be used to estimate their influence. The economics of the situation are also dynamic with the prices of cotton varying along with chemical and labour costs. The resulting model is not one which can easily be applied by most farmers and would have to be presented in the form of a look-up table to give a threshold for a particular stage of development and pest density. While this may be practical in countries where farmers have a good standard of education it would be nearly impossible to implement in countries where farmer education is poor.

The number of dynamic factors involved in the definition of EILs means that they require large amounts of data to define them and need to be calculated with many dimensions. The cost of collecting the necessary data and the calculation of the resulting equations is probably prohibitively expensive and indicates that a simplified robust approach that gives reliable results for the farmer and serves to reduce environmental impact at the same time is the sensible way to progress.

This section has tried to illustrate how factors other than just the cost of control and the damage caused by insects determine the control threshold. In the last example, assuming there are no other constraints, it can be seen that the problem of providing a reliable economic threshold for use in the field can be a complex process.

8.3. Objectives and Responses

The general objectives and forms of economic response to pest control problems have been reviewed several times (for example, Stern, 1973; Mumford and Norton, 1984; Reichelderfer *et al.*, 1984). Examples of economic approaches to specific types of pest problem have also been presented (Mumford and Norton, 1990; Mumford, 1993). Broader issues of environmental and ethical considerations have been discussed recently by Pimentel and Lehman (1993) and the broad subject of pest control decision making is covered by Norton and Mumford (1993).

An alternative view of economic thresholds was first presented by Headley (1972), who as an agricultural economist posed a very different question from that asked by entomologists. He and other economists (Hall and Norgaard, 1973; Feder, 1979) tried to answer the question, at what level of inputs should you stop applying pest control? This was answered using traditional marginal analysis techniques: a farmer should apply control inputs up to the point at which the marginal cost of a unit of input (normally pesticide) exceeded the marginal benefit of that input. In principle this is very similar to the entomological economic threshold approach, i.e. benefit must exceed cost. For the entomologist the key variable is the number of pests presented (and predicted), while considering only a single unit of input (the first). On the other hand, for the agricultural economist the key variable is the extent of management inputs, with the pest population considered as a fixed condition.

Several issues arise from this difference of viewpoint. First of all, the very nature of the decision problem appears to be determined by the discipline of the person advising the decision maker, either entomologist or economist. Entomologists know that pests vary and they naturally see that as the core element of the problem. They see the decision problem as a series of individual decisions about when to start applying control inputs based on the current level of pest populations. In fact, from an entomological point of view the control decision is a series of daily or weekly questions (determined by scouting schedules), should I spray tomorrow/soon?, based on each scouting result. Entomologists do not see control inputs as being endlessly divisible, in most cases for pesticides at least, only the dose on the label can be applied.

Economists, however, see management inputs as being continuously variable because that is their area of expertise. It is convenient for them to consider pest populations as a fixed condition, which may be reasonable if you are not scouting them.

Another approach would be to consider the objectives of the decision maker, rather than the expertise of the adviser as a basis for decision making. How does the farmer see the problem and what does he want to achieve? First of all, they are considering risk. This is a basic fault with both forms of economic threshold described above. The prediction of damage is not a definite curve, but a broad band of possible levels of damage. Control functions also vary, depending on application conditions, the stage of the pest, insecticide resistance, etc. A more realistic view of Fig. 8.5 could look like Fig. 8.11.

The lowest value for an economic threshold and the highest may cover a very wide range of populations. Further uncertainty should be added above and below these two values to account for sampling errors in scouting. Yet another problem arises in knowing where these curves are phenologically in the time dimension (see Fig. 8.6) and what the seasonal impact is (see Fig. 8.8). Once all this uncertainty is acknowledged the economic threshold may have very limited appeal to a farmer whose main aim is to avoid risk. Using a threshold may add risk, so he may find it is better to use a prophylactic treatment regardless of pest numbers.

That is an unduly pessimistic view of thresholds, however. Some pests,

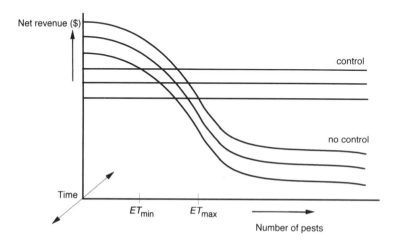

Fig. 8.11. Net revenue distributions with and without control.

crops, controls and environmental conditions are much easier to predict than others, so thresholds can reduce risks (Table 8.1).

On the basis of Table 8.1 it is clear that cotton bollworms fit many of the criteria for relatively certain threshold values. By contrast, autumn cereal aphids in northern Europe are not well suited to thresholds.

Decision makers can respond to pest problems in several ways. Thresholds are one, but they should not be seen as a universal solution. Alternative responses include calendar-based applications and standard operating procedures.

Calendar treatments reduce risk but may result in 'excessive' application of control inputs. This is much criticized in the case of pesticides, but widely accepted in the case of biological control. Two factors determine the usefulness of calendar treatments: costs of management and the certainty of threshold functions. If management is expensive or difficult then calendar

Table 8.1. Conditions that make threshold prediction more or less uncertain.

	Threshold uncertainty is reduced	Threshold uncertainty is increased
Prices	fixed	market varies
Type of damage	direct (i.e. feeding on the part of plant harvested)	indirect
		disease vectors
Crop growth	consistent (irrigation)	variable (rainfed)
Pest attack	short duration slow reproduction tied to crop stage	long duration fast reproduction independent of crop stage
	endemic pests later season	immigrant pests earlier season
Weather conditions	controlled (greenhouse) constant (irrigated crops, arid areas)	temperate areas
Scouting	cheap easy to detect damage or presence	expensive cryptic stages or damage
Control application	no pesticide resistance not affected by weather (adjuvants/stickers, or in dry climates)	variable resistance affected by weather
Natural enemies	very high or very low natural control, stable source nearby	variable natural control

treatments may be cost-effective substitutes for management effort. If threshold functions are uncertain, then the resulting thresholds will not meet the management objective of reducing risk.

Standard operating procedures are an alternative to calendar treatments that substitute a predetermined procedure for direct management effort. So, for example, a simple rule such as 'spray every time it looks like there will be a three day dry spell at least two weeks after the previous spray' or 'spray when your neighbour does' may give more cost-effective control than either calendar or threshold treatments without needing complex information or management.

The relative value of thresholds, calendar treatments and standard procedures can be compared using a pay-off matrix (Norton and Mumford, 1993). This would show the likely outcomes (based on historical evidence or expert opinion) of the various strategies under various levels of pest attack and under weighted average conditions (Table 8.2).

A low cost economic threshold is clearly the best overall strategy, but an expensive one is not attractive. A threshold based on uncertain values would need to be relatively conservative and may result in extra treatments, as in case E compared to case D in Table 8.2. While case E is still better than case C overall, it is worse 70% of the time and quite sensitive to management costs.

Considerable effort has gone into improving pest population and damage prediction and this may have reached its limit for some crop–pest complexes. As Table 8.1 implied, some situations are inherently uncertain, regardless of how good pest sampling is or how well we understand damage functions. Effort has also been put into improving the management part of the decision process, making it cheaper and more effective, for instance by using computers (Knight and Mumford, 1995).

8.4. Conclusions

The objective of this chapter has been to show the fundamental role of economic thresholds in the implementation of IPM programmes. Examples of how they are determined and the way in which they are used have been given along with examples where their use is not feasible. The conditions under which they can be used have been described and these can be used to assess whether determining economic thresholds will be appropriate for a particular case. The alternatives to economic thresholds, which are sometimes more suitable economically, if not environmentally, have also been listed. In some cases, these alternatives must be seen as being reasonable components of IPM. The temptation to use more economic thresholds with the old technical options must be avoided since this may still result in environmental damage, instead research into environmentally friendly inputs is essential. Alongside the development of ecological methods to make thresholds more precise, effort also needs to go into making decision-making

Table 8.2. An example of a pay-off matrix to show the relative value of different strategies.

	Low pest attack	Medium pest attack	High pest attack	Expected outcome
	probability = 0.2	probability = 0.5	probability = 0.3	(probability weighted)
A – No treatment	1000	600	200	560
B – Calendar ($200 sprays, 80% control)	800	720	640	712
C – Standard procedure ($100 sprays 70% control)	900	780	640	732
D – Economic threshold ($25 management, 90% control, $0, 100, 200 for control, respectively)	975	835	695	821
E – Economic threshold ($25 management, 90% control, $100, 200, 200 for control, respectively)	875	735	695	751
F – Economic threshold ($150 management, 90% control, $0, 100, 200 for control, respectively)	900	710	570	706

management input in IPM more cost-effective. The combination of precise economic thresholds, new solutions to pest control and a better understanding of the decision processes of the farmers will result in more cost-effective pest control with less environmental impact.

8.5. References

Bowden, J., Cochrane, J., Emmet, B. J., Minall, T. E. and Sherlock, P. L. (1983) A survey of cutworm attack in England and Wales and a descriptive population model of *Agrotis segetum. Annals of Applied Biology* 102, 29–47.

Feder, G. (1979) Pesticides, information, and pest management under uncertainty. *American Journal of Agricultural Economics* 61, 97–103.

Hall, D.C. and Norgaard, R.B. (1973) On the timing and application of pesticides. *American Journal of Agricultural Economics* 55, 198–201.

Headley, J.C. (1972) Defining the economic threshold. In: *Pest Control Strategies for the Future*. National Academy of Science, Washington, DC, USA, pp. 100–108.

Higley, L.G. and Wintersteen, W.K. (1992) A novel approach to environmental risk assessment of pesticides as a basis for incorporating environmental costs into economic injury levels. *American Entomologist* 38, 34–39.

Knight, J.D. and Mumford, J.D. (1995) Decision support systems in crop protection. *Outlook on Agriculture* 23, 281–285.

Mumford, J.D. (1993) The economics of integrated pest control in protected crops. *Pesticide Science* 36, 379–383.

Mumford, J.D. and Norton, G.A. (1984) Economics of decision making in pest management. *Annual Review of Entomology* 29, 157–174.

Mumford, J.D. and Norton, G.A. (1990) Economic analysis of stored grain pest management. *Proc. 5th Int. Working Conference on Stored-Product Protection*, Bordeaux, France, pp. 1913–1924.

Norgaard, R.B. (1976) The economics of improving pesticide use. *Annual Review of Entomology* 21, 15–60.

Norton, G.A. and Mumford, J.D. (eds) (1993) *Decision Tools for Pest Management*. CAB International, Wallingford, United Kingdom, 279pp.

Ochou, G.O. (1994) Decision making for cotton insect pest management in Côte d'Ivoire. Unpublished PhD thesis, University of London, 306pp.

Pedigo, L.P., Hutchins, S.H. and Welch, S.M. (1986) Economic injury levels in theory and practice. *Annual Review of Entomology* 31, 341–368.

Pimentel, D. and Lehman, H. (1993) *The Pesticide Question: Environment, Economics and Ethics*. Chapman & Hall, London. 441pp.

Reichelderfer, K.H., Carlson, G.A. and Norton, G.A. (1984) *Economic Guidelines for Crop Pest Control*. FAO Plant Production and Protection Paper 58, FAO, Rome, 93pp.

Ring, D.H., Benedict, J.H., Landivar, J.A. and Eddleman, B.R. (1993) Economic injury levels and development and application of response surfaces relating insect injury, normalized yield and plant physiological age. *Environmental Entomology* 2, 273–282.

Southwood, T.R.E. and Norton, G.A. (1973) Economic aspects of pest management strategies and decisions. In Geier, P.W., Clark, L.R., Anderson, D.J. and Nix, H.A. (eds) *Insects: Studies in Population Management*. Ecological Society of Australia, Canberra, Australia, pp. 168–184.

Stern, V.M. (1973) Economic thresholds. *Annual Review of Entomology* 18, 259–280.

Stern, V.M., Smith, R.F., van den Bosch, R. and Hagen, K.S. (1959) The integrated control concept. *Hilgardia* 29, 81–101.

Stewart, T.M. and Mumford, J.D. (1995) Pest and disease management in Hawkes Bay apple orchards: results of an advice givers survey, *New Zealand Journal of Crop and Horticultural Science* 23, 257–265.

Stone, J.D. and Pedigo, L.P. (1972) Development and economic-injury level of the Green Cloverworm on soybean in Iowa. *Journal of Economic Entomology* 65, 197–201.

9 Techniques in the Study of Insect Pollination

S.A. Corbet[1] and J.L. Osborne[2]
[1]*Department of Zoology, University of Cambridge, Downing Street, Cambridge CB2 3EJ, UK;* [2]*Department of Entomology and Nematology, Institute of Arable Crops Research, Rothamsted, Harpenden, Herts AL5 2JQ, UK*

9.1. Introduction

Whether the context is evolutionary ecology, conservation or agriculture, research on insect pollination often aims to quantify the costs and benefits of the insect/flower interaction. This may be done from the viewpoint of the insect, the plant, or the grower, in a currency that may be energy, progeny or cash. The net effect of the insect/flower interaction on the fitness of the partners interests ecologists as a determinant of selection acting over evolutionary time. It interests conservationists because they need to know whether maintenance of the interaction is a prerequisite for conservation of either partner. It interests growers because insects will not visit a flower consistently unless it yields profitable forage, and flower-visiting insects may not benefit the grower unless their visits augment yield. One objective of agricultural pollination studies is to identify cost-effective management options for the pollination of crops.

Ecological and conservation-related pollination studies often involve multispecific assemblages of plants and insects. In long-established semi-natural habitats that offer a diversity of alternative partners for both insect and plant, insects are unlikely to visit a plant species unless they can make a profit from it, and plants are likely to have evolved attributes that discourage visits from insect species that bring them no net benefit. In contrast, agricultural studies typically involve single species of crop plant and insect. Managed pollinators may share little of their evolutionary history with the crops on which they are imposed, and may have no access to alternative forage plants. Unless pollinators are selected judiciously for introduction to a crop, managing them may involve making the best of a misalliance.

A second difference between pollination studies of crops and wild flowers is a matter of scale. Ecological investigations often focus on small-scale interactions between individual insects and flowers, often with more detailed exploration of features such as insect foraging behaviour, floral reward and pollen flow (e.g. Harder, 1990; Young and Stanton, 1990; Wilson and Thomson, 1991; Davis, 1992). Agricultural studies are often on a larger scale, involving overall estimates of insect populations and seed yield in monocultures (e.g. McGregor, 1976; Free, 1993). Small-scale approaches have been little used in agriculture until recently (e.g. Carré *et al.*, 1994; Osborne, 1994; Willmer *et al.*, 1994; Vaissière, 1996; Michaelson-Yeates *et al.*, unpublished).

Social and solitary bees are probably among the most important pollen vectors of insect-pollinated flowers in both agriculture and conservation. They are the focus of this chapter.

9.2. Methods

Techniques used by pollination ecologists have been described in two recent books (Dafni, 1992; Kearns and Inouye, 1993). Instead of repeating them, we offer a more general framework for considering the role of insect/flower interactions in pollination, referring to Kearns and Inouye (1993) for details of methods.

9.2.1. Benefits and costs for pollinators

BENEFITS

Many insects that visit flowers collect floral products: nectar, pollen and, in special cases, oils. But some insects lay eggs, thermoregulate, hide, mate or hunt for prey in flowers. Careful observation is required to see what flower-visiting insects are doing. Not all insects that visit flowers forage for floral rewards, and not all those that forage are effective pollen vectors.

Nectar. Nectar consists largely of a solution of sugars in water (but see Inouye *et al.*, 1980). Other components such as amino acids or secondary metabolites are sometimes important to insect visitors, but nectar chemistry is commonly described in terms of sugar composition, especially the sucrose:hexose ratio. Much insect flower-visiting behaviour can be interpreted on the assumption that insects are maximizing their net rate of energy gain (or their gain:cost ratio) (Schmid-Hempel, 1987; Ydenberg *et al.*, 1994). The energy available to an insect at a given instant depends on the nectar sugar content in the flower, the standing crop. The potential rate of energy gain for a group of insects foraging on a patch of flowers depends on the rate of secretion of nectar sugar in the whole patch. Some flowers reabsorb sugars from nectar in contact with the nectary (Búrquez and Corbet, 1991). To understand the dynamics of energy gain by bees it is necessary to monitor not only standing crop, but also the rate of secretion and, if possible, reabsorption. All these variables change through the day and it is often useful to sample nectar at intervals of, say, 1 or 1.5 h from dawn until dusk. If bee counts and microclimate measurements are made regularly over the same dawn-to-dusk period, a very informative data set can be acquired (e.g. Corbet and Delfosse, 1984; Corbet *et al.*, 1995).

Methods of sampling nectar, and addresses of suppliers of equipment, are given by Kearns and Inouye (1993), Dafni (1992) and Prŷs-Jones and Corbet (1991).

The quantity of nectar sugar in a flower at a given time can be estimated from volume and solute concentration. Volume, v µl, is estimated from the length of the column of fluid in a slender disposable glass microcapillary tube used to withdraw nectar from flowers. For entomophilous flowers in temperate climates microcapillary pipettes of volume 0.2, 0.5, 1 or 5 µl are suitable. Translucent nylon tubing is cheaper and causes less damage to the flower, and makes it possible to monitor the rate of secretion over time by tracking the movement of the meniscus in a length of tubing wedged into a flower. Unless very fine, hydrophilic tubing can be found, this method can only be used for large flowers with high secretion rates.

Using pocket sugar refractometers, modified by the makers (Bellingham and Stanley, Tunbridge Wells) for small volumes of liquid, it is possible to measure solute concentration, $c\%$, on as little as 0.05 µl of nectar. Measurements should be made on individual flowers. An instrument that requires larger volumes may be adequate for hummingbird-pollinated flowers, but not for most temperate-zone entomophilous flowers. There is sometimes too little nectar in a flower for refractometric measurement. Pooling nectar samples from more than one flower is discouraged. Nectar is unevenly distributed among flowers; unvisited flowers may contain large volumes of dilute nectar, whereas recently-visited ones contain small volumes. Evaporation into dry air will concentrate a small droplet faster than a large droplet, and if volume correlates with concentration the sugar content calculated from the concentration of a pooled sample from ten flowers may be very different from the mean value calculated from the ten individual concentration measurements. Sugar content should be calculated for each individual flower, and not based on concentrations pooled or averaged from several flowers.

Sugar refractometers are generally calibrated in grams of sucrose per 100 g solution (i.e. mass per unit mass of solution). This is converted to mass of sugar per unit *volume* of solution by multiplying by the density, d, of the corresponding sucrose solution, taken from tables (e.g. from Lide, 1994) or an equation (e.g. see Bolten *et al.*, 1979; Kearns and Inouye, 1993), and the mass of sugar per flower, s µg, is given by

$$s = 1000(vcd/100)$$
$$s = 10vcd$$

Standing crop, measured by sampling, say, 10 or 20 flowers that are unprotected (open to bee visits) at each sampling time, represents the reward available to insect visitors whose flower-selecting behaviour resembles that of the research worker. Any ability the bees may have to select fuller flowers (e.g. Corbet *et al.*, 1984; Giurfa and Núñez, 1992) will increase the reward they receive above the estimated mean standing crop.

Secretion rate is estimated by carefully emptying 10 or 20 flowers, enclosing them in mesh bags that exclude insect visitors but have minimal effect on microclimate (e.g. nylon netting, 1.4 mm mesh, also known as bridal veil, available in fabric shops), and then removing the bags and sampling the accumulated nectar after a specified period (often 1 or 1.5 h). The same flowers may be rebagged and resampled after a further interval, or they may be discarded, in which case a fresh set of flowers emptied for standing crop measurements are bagged to allow nectar to accumulate over the next secretion interval.

By summing the secretion rates for successive 1-h periods over a day it is possible to calculate a cumulative secretion rate. This is often much higher than the secretion rate derived from flowers bagged for a full 24 h (Corbet and Delfosse, 1984). The difference has been attributed to reabsorption of nectar (Búrquez and Corbet, 1991). On the other hand the cumulative secretion rate in repeatedly-sampled flowers is sometimes *lower* than that of flowers bagged for 24 h. The difference may be due to reduction of secretion rate resulting from sampling damage to the flower (e.g. Corbet and Willmer, 1981).

If the nectar is too viscous or too sparse to run into a capillary and give a refractometer reading, sugar content per flower is sometimes measured colorimetrically (e.g. by the anthrone method, which requires heating and is therefore unsuitable for field use) on samples withdrawn from the flower with filter paper wicks (McKenna and Thomson, 1988) or by centrifuging (R.B. Roberts, 1979). Nectars rich in glucose, such as those of Brassicaceae, can be spotted on to a glucose test paper, and this colorimetric assay can be calibrated to allow surprisingly precise field estimation of sugar per flower (Búrquez, in Kearns and Inouye, 1993).

It is usually easy to see whether or not a bee takes nectar when it visits a flower, but it is harder to know how much it takes. A bee's rate of sugar gain is sometimes estimated as the product of the measured mean standing crop of nectar sugar in the flower patch and the number of flowers visited by a bee per unit time. This procedure may not be appropriate if the bee does not empty each flower (Hodges and Wolf, 1981), if occasional very full flowers sampled by chance distort the estimate of the mean standing crop, or if the bees select a different subset of flowers containing more (or less) nectar than those selected for nectar measurement (Corbet *et al.*, 1984). It is difficult to measure nectar content in a flower before and after a visit without affecting foraging, unless the nectar level can be seen from outside, for example through a translucent spur (Belavadi and Parvathi, 1993).

Pollen. In general, pollen is available in flowers from the time of anther dehiscence, which depends on species and weather (Percival, 1955; Corbet, 1990), until it is removed by insects or wind. The temporal pattern of pollen availability to flower-visiting insects depends on the schedule of flower development within a plant and the schedule of pollen release within a flower. The amount of pollen removed during each bee visit may be regulated by a dispensing or packaging mechanism (Harder and Thomson, 1989). Even when pollen is present in the dehisced anthers, pollen collection by insects is sometimes limited to a brief period in the day, perhaps because of humidity-related changes in the manipulability or the water content of the pollen (Corbet *et al.*, 1979; Corbet *et al.*, 1988).

The temporal pattern of pollen availability in flowers can be monitored by counting pollen grains in the laboratory with a microscope or a particle counter (Thomson, 1986; Vaissière, 1988, 1991; O'Rourke and Buchmann, 1991) or estimated more rapidly in the field in terms of the proportion of empty or dehisced anthers or some convenient index of anther fulness (Goodwin, 1986; Thomson and Thomson, 1992; Osborne, 1994).

It may be difficult to see whether or not bees are collecting pollen when they visit a flower unless they show distinctive behaviour, such as buzzing or scrabbling. When bees appear to be foraging only for nectar, their bodies often become dusted with pollen. This is groomed by the bee and either discarded or collected into the corbiculae. Bees with scopal pollen loads sometimes have large quantities of pollen on their body hairs (Osborne, 1994). Pollen loads are sometimes taken to indicate pollen collection, but the relationship between scopal pollen loads and pollen-foraging is not simple; some bees with scopal pollen loads may be individuals that are not currently collecting pollen but have failed to unload their scopae, and some bees without scopal pollen loads may be pollen collectors that have not yet collected enough pollen to produce visible pollen loads.

COSTS

The overall metabolic costs of a social bee colony include some elements incurred inside the nest and others incurred outside it, including many that depend on foraging activity, such as flight costs from the nest to the flower patch and from flower to flower in the patch, and the metabolic costs of endothermic maintenance of body temperature while not in flight.

Bees will not fly unless the thoracic temperature (T_{th}) exceeds a threshold, which can be achieved and maintained metabolically if the ambient temperature (T_a) is high enough (Stone and Willmer, 1989). Flight occurs at lower T_a when irradiance is high. Black globe temperature (T_g) integrates effects of radiation, convective cooling and T_a, and is often more useful than T_a as a predictor of flight activity. T_g is conveniently measured using a thermocouple embedded in a bee-sized sphere of Blue-Tack ® (Bostic, Leicester) painted matt black (Corbet *et al.*, 1993).

Overall foraging costs depend on the flight cost per unit time, the proportion of time spent in flight and the energetic cost of non-flight activities. Bees in flight are thought to maintain a steady high T_{th}, but some bees allow the thorax to cool down when they are on flowers, particularly male bees and particularly on flowers that are massed in compact inflorescences such as *Solidago* or *Sedum acre* (Heinrich, 1979; Corbet *et al.*, 1995). When this happens the mass specific rate of energy consumption falls (Silvola, 1984). But bees on flowers often retain flight-readiness by keeping T_{th} high – a condition sometimes called 'active rest' (Silvola, 1984). This elevates their energy consumption rate in proportion to the difference between T_{th} and T_a (or T_g). It is therefore useful to have some measure of the environmental temperature, as well as the proportion of foraging time spent in flight, in order to estimate costs of foraging (Corbet et al., 1995). The cost of a foraging trip depends on the rate of energy consumption while foraging on the patch plus the cost of the return flight between the nest and the patch. This commuting cost often remains unquantified when the bees' nest cannot be found.

THE SHIFTING BALANCE BETWEEN COSTS AND BENEFITS

Foraging rewards change with time, from hour to hour and day to day, depending on the balance between rates of replenishment and depletion of reward. Costs change too, depending on environmental temperature and the spatial distribution of reward. For nectar, shifts in the balance between costs and benefits of foraging can be expressed in terms of the common currency of energy. For pollen, costs and benefits are less well understood. There is no obvious common currency to relate costs to rewards, availability is probably not simply related to the number of grains per flower, and the demand may fluctuate depending on conditions in the nest.

Changes through the day in some of the costs and benefits of nectar foraging for a particular bee on a particular plant species can be represented graphically in the form of a competition box described elsewhere (Corbet et al., 1995). Briefly, values of the standing crop of nectar and the temperature at successive sampling times through the day are plotted in three-dimensional space in the form of a nectar trajectory, and the three-dimensional box containing the trajectory is divided into domains by three mutually perpendicular planes representing the bee's thresholds of energy gain per flower visit, environmental temperature and depth to nectar. When the nectar trajectory lies outside the domain accessible to and profitable for a given species of bee, the bee is not expected to forage. When the trajectory lies within the bee's domain, the bee may or may not visit the flowers depending on the availability of alternative forage nearby.

Because the cost/benefit balance changes with time it is unwise to evaluate an insect–flower relationship on the basis of a single survey or even a single dawn-to-dusk study; multiple dawn-to-dusk studies are desirable. The cost–benefit balance for each partner is also likely to vary spatially with habitat, region and climate. In so far as the outcome for each partner depends on local conditions, including the presence of robbers and other competitors for the floral rewards, most insect–flower interactions are, at best, conditional mutualisms (Thompson, 1994).

9.2.2. Benefits and costs for plants

BENEFITS

The most obvious benefit that plants may gain from the insect/flower inter-action is increased reproductive success due to pollen transfer. Ecological studies sometimes deal with both maternal and paternal reproductive success. In agriculture it is the former that is commonly of greater interest, although the fate of pollen of genetically manipulated organisms is attract-ing increasing attention.

In evolutionary terms, the reproductive success of a plant depends on the completion of a sequence of steps:

1. Transfer of pollen from an anther to a bee.
2. Maintenance of pollen viability in transit.
3. Transfer of pollen from the bee to a suitable stigma.
4. Germination.
5. Pollen tube growth to the base of the style.
6. Fertilization.
7. Seed development and maturation.
8. Survival of the seed to germination and establishment of a new plant capable of reproducing (see Osborne, 1994).

The seed yield of crops often depends on completion of steps 1–7, but steps 6 and 7 are usually the particular focus when maternal reproductive success is investigated in an agricultural context. Steps 1–7 are more likely to be examined in investigations of paternal reproductive success, which are con-sequently sometimes at a much finer scale (page 229) than investigations of maternal reproductive success (see below).

Maternal reproductive success. Does pollen transfer affect female repro-ductive success? The importance of pollination depends on the plant's breeding system. Insect pollen vectors are essential in many highly self-incompatible species, in which pollen from another plant is necessary for seed set, and they may increase seed yield or quality even in self-compatible forms (Free and Williams, 1976).

Is yield reduced by excluding insects? The contribution of natural insect pollination to female reproductive success is often estimated by comparing seed set of open-pollinated flowers with that of flowers from which insects have been excluded by bagging or caging (e.g. Brantjes, 1981). To allow for microclimatic effects of the bag or cage, such comparisons sometimes involve three treatments: tagged control flowers, caged or bagged flowers, and flowers open to insects but subjected to at least some of the micro-climatic effects of insect exclusion, e.g. wall-less cages (Free, 1993). In these experiments it is usual to bag the insect-free flowers before the buds open, and to tag flowers of the same stage at the same time for the unbagged treat-ments. Bagged flowers often set some seed by autopollination (Free, 1993), or by geitonogamy if the bag includes more than one flower. In an agricultural context, it is often useful to know the maximum yield increase that pollination management could achieve. This can be assessed by adding to the above experiment a fourth treatment, bagging with hand cross-pollination (Corbet *et al.*, 1991). The degree of self-incompatibility can be assessed by adding a fifth treatment, bagging with hand self-pollination.

The first four of these treatments can show whether seed set is reduced by exclusion of large insects and whether it is likely to be increased by enhanced pollination. They do not show which members of the current assemblage of flower-visiting insects contribute positively to seed set, or what species might enhance pollination if augmented in, or superimposed on, the existing assemblage.

Can yield be increased by bringing in honey bees? It is notoriously difficult to address this question experimentally in the open field by comparing crop yield with and without introduction of honey bees, because bees readily fly between plots. It is difficult to find the best compromise between the generality to be inferred from large-scale studies, in which the replicates are whole fields in a region, and the precise match with local circumstances that smaller-scale studies, comparing plots within a field, can provide. In a few studies hives placed at one end of a field have produced a gradient in bee density which could be associated with a gradient in crop seed yield (e.g. Fries and Stark, 1983), but when bees from a hive distribute themselves over a foraging area larger than an individual crop, local spatial patterns associated with hive introduction may become blurred, and other factors affecting yield may swamp field-to-field or year-to-year comparisons.

This is why effects of honey bees on yield are often quantified using field cages. Seed set or yield is compared in plots caged to exclude bees, plots caged with honey bees, and plots freely exposed to insect visitors. Some authors incorporate a wall-less cage treatment to control for the yield response to microclimatic effects of caging. Field-cage experiments have often shown higher yield in plots caged with honey bees than in open-pollinated plots or plots without bees (Free, 1993). The inference has sometimes been that if yield is increased by honey bees in a cage, yield could also be increased by bringing honey bee hives to the crop. In the absence of any other evidence, recommendations about the necessity for hive introduction, and the quantity of hives to be introduced to particular crops, have been based on such experiments (e.g. McGregor, 1976).

Problems of interpretion in field-cage experiments of this kind are increasingly recognized. Honey bees free in the countryside make choices among the forage plants within several hundred metres of the hive (Visscher and Seeley, 1982). Those choices often conform with the expectation that the bees will maximize the colony's net rate of energy gain or the net energy gain per unit cost. Honey bees caged on a crop have Hobson's choice – no alternative forage is available to them, their commuting costs between hive and crop are low, and they may work the crop even if its floral reward is small. Cage experiments can show whether very numerous honey bee visits could ever augment yield above natural levels, but they cannot tell us whether introduction of honey bee hives will do the same in the field. Bees unconfined by a cage may choose to forage elsewhere. Further, cage experiments cannot tell us whether superimposing free-ranging honey bees on the wild pollinator assemblage will increase or decrease yield. This important but challenging question has rarely been addressed.

The choice of size and material for the bags or cages for insect exclusion experiments must be a compromise. Large cages including several plants are easier to set up but they must be a compromise. Large cages including several plants are easier to set up but they allow cross-pollination by vectors (such as wind or small insects) inside the cage. Small cages holding an individual flower or plant are more time-consuming to use but make in-cage cross-pollination less likely. Paper or glassine bags used by plant breeders exclude windborne pollen and minute insects such as thrips, but probably modify the flower's microclimate in ways that may affect pollen performance or seed development. Coarser mesh such as mosquito netting or bridal veil has less effect on floral microclimate (Corbet and Delfosse, 1984; Ball *et al.*, 1992; Kearns and Inouye, 1993), but is not impermeable to pollen grains carried by wind or thrips. These more permeable bags are appropriate if the experiment is designed to assess the effect of excluding bees as pollen vectors.

These experimental comparisons should involve whole plants, rather than individual flowers or branches of a plant, because the treatment of one branch may influence resource allocation, and therefore seed yield, on another (Stephenson, 1981).

Pre-dispersal seed predation (damage by seed parasites) may be a major determinant of reproductive success or yield, affecting the evolution or economic consequences of the pollination process. It is an intrinsic cost in a few tightly-coevolved plant–pollinator mutualisms (Thompson, 1994), but is more often considered separately from the pollination process.

Paternal reproductive success. It is much more difficult to estimate male reproductive success of a plant than female reproductive success because the pollen grains are small, mobile and hard to track. Tracing marked or dyed grains or pollen mimics (Handel, 1976; Waser and Price, 1982; Thomson *et al.*, 1986; Eguiarte *et al.*, 1993), or even individually marked pollinia (Nilsson *et al.*, 1992) can give some information on spatial patterns of pollen dispersal, but these may be poor predictors of patterns of gene dispersal, for example if the probability of successful fertilization varies with the distance apart of the two parent plants (Price and Waser, 1979). Patterns of gene dispersal are examined directly by other techniques such as those involving genetic attributes (Handel, 1982; Meagher, 1986; Arnold *et al.*, 1992; Carré *et al.*, 1993; Kearns and Inouye, 1993; Scheffler *et al.*, 1993; Michaelson-Yeates *et al.*, 1997).

Transfer of pollen from a flower to a bee can be confirmed by direct observation and quantified by grain counts on the bee's body (Beattie, 1971). In-transit viability may be compromised by microclimatic conditions

(Corbet, 1990) or bee-derived chemical contamination (Vaissière *et al.*, 1996). It can be assessed by viability tests *in vitro* (e.g. Mesquida and Renard, 1989) or germination tests *in vivo*. To see how much pollen a bee of a given type transfers on to the stigma, a flower that is known to be pollen free (either by inspection *in situ*, or because it has been emasculated and bagged from the bud stage onwards) is exposed, allowed to receive one (or two, three …) visit(s) by a known insect visitor, and then inspected for a count of pollen grains on the stigma. The grains can sometimes be counted *in situ* (e.g. A.V. Roberts, 1979), but if the grains are small counting may require staining and microscopy (Kearns and Inouye, 1993). Transport to the laboratory for counting requires care. In the minutes after arrival on the stigma some grains become firmly attached to the stigma and others do not. The proportion of grains displaced by disturbance during transport will depend on the stigmatic pollen load, the time since deposition, the provenance of the pollen and the conditions of transport. Some of the grains deposited on a stigma may be removed again by a subsequent insect visitor (Osborne, 1994). The importance of secondary redeposition of such pollen on to a fresh stigma remains poorly understood.

Processes on and in the gynoecium (germination, pollen tube growth and fertilization) are assessed by microscopical examination. Fluorescent staining of pollen, for example with aniline blue, can reveal detail clearly (Stoddard, 1986a,b; Davis, 1992). Sometimes it is necessary to macerate or clear opaque stigmatic tissue (Kearns and Inouye, 1993).

Seed development or maturation is often interrupted by abortion. In so far as the probability of abortion of a zygote depends on the provenance of its paternal pollen (Stephenson, 1981), this is relevant to pollination studies in which mature seed yield is scored as the outcome.

COSTS

A plant's pollination costs may compete with seed development for resources from a common pool. It may be useful to separate pollination costs into the costs of the pollination apparatus such as the corolla or the pollen, comparable with a non-returnable deposit, and the costs of nectar, which are sometimes paid out on a 'sale or return' basis. Nectar sugar remaining in unvisited flowers may be reabsorbed in some species, as experiments with labelled sugars have shown (Búrquez and Corbet, 1991). In theory, if one ramet of a modular plant received disproportionately large numbers of insect visits, it might maintain nectar flow and support subsequent seed development by drawing on assimilate reabsorbed in the nectar of unvisited flowers on sister ramets. This possibility remains to be explored.

The estimation of plant pollination costs, in a currency of energy or seed number, is complicated by the capacity of plants to reallocate assimilate in time and space (Southwick, 1984; Pyke, 1991; Ashman and Baker, 1992). In short-lived annuals, it is appropriate to relate immediate nectar production to the concurrent rate of assimilation, but in perennial herbs and woody plants assimilate can be stored and later remobilized. Here the relation between assimilation and nectar secretion or seed production is less susceptible to analysis in a short-term study; a steady assimilation rate may be associated with biennial or irregular peaks of seed production, as exemplified by apple or masting fruits, respectively. Similarly, spatial patterns of carbohydrate reallocation depend on plant structure and growth form. Green tissues associated with the flower probably produce much of the assimilate destined for nectar and seeds in some species (e.g. Carnowski, 1952), but flowers that bloom during a leafless period must derive nectar sugars from reserves elsewhere.

THE SHIFTING BALANCE BETWEEN COSTS AND BENEFITS

The balance between resource limitation and pollination limitation as constraints on seed yield is a dynamic one (Campbell and Halama, 1993; Casper and Niesenbaum, 1993); it may change through time and space and it is not always clear which dominates in any particular case. Thus, enhanced pollen transfer may bring no benefit in an unthrifty crop where resources limit seed yield, but the same crop might show a yield benefit from enhanced pollination if its nutrient or water status were improved. Commonly, pollinator limitation is assessed at a single resource level, and resource limitation is assessed at a single pollination level. A three-dimensional plot showing the relation between resource level, pollination level and yield would often be more useful (Corbet, 1996).

9.2.3. Benefits and costs for growers

So far this chapter has dealt with both wild plants and crops, but here we focus on the benefits and costs of managing the insect/flower interaction in field and orchard crops, in horticultural production of cultivated and wild flowers, in glasshouse crops and in plant breeding.

The aim of pollinator management is not always maximization of seed yield; sometimes it is to synchronize seed set reducing wastage at harvest, to improve seed or fruit quality, or even to avoid pollination or to identify and control patterns of distribution of pollen from a particular source. Management may include introduction of bees to a crop, or habitat management to increase the probability that wild bees will pollinate a crop, or an integrated programme involving both elements.

A case can be made for management of a pollination system wherever the increase in profit is expected to exceed the cost of management, or where management is necessary to reduce the risk of genetic contamination by genetically manipulated organisms. It is the role of a crop pollination ecologist to assess and make clear to the grower the benefits and potential costs of management of the pollination system.

BENEFITS

In so far as the role of insects in pollination fluctuates through time and space (see above), the pollinating value of the existing assemblage of insect visitors, and the potential effects of changing it, will depend on local circumstances. The benefits to the grower of pollinator management will depend on the local circumstances including location and weather conditions, as well as on the species of crop. Small-scale experiments replicated locally give better resolution in time and space than large-scale experiments that are too costly and time consuming for local replication. Multiple experiments of the type described in Section 9.2.2 could be used to investigate the maximal potential benefit that augmented pollination could bring in each local situation, but it is more difficult to discover how nearly that maximum is achieved by various different management procedures. It is necessary to enquire how the proposed management will affect the numbers of bee visits, and how a change in the numbers of visits by a given species of bee will affect yield. Experimental approaches to unravelling the pollinating role of individual species of bee are considered in Section 9.2.2.

An increase in seed or fruit mass or number resulting from augmented pollination is most obvious in completely self-incompatible crops, but partially or fully self-compatible crops may also benefit from augmentation of bee visits if these increase the level of self-pollination (Free and Williams, 1976). Even if pollination management does not increase overall seed yield, it may improve the synchrony of seed or pod set so that more seed can be harvested. In oilseed rape a large population of bees on the crop early in the flowering period leads to more synchronous early pod set, reducing problems of pod shatter and lodging at harvest (Williams *et al.*, 1987). To establish such effects, bee numbers are manipulated, and pollinator levels and seed set are scored, at intervals over the flowering season.

In some crops fruit size depends on seed number. In passionfruit (Akamine and Girolami, 1959), kiwifruit (Hopping, 1976; Pyke and Alspach, 1986), tomato (Verkerk, 1957; Banda and Paxton, 1991) and melon (McGregor and Todd, 1952; Winsor *et al.*, 1987; Subirana, 1993) large numbers of seeds are required for a fruit of marketable size. Extra bee visits may therefore improve the quality of the fruit, perhaps bringing it above some commercial quality threshold. Many fruits of this type are high-value crops grown in glasshouses or polythene tunnels, inaccessible to wild pollinators. Hand pollination or pollinator introduction is therefore required. It is in crops of this kind that the need for pollination management is often most obvious to growers.

In crops grown in such enclosures, the economic benefit due to pollinator introduction can be assessed by comparing crops with and without introduced pollinators. In the field, where pollination management involves supplementing or augmenting wild bee populations rather than introducing pollinators into a bee-free area, assessing the benefits of management requires comparison of seed set due to the natural assemblage of pollinators with that in the managed system. This comparison is not straightforward. In a small-scale plot experiment the managed and unmanaged sites are close together and bees can readily fly between them, so that the proximity of the managed site may affect pollination in the unmanaged site. In a large-scale field experiment the comparison is likely to be confounded by site-to-site differences in growing conditions, microclimate or the insect community, and adequate replication is difficult for reasons of scale.

Estimates of the overall economic value of domesticated honey bees as crop pollinators, in the United States of America (Robinson *et al.*, 1989), in Canada (Winston and Scott, 1984) and in Europe (Borneck and Merle, 1989), are derived from the estimated economic value of the seed produced, multiplied by a factor representing the proportional contribution of honey bees to seed yield, largely inferred from cage studies. In this way Robinson *et al.* (1989) calculated that about 31% of the value of commercial crops in the USA was attributable to honey bees. For insect-pollinated crops in Europe, with a total annual market of 65,000 million ecus, Borneck and Merle (1989) calculated that 5000 million ecus were attributable to pollination by domestic honey bees. Robinson *et al.* (1989) showed that their estimate of the yield benefit attributable to introduction of honey bee colonies in the USA far outweighed the cost to growers of introducing those colonies. However, the assumptions on which these estimates necessarily rest have yet to be fully substantiated. In particular, they may change when account is taken of the background pollination service provided by local assemblages of wild bees.

COSTS

Pollination management commonly involves managing the habitat to augment the existing assemblage of wild pollinator species, introducing extra pollinators in artificial domiciles, or employing a combination of both strategies. The choice will depend on the context, the resources available and the cost.

Habitat management for the maintenance of substantial populations of bees involves meeting their requirements: suitable forage plants to provide nectar and pollen throughout the season; nesting sites; and in some cases, for example where bumble bees are being considered, also sites for overwintering and courtship. Sources of information on forage and nesting requirements include Falk (1991) and Westrich (1990a, b) for solitary bee species, and for British bumble bee species Fussell and Corbet (1992a) for forage plants, Fussell and Corbet (1992b) for nesting places, and Fussell and Corbet (1992c) for male patrolling sites used in courtship. Hibernation sites have received less attention (but see Sladen, 1912; Skovgaard, 1936; Alford, 1975).

Habitat management for bees may simply involve protecting uncultivated areas such as long-term set-aside, hedgerows, field margins and woodland edges in which destructive activities such as ploughing, spraying and mowing are minimized, so that a floristically diverse sward of perennial herbaceous species can develop providing forage and nesting places. Where such a sward does not develop naturally, the more expensive option of sowing wild flower mixtures or bee forage crops may be appropriate (Smith *et al.*, 1993; Corbet *et al.*, 1994). Indirect costs may include an increased weed problem on unsprayed and unploughed areas, but this is expected to decrease over the first few years as perennial plant species out-compete the annual weeds (Roebuck, 1987; Smith and Macdonald, 1992).

It is difficult to evaluate the impact of habitat management on local bee populations and communities, because the bees' foraging range is large in relation to the size of an experimental plot convenient for replication, and because there is as yet no reliable technique for the comprehensive mapping of bumble bee nests in an area. The impact of management on the flower-visiting behaviour of potential pollinators and on forage resources available is easier to estimate. An example of successful analysis is described by Feber *et al.* (1994).

If habitat management cannot provide the crop with adequate pollina-tors, or where no local bee fauna exists, as in glasshouses, it may be appropriate to introduce selected pollinator species reared elsewhere. A pol-lination fee may be payable by the grower, but it may not reflect the true cost of providing the pollinators. Beekeepers sometimes charge little or nothing for bringing hives to an insect-pollinated crop because they expect an increased yield of honey. In contrast, providers of bumble bees for polli-nation of high-value glasshouse crops charge a high fee that reflects the potential economic gain resulting from their use rather than the cost of pro-duction. Other bees managed on a large scale include alfalfa leafcutter bees, *Megachile rotundata*, the management of which brings economic benefits in the form of saleable bees, as well as increased seed yield (Richards, 1987).

Honey bees are the most easily managed pollinators and the most wide-ly available, but they are not necessarily the most effective. The success of a pollinator introduction depends on the match between the crop's flowers and the introduced bee's foraging requirements (Corbet *et al.*, 1995), and on the impact of the introduced bees on the background pollination service performed by the local wild bees (Corbet, 1996). Introduction of an unsuit-able species to a crop may reduce the effectiveness of existing pollinators to an extent that outweighs any benefit from the introduction; the newcomer may not visit the crop, or if it does, it may displace a more effective wild pollinator or reduce its efficacy. Wilson and Thomson (1991) drew attention to the possibility that seed yield might be reduced, rather than increased, by bringing extra bees to a crop, and they showed that a bee species may reduce the overall rate of pollen transfer by a more effective pollinator by depleting pollen and reducing the per-visit pollen transfer rate of the more effective pollinator.

THE SHIFTING BALANCE BETWEEN COSTS AND BENEFITS: WHICH MANAGEMENT STRATEGY?

If the pollination provided by the local pollinator assemblage is close to the maximum that can be achieved by hand cross-pollination, the costs and risks associated with the introduction of managed pollinators can be avoided; all that is necessary is to ensure that the habitat continues to support a flour-ishing community of wild bees. If the pollination provided by the local assemblage is inadequate, and suitable pollinating species are present but their numbers are too low or too variable over space and time for reliable pollination, habitat management may be considered. If suitable species are absent, or cannot be augmented by habitat management, pollinator intro-duction may be necessary.

The relative costs and benefits of these alternative management options depend on attributes of the crop and the local environment, such as the value of the crop, the nature of the existing pollinator community, the habi-tat potential of the surrounding land and the cost, suitability and availability of pollinators for introduction to the crop. Wise choice will require information on the pollination system of the crop, its requirements for bee pollination, the most suitable pollinating species and the most appropriate way to increase their numbers, whether by habitat management or by the introduction of managed bees.

The decision will depend in part on how much risk the grower is pre-pared to take. If an appropriate pollinator species is successfully introduced, the initial cost of introduction is balanced by a lowered risk of poor seed yield. Habitat management often involves lower initial costs but less pre-dictable yield, because the numbers of wild populations may fluctuate.

Before introducing a bee species to a crop it is important to establish its likely effects on overall pollen transfer. Effects on the pollination of local wild flowers and the survival of local wild bees should also be taken into account. An investigation of the background pollination system of the crop should be undertaken, together with a detailed comparison of the pollinat-ing effects of different potential pollinator species (Section 9.2.2). The list of bee species that can be managed effectively is growing year by year (Torchio, 1994). In selecting candidate species for introduction from this list, pollination managers must consider not only the direct economic costs to the grower in terms of management costs and pollination fees, but also the biological costs to the crop and to the local assemblage of pollinators. Unless the introduced bee can forage profitably on the crop, it will desert the crop for more attractive alternative forage. This means the bee must suit the crop in terms of the temperature threshold for foraging flight, tongue length and energetic requirements. Corbet *et al.* (1993) deal with methods for identify-ing temperature thresholds, and Corbet *et al.* (1995) outline a graphical method for comparing the attributes of a given bee species with those required for a given crop in a given climate.

The long-term biological costs of an ill-considered pollinator introduc-tion may be very great, far exceeding any short-term local benefit. Introduction of a species or genotype of bee to new areas may disrupt local pollinator assemblages, contaminate local races genetically, or introduce parasites and pathogens that will affect local bees. To introduce pollinators without first evaluating the risks would be irresponsible, but a substantial research programme would be required for that evaluation.

9.3. Benefits and Costs of Pollination Research

Growers know that resource availability sometimes limits yield, and they commonly remedy this by fertilization or irrigation. The possibility that pollination may limit yield receives less attention. This constraint is expected to become increasingly important as honey bee numbers decline (Williams *et al.*, 1991) and wild bee communities are eroded (Westrich, 1996). Pollination research can highlight the need for pollination management on certain crops with unstable yields, and can help growers to select appropriate pollinating species and management options, and to target their resources effectively. In addition to its academic value to students of ecology and evolution, research on pollination brings practical benefits to plant breeders and those responsible for managing gene flow of genetically manipulated organisms, as well as conservationists concerned with wild flowers and wild bees.

The tactics of pollination research are now well documented, but the strategies remain poorly developed (Torchio, 1990). More standardized experimental approaches to the evaluation of the role of natural pollinator assemblages and the costs and benefits of intervention could facilitate the production of a sound scientific basis for decisions on pollination management. A better research base could help to reduce the probability of irreversible damage due, for example, to pollen flow from genetically manipulated plant species or competitive displacement of wild species by introduced pollinators. Until we have a clearer idea of the extent to which pollination limits seed yield in a wide range of crops and wild flowers, it is difficult to quantify the economic benefits that will accrue from soundly based pollination management, but if global estimates of economic benefits of honey bee pollination are anywhere near the mark, small improvements in the efficacy of pollination will result in economic gains far greater than the costs of research.

9.4. References

Akamine, E.K. and Girolami, G. (1959) Pollination and fruit set in the yellow passion fruit. *Hawaii Agricultural Experiment Station Technical Bulletin* 59, 44pp.

Alford, D.V. (1975) *Bumblebees.* Davis-Poynter Ltd, London, 352pp.

Arnold, M.L., Robinson, J.J., Buckner, C.M. and Bennet, B.D. (1992) Pollen dispersal and interspecific gene flow in Louisiana irises. *Heredity* 68, 399–404.

Ashman, T.-A. and Baker, I. (1992) Variation in floral sex allocation with time of season and currency. *Ecology* 73, 1237–1243.

Ball, S.T., Campbell, G.S. and Konzak, C.F. (1992) Pollination bags affect wheat spike temperature. *Crop Science* 32, 1155–1159.

Banda, H.J. and Paxton, R.J. (1991) Pollination of greenhouse tomatoes by bees. *The 6th International Symposium on Pollination, Tilburg, The Netherlands, August 1990. Acta Horticulturae* 288, 194–198.

Beattie, A.J. (1971) A technique for the study of insect-borne pollen. *Pan-Pacific Entomologist* 47, 82.

Belavadi, V.V. and Parvathi, C. (1993) Landing and departure rules used by honeybees on cardamom (*Elettaria cardamomum* Maton) In: Veeresh, G.K, Uma Shaanker, R. and Ganeshaiah, K.N. (eds) *Proceedings of the International Symposium on Pollination in Tropics, 1993, India*. IUSSI Indian Chapter, Bangalore, pp. 73–76.

Bolten, A.B., Feinsinger, P., Baker, H.G. and Baker, I. (1979) On the calculation of sugar concentration in flower nectar. *Oecologia* 41, 301–304.

Borneck, R. and Merle, B. (1989) An attempt to evaluate the economic value of the pollinating honeybee in European agriculture. *Apiacta* 24, 33–38.

Brantjes, N.B.M. (1981) Ant, bee and fly pollination in *Epipactis palustris* (L.) Crantz (Orchidaceae) *Acta Botanica Neerlandica* 30, 59–68.

Búrquez, A. and Corbet, S.A. (1991) Do flowers reabsorb nectar? *Functional Ecology* 5, 369–379.

Campbell, D.R. and Halama, K.J. (1993) Resource and pollen limitations to lifetime seed production in a natural plant population. *Ecology* 74, 1043–1051.

Carnowski, C. von (1952) Untersuchung zür Frage der Nektarabsonderung. *Zeitschrift für Bienenforschung* 1, 171–173.

Carré, S., Badenhausser, I., Tasei, J.N., Le Guen, J. and Mesquida, J. (1994) Pollen deposition by *Bombus terrestris* L, between male-fertile and male-sterile plants in *Vicia faba* L. *Apidologie* 25, 338–349.

Carré, S., Tasei, J.N., Le Guen, J., Mesquida, J. and Morin, G. (1993) The genetic control of seven isozymic loci in *Vicia faba* L. Identification of lines and estimates of outcrossing rates between plants pollinated by bumble bees. *Annals of Applied Biology* 122, 555–568.

Casper, B.B. and Niesenbaum, R.A. (1993) Pollen versus resource limitation of seed production: a reconsideration. *Current Science* 65, 210–214.

Corbet, S.A. (1990) Pollination and the weather. *Israel Journal of Botany* 39, 13–30.

Corbet, S.A. (1996) Which bees do plants need? *Proceedings of a symposium 'Conserving Europe's Bees', London, April 1995*. International Bee Research Association and Linnean Society of London.

Corbet, S.A. and Delfosse, E.W. (1984) Honeybees and the nectar of *Echium plantagineum* L. in south-eastern Australia. *Australian Journal of Ecology* 9, 125–139.

Corbet, S.A. and Willmer, P.G. (1981) The nectar of *Justicia* and *Columnea*: composition and concentration in a humid tropical climate. *Oecologia* 51, 412–418.

Corbet, S.A., Unwin, D.M. and Prŷs-Jones, O.E. (1979) Humidity, nectar and insect visits to flowers, with special reference to *Crataegus, Tilia* and *Echium*. *Ecological Entomology* 4, 9–22.

Corbet, S.A., Kerslake, C.J.C., Brown, D. and Morland, N.E. (1984) Can bees select nectar-rich flowers in a patch? *Journal of Apicultural Research* 23, 234–242.

Corbet, S.A., Chapman, H. and Saville, N. (1988) Vibratory pollen collection and flower form: bumble-bees on *Actinidia, Symphytum, Borago* and *Polygonatum*. *Functional Ecology* 2, 147–155.

Corbet, S.A., Williams, I.H. and Osborne, J.L. (1991) Bees and the pollination of crops and wild flowers in the European Community. *Bee World* 72, 47-51.

Corbet, S.A., Fussell, M., Ake, R., Fraser, A., Gunson, C., Savage, A. and Smith, K. (1993) Temperature and the pollinating activity of social bees. *Ecological Entomology* 18, 17–30.

Corbet, S.A., Saville, N.M. and Osborne, J.L. (1994) Farmland as a habitat for bumble bees. In: Matheson, A. (ed.) *Forage for Bees in an Agricultural Landscape.* International Bee Research Association, Cardiff, pp.35–45.

Corbet, S.A., Saville, N.M., Fussell, M., Prŷs-Jones, O.E. and Unwin, D.M. (1995) The competition box: a graphical aid to forecasting pollinator performance. *Journal of Applied Ecology* 32, 707–719.

Dafni, A. (1992) *Pollination Ecology. A Practical Approach.* Oxford University Press, Oxford, 250pp.

Davis, A.R. (1992) Evaluating honey bees as pollinators of virgin flowers of *Echium plantagineum* L. (Boraginaceae) by pollen tube fluorescence. *Journal of Apicultural Research* 31, 83–95.

Eguiarte, L.E., Búrquez, A., Rodríguez, J., Martínez-Ramos, M., Sarukhán, J. and Piñero, D. (1993) Direct and indirect estimates of neighborhood and effective population size in a tropical palm, *Astrocaryum mexicanum*. *Evolution* 7, 75–87.

Falk, S. (1991) *A Review of the Scarce and Threatened Bees, Wasps and Ants of Great Britain.* Nature Conservancy Council, Peterborough, 344pp.

Feber, R., Smith, H. and Macdonald, D.W. (1994) The effects of field margin restoration on the meadow brown butterfly, *Maniola jurtina*. In: N. Boatman (ed.) *Field Margins: Integrating Agriculture and Conservation.* Monograph no. 58, British Crop Protection Council, Farnham, pp.295–300.

Free, J.B. (1993) *Insect Pollination of Crops.* Academic Press, London, 684pp.

Free, J.B. and Williams, I.H. (1976) Pollination as a factor limiting the yield of field beans (*Vicia faba* L.) *Journal of Agricultural Science (Cambridge)* 87, 395–399.

Fries, I. and Stark, J. (1983) Measuring the importance of honeybees in rape seed production. *Journal of Apicultural Research* 22, 272–276.

Fussell, M. and Corbet, S.A. (1992a) Flower usage by bumble-bees: a basis for forage plant management. *Journal of Applied Ecology* 29, 451–465.

Fussell, M. and Corbet, S.A. (1992b) The nesting places of some British bumble bees. *Journal of Apicultural Research* 31, 32–41.

Fussell, M. and Corbet, S.A. (1992c) Observations on the patrolling behaviour of male bumblebees (Hym.). *Entomologist's Monthly Magazine* 128, 229–235.

Giurfa, M. and Núñez, J.A. (1992) Honeybees mark with scent and reject recently visited flowers. *Oecologia* 89, 113–117.

Goodwin, R.M. (1986) Kiwifruit flowers: anther dehiscence and daily collection of pollen by honeybees. *New Zealand Journal of Experimental Agriculture* 14, 449–452.

Handel, S.N. (1976) Restricted pollen flow of two woodland herbs determined by neutron-activation analysis. *Nature* 260, 422–423.

Handel, S.N. (1982) Dynamics of gene flow in an experimental population of *Cucumis melo* (Cucurbitaceae). *American Journal of Botany* 69, 1538–1546.

Harder, L.D. (1990) Behavioural responses by bumble bees to variation in pollen availability. *Oecologia* 85, 41–47.

Harder, L. D. and Thomson, J. D. (1989) Evolutionary options for maximizing pollen dispersal of animal-pollinated plants. *American Naturalist* 133, 323–344.

Heinrich, B. (1979) *Bumblebee Economics.* Harvard University Press, Cambridge, 245pp.

Hodges, C.M. and Wolf, L.L. (1981) Optimal foraging in bumblebees: why is nectar left behind in flowers? *Behavioral Ecology and Sociobiology* 9, 41–44.

Hopping, M.E. (1976) Effect of exogenous auxins, gibberellins and cytokinins on fruit development in Chinese gooseberry (*Actinidia chinensis* Planch). *New Zealand Journal of Botany* 14, 69–75.

Inouye, D.W., Favre, N.D., Lanum, J.A., Levine, D.M., Meyers, J.B., Roberts, F.C.,

Tsao, F.C. and Wang, Y. (1980) The effects of nonsugar nectar constituents on nectar energy content. *Ecology* 61, 992–996.

Kearns, C.A. and Inouye, D.W. (1993) *Techniques for Pollination Biologists*. University Press of Colorado, Colorado, 583pp.

Lide, D.R. (ed.) (1994) *CRC Handbook of Chemistry and Physics*. CRC Press, Boca Raton, Florida.

McGregor, S.E. (1976) *Insect Pollination of Cultivated Crops*. ARS, US Department of Agriculture, Agriculture Handbook 496, Washington, DC, 411pp.

McGregor, S.E. and Todd, F.E. (1952) Canteloupe production with honeybees. *Journal of Economic Entomology* 45, 43–47.

McKenna, M. and Thomson, J.D. (1988) A technique for sampling and measuring small amounts of floral nectar. *Ecology* 69, 1306–1307.

Meagher, T.R. (1986) Analysis of paternity within a natural population of *Chamaelirium luteum*. I. Identification of most-likely male parents. *American Naturalist* 128, 199–215.

Mesquida, J. and Renard, M. (1989) Étude de l'aptitude a germer in vitro du pollen de colza (*Brassica napus* L.) recolté par l'abeille domestique (*Apis mellifera* L.). *Apidologie* 20, 197–205.

Michaelson-Yeates, T.P.P., Marshall, A.H., Williams, I.H., Carreck, N.L. and Simpkins, J.K. (1997) The use of isoenzyme markers to determine gene flow mediated by different bee species in an outbreeding plant species (*Trifolium repens* L.). *Journal of Apicultural Research* (in press).

Nilsson, L.A., Rabakonandrianina, E. and Pettersson, B. (1992) Exact tracking of transfer and mating in plants. *Nature* 360, 666–668.

O'Rourke, M.K. and Buchmann, S.L. (1991) Standardized analytical techniques for bee-collected pollen. *Environmental Entomology* 20, 507–513.

Osborne, J.L. (1994) Evaluating a pollination system: *Borago officinalis* and bees. Unpublished PhD thesis, University of Cambridge.

Percival, M.S. (1955) The presentation of pollen in certain angiosperms and its collection by *Apis mellifera*. *New Phytologist* 54, 353–368.

Price, M.V. and Waser, N.M. (1979) Pollen dispersal and optimal outcrossing in *Delphinium nelsoni*. *Nature* 277, 294–297.

Prŷs-Jones, O.E. and Corbet, S.A. (1991) *Bumblebees*. Richmond Publishing Co. Ltd, Slough, 92pp.

Pyke, G.H. (1991) What does it cost a plant to produce floral nectar? *Nature* 350, 58–59.

Pyke, N.B. and Alspach P.A. (1986) Inter-relationships of fruit weight, seed number and seed weight in kiwifruit. *New Zealand Agricultural Science* 20, 153–156.

Richards, K.W. (1987) Alfalfa leafcutter bee management in Canada. *Bee World* 68, 168–178.

Roberts, A.V. (1979) The pollination of *Lonicera japonica*. *Journal of Apicultural Research* 18, 153–158.

Roberts, R.B. (1979) Spectrophotometric analysis of sugars produced by plants and harvested by insects. *Journal of Apicultural Research* 18, 191–195.

Robinson, W.S., Nowogrodski, R. and Morse, R.A. (1989) The value of honeybees as pollinators of US crops. *American Bee Journal* 129, 411–423 and 477–487.

Roebuck, J.F. (1987) Agricultural problems of weeds on the crop headland. In: Way, J.M. and Greig-Smith, P.W. (eds) *Field Margins*. British Crop Protection Council Monograph No. 35, Thornton Heath, pp. 11–22.

Scheffler, J.A., Parkinson, R. and Dale, P.J. (1993) Frequency and distance of pollen dispersal from transgenic oilseed rape (*Brassica napus*). *Transgenic Research* 2, 356–364.

Schmid-Hempel, P. (1987) Efficient nectar collection by honeybees. I. Economic models. *Journal of Animal Ecology* 56, 209–218.

Silvola, J. (1984) Respiration and energetics of the bumblebee *Bombus terrestris* queen. *Holarctic Ecology* 7, 177–181.

Skovgaard, O.S. (1936) Rødkløverens Bestøvning, Humlebier og Humleboer. *Det Kongelige Danske Videnskabernes Selskabs Skrifter Naturvidenskabelig og Mathematisk Afdeling. 9, Raekke* 6, pp. 1–140.

Sladen, F.W.L. (1912) *The Humble-bee, its Life History and How to Domesticate It.* Macmillan, London, 283pp.

Smith, H. and Macdonald, D.W. (1992) The impacts of mowing and sowing on weed populations and species richness of field margin set-aside. In: Clarke, J. (ed.) *Set-aside.* British Crop Protection Council Monograph No. 50, Bracknell, pp. 117–122.

Smith, H., Feber, R.E., Johnson, P.J., McCallum, K., Plesner Jensen, S., Younes, M. and Macdonald, D.W. (1993) *The Conservation Management of Arable Field Margins.* (English Nature Science no. 18) English Nature, Peterborough, 455pp.

Southwick, E.E. (1984) Photosynthate allocation to floral nectar: a neglected energy investment. *Ecology* 65, 1775–1779.

Stephenson, A.G. (1981) Flower and fruit abortion: proximate causes and ultimate functions. *Annual Review of Ecology and Systematics* 12, 253–279.

Stoddard, F.L. (1986a) Floral viability and pollen tube growth in *Vicia faba* L. *Journal of Plant Physiology* 123, 249–262.

Stoddard, F.L. (1986b) Pollination and fertilization in commercial crops of field bean (*Vicia faba* L.). *Journal of Agricultural Science (Cambridge)* 106, 89–97.

Stone, G.N. and Willmer, P.G. (1989) Warm-up rates and body temperature in bees: the importance of body size, thermal regime and phylogeny. *Journal of Experimental Biology* 147, 303–328.

Subirana, M. (1993) Efficacité pollinisatrice comparée du bourdon, *Bombus terrestris* L., et de l'abeille domestique, *Apis mellifera* L., sur le melon, *Cucumis melo* L., cultivé sous abris isolés. Unpublished Diplôme d'Etudes Approfondies Thesis, Université d'Aix, Marseilles III.

Thompson, J.N. (1994) *The Coevolutionary Process.* University of Chicago Press, Chicago, 376pp.

Thomson, J.D. (1986) Pollen transport and deposition by bumble bees in *Erythronium*: influences of floral nectar and bee grooming. *Journal of Ecology* 74, 329–341.

Thomson, J.D. and Thomson, B.A. (1992) Pollen presentation and viability schedules in animal-pollinated plants: consequences for reproductive success. In: Wyatt, R. (ed.) *Ecology and Evolution of Plant Reproduction.* Chapman & Hall, London, pp. 1–24.

Thomson, J.D., Price, M.V., Waser, N.M. and Stratton, D.A. (1986) Comparative studies of pollen and fluorescent dye transport by bumble bees visiting *Erythronium grandiflorum. Oecologia* 69, 561–566.

Torchio, P.F. (1990) Diversification of pollination strategies for U.S. crops. *Environmental Entomology* 19, 1649–1656.

Torchio, P.F. (1994) The present status and future prospects of non-social bees as crop pollinators. *Bee World* 75, 49–53.

Vaissière, B.E. (1988) A novel method for quantification of pollen production and pollen loads on bees and stigmas using a Coulter Counter. *American Bee Journal* 128, 810.

Vaissière, B.E. (1991) Honey bees, *Apis mellifera* L. (Hymenoptera: Apidae), as pollinators of upland cotton, *Gossypium hirsutum* L. (Malvaceae), for hybrid seed production. Unpublished PhD Thesis, Texas, A&M University.

Vaissière, B.E. (1996) Pollen flow and pollination efficiency in entomophilous systems: a case study with bumble bees, honey bees and a monoecious crop in enclosures. *Proceedings of a symposium 'Conserving Europe's Bees', London, April 1995.* International Bee Research Association and Linnean Society of London.

Vaissière, B.E., Malaboeuf, F. and Rodet, G. (1996) Viability of cantaloupe pollen carried by honeybees, *Apis mellifera*, varies with foraging behavior. *Naturwissenschaften* 83, 84–86.

Verkerk, K. (1957) The pollination of tomatoes. *Netherlands Journal of Agricultural Science* 5, 37–54.

Visscher, P.K. and Seeley, T.D. (1982) Foraging strategy of honeybee colonies in a temperate deciduous forest. *Ecology* 63, 1790–1801.

Waser, N.M. and Price, M.V. (1982) A comparison of pollen and fluorescent dye carry-over by natural pollinators of *Ipomopsis aggregata* (Polemoniaceae). *Ecology* 63, 1168–1172.

Westrich, P. (1990a) *Die Wildbienen Baden-Württembergs, Spezieller Teil.* Eugen Ulmer, Stuttgart, 431pp.

Westrich, P. (1990b) *Die Wildbienen Baden-Württembergs, Allgemeiner Teil.* Eugen Ulmer, Stuttgart, 972pp.

Westrich, P. (1996) Considering the needs of our native bees: the problems of partial habitats. *Proceedings of a symposium 'Conserving Europe's Bees', London, April 1995.* International Bee Research Association and Linnean Society of London.

Williams, I.H., Martin, A.P. and White, R.P. (1987) The effect of insect pollination on plant development and seed production in winter oilseed rape (*Brassica napus* L.). *Journal of Agricultural Science (Cambridge)* 109, 135–139.

Williams, I.H., Corbet, S.A. and Osborne, J.L. (1991) Beekeeping, wild bees and pollination in the European Community. *Bee World* 72, 170–180.

Willmer, P.G., Bataw, A.A.M. and Hughes, J.P. (1994) The superiority of bumblebees to honeybees as pollinators: insect visits to raspberry flowers. *Ecological Entomology* 19, 271–284.

Wilson, P. and Thomson, J.D. (1991) Heterogeneity among floral visitors leads to discordance between removal and deposition of pollen. *Ecology* 72, 1503–1507.

Winsor, J.A., Davis, L.E. and Stephenson, A.G. (1987) The relationship between pollen load and fruit maturation and the effect of pollen load on offspring vigor in *Cucurbita pepo. American Naturalist* 129, 643–656.

Winston, M.L. and Scott, C.D. (1984) The value of bee pollination to Canadian apiculture. *Canadian Beekeeping* 11, 134.

Ydenberg, R.C., Welham, C.V.J., Schmid-Hempel, R., Schmid-Hempel, P. and Beauchamp, G. (1994) Time and energy constraints and the relationships between currencies in foraging theory. *Behavioural Ecology* 5, 28–34.

Young, H.J. and Stanton, M.L. (1990) Influences of floral variation on pollen removal and seed production in wild radish. *Ecology* 71, 536–547.

10 Techniques to Evaluate Insecticide Efficacy

G.A. Matthews

International Pesticide Application Research Centre, Imperial College of Science, Technology and Medicine, Silwood Park, Ascot, Berkshire SL5 7PY, UK

10.1. Introduction

The pesticide industry expanded very considerably in the 1950s following the introduction of organochlorine and organophosphate insecticides. Chemical control became the dominant way of checking insect pest populations of agricultural and medical importance. While the trend is now to adopt integrated pest management, a large number of chemicals continue to be screened each year by the multinational agrochemical companies in their search for new insecticides. Health and safety regulations have to be followed as part of good laboratory practice.

In contrast to the earlier development of neurotoxins, development has also included chemicals which interfere with insect development, in particular the insect growth regulators and the integration of conventional insecticides with pheromones to 'lure and kill' insect pests. Attention has also been directed at screening plant extracts to see if there are other molecules, like the pyrethrins, which can be extracted as insecticides. In particular extracts of the neem tree (*Azadirachta indica*) have received much attention. The evaluation of other biological agents as insecticides has focused notably on bacterial, fungal and viral pathogens. Another area for evaluating insecticidal activity has been initiated by the development of genetically engineered plants. Initially interest has been on the gene for *Bacillus thuringiensis* endotoxin, but as with other insecticides there is the likelihood of selection of pests resistant to Bt. However, there is considerable potential for other genes conferring pest resistance to be used.

In addition to assessing activity against pest species, the need to provide an environmental impact assessment has increased the evaluation of insecticides against beneficial and other non-target species (Hassan, 1977, 1985; Carter, 1992; Cooke, 1993; Dohman, 1994). The extremely high investment needed to market any new insecticide has also led to the use of predictive

© CAB INTERNATIONAL 1997. *Methods in Ecological and Agricultural Entomology* (D.R. Dent and M.P. Walton eds)

models in an attempt to reduce the number of chemicals that are retained after initial screening for more detailed assessments.

Earlier reviews of the techniques for assessing insecticides, in the laboratory, were given by Shepherd (1958) and Busvine (1971). Ford and Salt (1987) review the studies on the transfer of insecticides from plant surfaces to insects.

10.2. Laboratory Evaluation

The initial evaluation by agrochemical companies is usually confined to a limited number of species that are easily reared in the laboratory and represent major insect pests. A primary screen is to select only those compounds which have sufficient activity to warrant further consideration. A secondary screen is then used to define the level of activity of each compound, especially in comparison with close analogues, and ascertain whether the compound has specific features that justify more detailed evaluation (Giles, 1989). Subsequent tests by government agencies and universities will extend the range of test species and be concerned with pests in a particular crop or area. The problem is how to predict with a high level of certainty and as early in the screening programme as possible how successful an insecticide will become commercially.

The choice of a technique to assess the efficacy of an insecticide should be chosen in relation to the life cycle of the pest and be aimed at a susceptible stage. Use of eggs, immature larval or nymphal stages or adult insects will depend on which is most exposed to an insecticide treatment under practical conditions. Internal feeders such as lepidopterous stem borers or bollworms are exposed to sprays as adults, eggs and first instar larvae. Later instars are protected inside the plant, so an insecticide is needed to affect the adults, reduce oviposition, be ovicidal or control the first instar larvae before they penetrate into the plant. An egg dipping test can be used if the first instar larva eats its way out of the bollworm egg (Matthews, 1966). Walker (1960) also used first instar larvae in his tests with the maize stem borer *Busseola fusca*. A series of tests may be required to investigate the effectiveness of different formulations by topical, residual contact and ingestion of an insecticide (Thacker *et al.*, 1992).

The importance of knowing as much as possible about the behaviour of the insect in designing evaluation methods was again exemplified by the results of limited exposure tests of contact activity reported by Leonard *et al.* (1994). The larvae of the test species (*Heliothis virescens*) walk on plant surfaces exposed to sprays for a limited period, so a 2–4 h exposure gave more realistic results in contrast to a 24 h exposure on the treated substrate. This suggested that the experimental insecticide would not be so effective under field conditions as a standard pyrethroid.

10.3. Factors which Affect the Evaluation of an Insecticide

Standardization of test methods is very important, especially as results of tests carried out on different occasions may need to be compared. These factors include intrinsic differences between groups of test insects, changes in the environment and experimental variables, such the formulation and method of application.

10.3.1. Intrinsic factors

SPECIES SPECIFICITY

Initially one species of a particular insect order, family or genus is selected for the screening of insecticides, and although an insecticide may have shown activity to this test species, it may not be equally effective against another related or similar species. Differences may be due to the behaviour of the insects or rate of uptake and penetration through the cuticle as much as to the insect's ability to metabolize the active ingredient and excrete it. Two lepidopterous pests of cotton, *Helicoverpa armigera* and *Diparopsis castanea*, exhibited contrasting susceptibilities to DDT and carbaryl (Matthews, 1966), so that each needed to be used separately depending on the pest population in the crop.

STAGE SPECIFICITY

The susceptibility of a pest to a particular insecticide will change according to age, size, physiology and behaviour of the individual stages of development. Often the later immature stages are more tolerant to certain insecticides, for example first instar *Schistocerca gregaria* LD_{50} for dieldrin was $0.85 \ \mu g \ g^{-1}$, but was $1.9 \ \mu g \ g^{-1}$ at the fifth instar at $25°C$ (MacCuaig, 1983). Hence, the need to control the young stages as much as possible to minimize the total amount of insecticide that has to be applied, reduce selection for resistance and avoid unnecessary contamination of the environment. Higher dosages may also prove to be uneconomic. However, the oral toxicity of diflubenzuron increased with increasing size from second to sixth instar larvae of *Choristoneura fumiferana* (Granett and Retnakaran, 1977). The toxicity may not change between the second and sixth instars, thus in tests with DDT and malathion, toxicity (LD_{50}) expressed as $\mu g \ g^{-1}$ to *Mamestra brassicae* larvae did not change significantly whether the insecticide was injected, applied in food or applied topically (Aikens and Wright, 1985).

Tolerance to an insecticide may also vary within an instar, thus in studies with the insect growth regulators, teflubenzuron and flufenoxuron, Fisk and Wright (1992a) showed that while the age distribution within an instar had little influence on the end point mortality, it can have a major impact on the speed of response. Thus, no response was observed when one-day-old third instar *Spodoptera* were treated until 48 h later, whereas two-day-old larvae showed some response after 24 h and considerably more were affected in 24 h, when the non-feeding pre-moult stage was treated.

Age of adult moths has been considered one of the problems in resistance studies using treated vials, so this test is unlikely to be used to monitor accurately resistance frequencies of field populations in Australia (Gunning, 1993).

SEX

Female insects often have a greater tolerance to insecticides, when tests are corrected for any differences in body weight. This may possibly be due to differences in their metabolism, so bioassays should be with all of one sex or equal proportions of each sex. The tolerance of male moths alone was assessed when developing a 'lure and kill' technique (De Souza *et al.*, 1992).

SIZE

Dosages of a drug are often related to body weight, although in many situations the dosage may be proportionately smaller with larger animals. There is no doubt that the larger insect pests do require more insecticide to be killed. In Australia, it was shown that an application rate designed to kill third instar larvae would also kill smaller larvae of resistant populations (Daly *et al.*, 1988).

RESISTANCE

Ismail and Wright (1991) showed that with a leaf-dip bioassay the end point mortality of *Plutella xylostella* varied from 9 to 17 days, being least with a laboratory-susceptible strain, compared to a field-collected strain resistant to certain insecticides. Therefore, if the end point is too soon, it may underestimate mortality and exaggerate the level of resistance.

10.3.2. Extrinsic factors

TEMPERATURE

Laboratory tests should be done at a constant temperature to facilitate comparison of data from tests carried out on separate occasions. Many tests are carried out at 25°C, but ideally a temperature similar to that likely to be encountered under field conditions should be selected. Most insecticides will be more effective at higher temperatures, for example in a tropical crop, but there are insecticides, such as the pyrethroids (Miller and Salgado, 1985), which have a negative temperature coefficient, whereby if treated insects are retained on a clean surface, more will die at a lower temperature. This effect is masked if insects are retained on treated surfaces as greater activity at a higher temperature will increase the amount picked up by the insects.

HUMIDITY

Under natural field conditions humidity will fluctuate significantly each day and affect the behaviour and survival of insects. Many insects avoid being in direct sunlight and remain on the undersurface of leaves or low within a crop canopy where the air is more humid. Provision of constant humidity is expensive, but retention of treated insects may require potted plants or a source of moisture to ensure the insects are not desiccated. Brodsgaard (1994) used air bubbled through deionized water to ventilate a glass cell and regulate the humidity in insecticide tests with thrips. Humidity can be controlled to some extent in small enclosed containers by using suitable saturated salt solutions (Stokes and Robertson, 1949), but total enclosure is inappropriate if the insecticide has any vapour action.

FOOD

The response of a test insect may be affected by whether or not it had recently fed. The amount and quality of food available to an insect culture will also affect growth and the ability to metabolize an insecticide. Feeding and time of treatment should therefore be standardized, and in some cases preference has been given to artificial diets as the quality of natural food may vary seasonally. Adequate provision of food is particularly important where undue mortality would occur in its absence. The toxicity of abamectin to fourth instar *Plutella xylostella* varied in relation to the cabbage cultivar on which the larvae were fed and the test method used (Abro and Wright, 1989).

POPULATION DENSITY

If a large number of insects are retained on a treated surface, mutual disturbance could increase activity and subsequent pick-up of insecticide. In some species cannibalism can occur and necessitate individuals being kept separate in culture as well as after a test treatment. Population density may be important in tests with natural enemies, although in studies on possible sublethal effects of an insect growth regulator teflubenzuron, Furlong *et al.* (1994) were unable to detect any significant changes in parasitism by *Diadegma semiclausum* or *Cotesia plutella* due to different densities of the second instar *Plutella xylostella*. However, changes in parasitism might be expected to occur with behaviourally active compounds, such as a pyrethroid.

ILLUMINATION

Most laboratory cultures are maintained with a set daylength, such as a 16:8 h light:dark cycle, in case diurnal rhythms affect tolerance to an insecticide. Ideally, all the insects should be treated at the same time of day. In some experimental work a low light intensity is required, for example in studies on aerosols to control mosquitoes, where the conditions should encourage flight activity.

10.3.3. Experimental factors

The choice of experimental technique will depend on the objective of the evaluation. Topical application of an insecticide may provide a more precise

dose with which to measure intrinsic toxicity, while tests using treated surfaces or direct spraying of insects will represent more closely what may
happen under field conditions. Measurement of intrinsic toxicity may be
required to compare different insecticides, but a range of other tests will be
required to establish the formulation and dosage rate for field use. This will
involve both laboratory orientated and semi-field tests.

The design of these tests needs to be standardized for comparisons of
different experiments, especially in terms of the number of insects used, the
number of replicates, the duration of the test and the method of determining
the response. An untreated control is needed to establish whether the
method of handling the insects has any effect on mortality. The data can be
corrected for a low level of mortality in the control, but natural mortality
should be kept to an absolute minimum. Apart from an untreated control it
is sometimes necessary to include another standard treatment. This may be
to check that the solvent used in a formulation without any active ingredient does not adversely affect the insects. In some cases it is important to
include a previously recommended insecticide to provide a reference point
to compare the new insecticides being evaluated as the susceptibility of a
batch of test insects may change with time.

When an insect population is suspected of developing resistance to an
insecticide, rather than do tests with a series of dosages to establish an LD_{50}
value, it is usual to assess the mortality obtained with a discriminating dose,
based on an expected mortality of > 90% in susceptible populations.

Mortality is not always easy to assess as there can be apparent mortality
or 'knockdown' from which an insect may recover some hours later. This
effect has been noted in particular with pyrethroids. A rapid response or
'knockdown' is needed in some pest situations, such as the control of vectors of disease, provided there is no recovery. Special criteria may be used to
determine whether the insects are still alive, thus the test insects may be
placed in a circular arena and only those that can escape are counted as still
alive. In contrast to the need for a quick response, end point mortality with
some of the newer types of insecticide, such as acylureas, may not occur
until after 120 h (Fisk and Wright, 1992b).

10.3.4. Experimental techniques

TOPICAL TESTS

A precise dose of insecticide is applied directly to a specific part of the outer
surface of an insect. The choice of volume of liquid, carrier liquid and site of
placement does vary between reports; for example 1 µl in acetone was
applied to the dorsal surface of the thorax (Elzan et al., 1994), whereas De
Souza et al. (1992) applied 0.78 µl in ethyl methyl ketone specifically to the
mesoscutum; other sites may be on the abdomen or in some resistance studies, topical application is to the left eye of moths (Forrester et al., 1993).

Small micropipettes can be used to dispense uniform volumes
(MacCuaig and Watts, 1968), but usually a pesticide dosage is applied using

a precision glass syringe fitted with a very fine hypodermic needle. This is attached to a micrometre to control the movement of the plunger, so that repeated regular dosing is possible. Forrester *et al.* (1993) used a small hand-held repeating dispenser with a 50 µl microsyringe for topical treatments in resistance studies. One of the earlier designs of manually operated microm-eter has a cylinder with five rings each of which has a different number of equally spaced depressions. The dose is regulated by the position of a spring-loaded ball that fits against the appropriate ring that is selected to adjust the amount applied from 0.25 up to 5 µl (Arnold, 1967). A more expensive version was designed to be electrically operated (Arnold, 1965) and provide doses in the range 0.1 to 1.0 µl. The latest version is electronical-ly controlled to deliver from 0.1 to 100 µl droplets. Other adaptations include the use of a twin-fluid nozzle to provide an air flow to detach a spray droplet from the needle and impact it on an insect (Hewlett, 1954, 1962) and enable small insects such as aphids to be treated (Needham and Devonshire, 1973).

Another specialized technique for assessing the effect of insecticides applied as an aerosol to control tsetse flies was developed using a vibrating orifice droplet generator (Berglund and Liu, 1973) to deliver an aerosol into a low-speed wind tunnel. The highly volatile carrier in the insecticide for-mulation evaporated so that droplets (10–25 µm diameter) of field composition could be collected on silk threads (*c.* 2 µm diameter) and subse-quently transferred to tsetse flies (Johnstone *et al.*, 1989 a, b).

SYSTEMIC ACTIVITY

Special tests are needed to assess whether an insecticide is systemically transferred through the xylem to the leaves to control sucking pests, espe-cially aphids and whiteflies. In initial trials, a seedling can be removed from soil and the cut tap root immersed into a tube containing pesticide solution, so that the effect of adsorption of the pesticide on soil particles is removed. Test insects can be caged on the leaves. Ultimately a formulation such as a granule or seed treatment of a new insecticide has to be assessed with pot-ted plants to ensure that there is adequate transfer of the active ingredient to the roots. A recent trial compared the activity of a nitromethylene insecticide (imidacloprid) seed treatment with granular insecticides for the protection of maize from false wireworm (*Somaticus* spp.) damage (Drinkwater, 1994). Uniformity of the potting soil is important to ensure different trials on dif-ferent occasions can be compared.

RESIDUAL FILMS

A simple method of assaying a residual deposit of insecticide is to prepare a series of dilutions in a suitable carrier and pipette equal volumes of each dilution on a set of filter papers placed on a non-absorbent surface, such as a petri dish. Glass equipment may be used provided it is adequately cleaned with an appropriate solvent before use, but preference is often given to dis-posable plastic equipment. Sometimes the insecticide solution is placed directly on the glass surface, for example in the tarsal plate test for moths (Forrester *et al.*, 1993). In tests with both pests and parasitoids (Plapp and Vinson, 1977) and in resistance studies (Plapp *et al.*, 1987), insecticides dissolved in acetone have also been pipetted (usually 0.5–1.0 ml) into 20 ml glass scintillation counting vials, manually rotating the vials on their sides until the solvent has evaporated. Insects can be placed on the treated surfaces after sufficient time has been allowed for any solvent to volatilize and prevented from escaping by covering with an untreated lid, preferably ventilated, using open metal rings treated with Fluon (polytetrafluoroethylene) or the circumference of the container may be treated with Fluon, depending on the insect, the stage of development and its behaviour.

In some tests the surface may be treated by dipping. A slide dip technique (Voss, 1961; Dittrich, 1962) in which a strip of double-sided adhesive tape is attached to a microscope slide, has been used to evaluate acaricides. Using a fine brush, 50 mites were transferred to the adhesive surface and stuck dorsal side down in 5 rows of 10 mites. The slide was dipped for 5 s in serial dilutions of acaricide before being drained and retained at 27°C and 95%RH. Mortality was recorded at 24, 48 and 72 h. The technique has also been used to assess effects of insecticides on predaceous mites (Croft *et al.*, 1976), and modified for use with aphids (Kerns and Gaylor, 1992). In another simple test individual *Diparopsis castanea* eggs were dipped in commercial formulations diluted in distilled water and because the first instar larva has to eat its way from the egg, efficacy was assessed in terms of whether the first instar larvae were able to hatch from the egg (Matthews, 1966).

Bishop and Grafius (1991) describe a test in which Whatman No. 1 9 cm filter paper was dipped for 5 s in formulated insecticide dilutions and then air-dried before placing on the bottom of a disposable petri dish. Instead of an artificial surface, Elzan *et al.* (1992), using a camel hair brush, transferred five neonate (<1-day-old) larvae of tobacco budworm into a small plastic cup over which a leaf, previously dipped in a distilled water solution of insecticide for 20–30 s , was placed. The leaf was then covered with wet cotton wadding and a wax-coated paper lid to prevent desiccation. Leaf dipping has been selected for resistance studies with diamondback moth (Tabashnik and Cushing, 1987) and bollworms (Perrin and Löwer, 1994). As suggested earlier the choice of method should be to simulate as near as possible the field situation, but with greater uniformity to allow more valid comparisons between treatments to be made. Where different test methods have been compared, for example in tests with the diamondback moth, similar resistance ratios have been obtained with topical, leaf residue and filter paper methods (Zhao and Grafius, 1993), but the test method might influence actual toxicity, for example abamectin was much more toxic on glass compared with a leaf surface against *Cotesia plutella* (Fauziah, 1990). In comparisons by Perrin and Löwer (1994), the adult vial test was good where a rapid result was required, whereas the topical application tests using an accurate dose to each insect was less suitable where facilities are limited.

The leaf dipping test with second instar larvae was the simplest and easiest to use provided there was a good supply of uncontaminated leaves.

The period of exposure is often 24 h, but shorter exposure periods are often more relevant. An exposure of 1 min is used in certain tests with disease vectors, such as mosquitoes, to simulate the short period that they may be exposed to a surface deposit in the field. In the development of a lure and kill technique of combining an insecticide with a pheromone, the exposure was limited to only 10 s when 3-day-old *Spodoptera littoralis* moths were walked over a treated surface to mimic moths just touching an insecticide with their tarsi (De Souza *et al.*, 1992)

TECHNIQUES OF SPRAYING SURFACES

A number of systems have been devised to treat a surface with a consistent measured dosage of insecticide. Early work by Potter (1941, 1952) led to the development of a spray tower designed to minimize air turbulence and reduce the amount deposited on the sides of the tower. A twin-fluid nozzle was used mounted centrally at the top of an open-ended metal tube, so that the spray fell vertically and was deposited on a horizontal plate. A typical example of its use was spraying fertile bollworm eggs of filter paper discs to expose first instar larvae to deposits on hatching (Gunning, 1993). Edwards *et al.* (1994) covered sections of a leaf surface with wax before spraying to produce a checkerboard-like discontinuous residue on the leaf. Recently a computer controlled spraying apparatus has been designed using a similar twin-fluid nozzle, the output from which can be by adjustment of nozzle volume control needle and/or the air pressure (Fig. 10.1). The compact unit

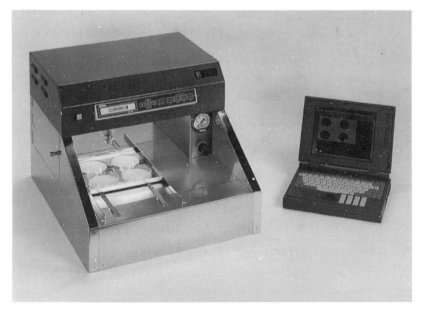

Fig. 10.1. Computer controlled spraying apparatus.

can be used inside a fume cupboard and operated through a software package, such as Microsoft Paintbrush (Arnold, personal communication). Morgan and Pinniger (1987) described a small sprayer incorporating a car windscreen-wiper assembly to move the spray nozzle to give an even deposit on surfaces up to 27 cm in diameter.

In most tests in a Potter tower or equivalent, the spray volume is comparable with field applications of around 200 l ha^{-1}, but the droplets produced by the twin-fluid nozzle are in the aerosol and mist size range (<100 μm diameter) compared with the wider droplet spectra produced by conventional hydraulic nozzles (Fig. 10.2) (Matthews, 1994). In order to assess the importance of droplet size, Munthali and Scopes (1982) adapted a twin-fluid nozzle described originally by Uk (1978) and adapted by Coggins and Baker (1983) so that monosized droplets could be placed on leaf surfaces. A fluorescent tracer was added to the spray so that the deposition of individual droplets could be seen under ultraviolet light. A range of droplet densities was achieved by moving the leaf manually under the spray for different periods of time. The equipment is useful in studying the factors which influence the number of droplets required on a treated surface (Adams et al., 1987). Similar experiments have used a Berglund and Liu droplet generator modified by removing the air column and using orifice plates with larger apertures to produce uniform droplets in the 50–500 μm diameter range (Reichard et al., 1987). Voltage to a piezo-electric disc has

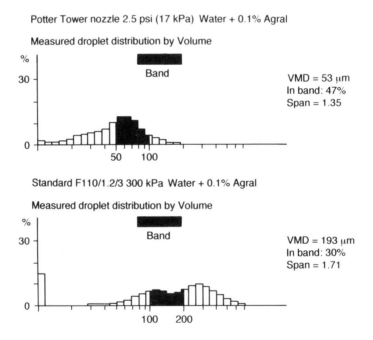

Fig. 10.2. Histograms to compare droplet spectra from airshear nozzle used in Potter tower and from fan nozzle used in field experiments.

been used to pulse liquid from a chamber through an fine orifice (Young, 1986; Reichard, 1990; Womac *et al.*, 1992) to produce monosized droplets as small as 60 μm in diameter. In tests with *Trichoplusia ni,* 100 μm droplets of a pyrethroid were more toxic than larger droplets (Hall and Thacker, 1994). Linked with appropriate software, commercial equipment is now available to produce either a single or series of droplets on demand (Thacker *et al.*, 1995). Well-filtered solutions are needed when the smallest apertures are used.

In some studies an electrically driven spinning disc has been used to produce a narrow range of droplet sizes. Rotary devices have been used for some time, for example one device, described by Rayner and Halliburton (1955), was modified by Sundaram *et al.* (1991) so that a rotating point was fed from a microsyringe. Control of the flow rate and disc speed enables droplet size to be adjusted, and if the disc is shrouded so that spray is emitted through a slit, the droplets can fall as a fan-shaped curtain on the target surface. Droplets from about 40 μm diameter upwards can be produced. The target can be mounted on a moving belt or trolley. Bateman (1994) used a spinning disc to apply oil formulations of *Metarhizium flavoviride* to locusts.

SPRAY CHAMBERS

One of the problems confronting those evaluating insecticides is how, on the basis of the laboratory data, the effects in the field can be predicted. Field investigations inevitably take a long time from planning to final analysis and due to changes in weather and variability in the incidence of insects, the results may be inconclusive. Therefore, the intermediate step between laboratory screening to field trials is crucial to study a range of factors, such as dosage rate, formulation changes and application parameters to limit the number of treatments that have to be taken through to the later stages of evaluation under practical field conditions (Fig. 10.3). In this section a number of spray chambers, designed to evaluate insecticides in a way which simulates field conditions, are described. Health and safety regulations require the whole area in which the spray is applied to be enclosed in a chamber so that the operator is not exposed to the insecticide, and the treated area can be subsequently washed down to remove deposits from the inner surfaces of the chamber.

The simplest spray chambers have a nozzle, such as a twin-fluid atomizer mounted to direct spray towards a plant placed on a turntable so that the foliage is completely wetted ('runoff' spray). This does not provide a very accurate simulation of field application so the trend has been to have a gantry along which one or more nozzles may be traversed across potted plants. The linear transporter of one type of spray chamber (the 'Mardrive' track sprayer) has a sealed tube in which a small polymer-bonded slug referred to as a 'mole' is pushed to and fro by compressed air. A shuttle mounted on rollers is moved along the tube by a set of permanent magnets, which keep in step with the 'mole' by magnetic coupling. Spray can be applied with one or more conventional hydraulic nozzles or the sprayer adapted to use a spinning disc. Fisk and Wright (1992b) used hollow cone (Conejet 6) nozzles to treat maize and Chinese cabbage, but in their simulation of an ultra-low volume spray they mounted a spinning disc nozzle (Micro Ulva) with the use of a ventilation fan to disperse spray droplets down the chamber over plants with and without larval infestation.

A larger spray chamber incorporating a wind tunnel (12.6 m long, 3.6 m wide and 2.7 m high) was designed with a ceiling mounted single-axis beam carrying a chain-driven module on which nozzles can be mounted (Hislop, 1989). A 3 kW electric motor with a 4 kW variable speed motor controller allows continuous adjustment of speed of travel between 0.5 and 6.0 m s^{-1} in either direction along the track. Sensors are used to measure the actual speed to an accuracy of 0.01 m s^{-1} over a 5 m section in the area used for spray application. Wind speeds along the chamber can be varied between 1 and 10 m s^{-1} so that sprays can be applied either with or against the direction of the wind. This unit has been used for a number of spray application studies, including assessment of airborne drift with different nozzles and airspeeds (Miller *et al.*, 1993). The study of airborne drift is crucial in relation to effects of insecticides on natural enemies in hedgerows, which can filter some of the spray drift. In particular spider webs are efficient collectors of spray (Samu *et al.*, 1992). A tunnel similar to that described by Hislop, arranged with different types of nozzles and the capability of applying electrostatic sprays, has been used for studies with semiochemicals and insecticides (Pye, personal communication). One example is reported with a mycopesticide *Metarhizium flavoviride* applied to a non-target organism – bees (Ball *et al.*, 1994). Analysis of some aerial spray operations in forests has been simulated using potted trees in a spray chamber (Sundaram, 1991).

The nozzle used in the spray chamber needs to be selected carefully. The choice is often dictated by the need to apply the same volume rate as used in the field, but, if the speed of travel in the laboratory spray chamber is less, a smaller orifice nozzle may be used which applies a different droplet spectrum (Matthews, 1994). In contrast to measurements of spray distribution under static conditions, the pattern will also be influenced by the speed of travel, with the smallest droplets entrained in the air vortices created by the spray.

An important aspect of laboratory studies is to have plants that mimic field conditions. Often plants grown in glasshouses are used, but it has been argued that because of the differences in temperature and other factors, the surface characteristics of leaves are different to those grown outside. In the spray chamber described by Hislop (1989) it was possible to use cereal crops grown in trays outdoors so that the leaf surfaces resembled field plants in terms of wettability, retention of spray droplets and uptake of a pesticide.

A more specialized spray cabinet was devised to assess the effects of a simulated aerial spray on several generations of a population of whiteflies on cotton plants (Rowland *et al.*, 1990). A spinning disc sprayer was fitted above a cage in which cotton plants were maintained, so that the response of susceptible and resistant populations to insecticides could be compared (Sawicki *et al.*, 1989). Sprays could be applied with different droplet sizes and by using an endoscope the numbers of adult whiteflies could be assessed on leaves without interfering with the plants or insects. The equipment also allowed the effect of sprays on mixed populations of whiteflies and natural enemies to be studied.

Space treatment with insecticide droplets <50 µm diameter may be used to control flying insects in glasshouses, animal houses and warehouses, and in some circumstances outdoors, for example in forestry and mosquito control. Tests have been devised especially to assess the efficacy of pressure packs used for fly control. In a typical standard specification, based on studies by Goodwin-Bailey *et al.* (1957), an operator continuously discharges a pressure pack while walking down the long axis of a test-room of about 50 m³ capacity, but smaller chambers are used. The room is uniformly illuminated by fluorescent lights to provide a light intensity of 108 mcd when measured 1 m above the floor. A temperature of 27±1°C and relative humidity of 50±5% should be maintained and the room ventilated between tests by a fan displacing air at a rate of at least 10 m³ min⁻¹. The floor should be covered with a new layer of absorbent paper for each test and deposition on the walls and ceiling avoided, positioning the nozzle at least 1 m from any surface. The pressure pack is weighed before and after spraying to determine the amount discharged. A test may involve the release of 500 flies at floor level. To avoid entering the area treated, some users have a 25 m³ room with transparent panels inside a larger room so that the number of flies 'knocked down' on a grid demarcated on the floor can be assessed at 2 min intervals without entering the room. Later, after adequate ventilation, the flies are collected and placed in clean containers with food so that total mortality can be recorded after 24 h.

Individual clear polythene chambers (2250 × 2000 × 2250 mm) with a plastic zip fitted centrally in the front to allow access, have been used to test space sprays (Learmount, 1994). A mesh is fitted in the back wall to allow ventilation, but is covered by a black plastic sheet during the entry of spray from an airbrush nozzle operated at about 2.5 bar pressure and directed through the zipper.

10.4. Statistical Tests

Techniques to analyse the results of laboratory bioassays were developed originally by Bliss (1935) and subsequently refined by Finney (1971). Hewlett and Plackett (1979) pointed out that transformation of mortalities into logits rather than probits was more easily done with modern programmable calculators. A number of computer software packages have a facility for analysis of bioassays, but often lack the ability of the original techniques to adjust for heterogeneity in the data. Robertson and Preisler (1992) give an interesting discussion of the design of bioassays and their analysis. In particular they discuss the analysis where time of response is a factor. This is particularly important where biological agents such as entomopathogens are used and mortality may occur several days after treatment (Prior *et al.*, 1996).

The use of the formula devised by Abbott (1925) to correct for control mortality has been standard practice, but there are modifications to correct for skewness of data and provide confidence limits (Gart and Nam, 1988; Rosenheim and Hoy, 1989; Koopman, 1994).

10.5. Field Trials

10.5.1. Small-scale trials

The very high cost of field trials indicates that there may be situations where some small-scale investigations are needed to define more precisely what treatments need to be included in a formal field trial. It is possible to grow most plants to a reasonable size in pots and treat these with a small-scale plot sprayer. Such plants can be artificially infested before or after treatment. These trials can investigate a series of dosages and effects of rainfall on different formulations or adjuvants. Using an artificial rainfall simulator, Taylor and Matthews (1986) showed that an oil adjuvant based on rapeseed oil increased rainfastness of an insecticide on brassica foliage. In studies with transgenic cotton, plant material sprayed in the field was moved to the laboratory to assess the effects on *Heliothis virescens* (Jenkins *et al.*, 1993).

In locust control studies with *Metarhizium flavoviride*, Bateman *et al.* (1992) used the decrease in numbers of droplets deposited at different distances downwind from a line source of spray to do a field bioassay with laboratory bred *Schistocerca gregaria* placed on foliage and recaptured after spraying for subsequent mortality assessments.

10.5.2. Field trials

Typically, having selected a particular insecticide, formulation and probable dosage from the small-scale studies, its performance is investigated in a field trial, which consists of a small number of replicated treatments in a randomized block design. An unsprayed 'check' plot should be included to show the extent to which yields can be improved. However, the usefulness of untreated plots in these trials has been questioned, because they tend to yield more than larger untreated areas not adjacent to treated plots (Reed, 1972). Alternatively the 'check' can be another standard insecticide treatment.

Every effort should be given to collecting adequate insect and plant growth/crop data throughout the trial to assess the contribution of insect control to any improvement in yields. This means that plots need to be sufficiently large so that there is enough area within the peripheral guard section to allow sampling over an extended period without damage to the plants. Access paths between plots are important to facilitate routine sampling. In addition to sampling of pest species, information on natural enemies should also be obtained if possible, but as detailed sampling of these is often more difficult, such studies may be better on large-scale trials. The operators should be allocated at random to the sprayers and treatments on each spray date so that the results are not biased in any way due to the efficiency of a particular person. Another aspect, often neglected, is a check on the insecticide distribution in relation to where the insect is on the plants. This can be checked in supplementary trials, for example by using water-sensitive cards or some other sampling technique (Cooke and Hislop, 1993). In large-scale aerial spray trials against the tsetse flies in Africa, Johnstone *et al.* (1989a) studied the collection efficiency of aerosol droplets on different size cylinders to simulate the resting sites of flies in the natural environment to assist interpretation of spray monitoring in the field.

Apart from considerations of sampling, choice of plot size is important in relation to the mobility of pests and spray droplets. Too small a plot may result in droplets drifting across neighbouring plots, even if a shield is held along the downwind edge during treatment. A shield, if used, should be fairly porous and act as an efficient filter, otherwise the air will be deflected over the top of the shield carrying the smallest airborne droplets, most prone to drift. The extent to which downwind drift is a factor will also depend on the method of applying the insecticide. Clearly quite small plots are acceptable if there is a placement technique such as granule application or seed treatment. Where knapsack sprayers were used with nozzles placed in the inter-row, plots of 10 × 10 m were satisfactory with a range of cotton pests, but if the nozzle is held above the crop canopy, and drift can occur, either a larger plot should be considered or a wider guard area used. Where downwind displacement of the spray is expected as part of the technique, for example when hand-carried spinning disc sprayers are used, plots 30 × 30 m are required. Square plots are preferred to long rectangular plots as the effect of spray drift across the latter will be more pronounced if there is a cross-wind.

Where mobility of insects is concerned, Joyce and Roberts (1959) had demonstrated that jassids moving over long distances could have an inter-plot effect even when 4.2 ha plots were separated by 450 m. In assessments of insecticides to control locusts, the minimum plot size recommended using ground equipment is 1 ha, whereas aerial spraying always demands large plots as spray deposition is less controlled. Where techniques, such as 'lure and kill' and the use of viruses, are being examined, area-wide treatments are essential, for example when treating alternate hosts to reduce the population of the cotton bollworm and tobacco budworm (Bell and Hayes, 1994).

Even when a field trial has been conducted, it is preferable to assess the proposed recommended treatment on a large-scale farm trial, where the replicates are the individual farms. Such trials need to be kept as simple as possible and may only consist of two treatments: the farmer's normal practice and the treatment that research suggests will be an improvement. Plots should be as large as possible. A series of farm-scale extension trials in Central Africa initially compared untreated with knapsack sprayed cotton and later compared knapsack treatments with a ULV technique (Tunstall and Matthews, 1966; Matthews, 1973).

10.5.3. Design of trials

Before embarking on any field trial, its aims and objectives need to be clearly defined and advice on the trial design should be sought. While there are many statistical books, such as Gomez and Gomez (1976) and Pearce (1976), discussion with a biometrician at the outset is invaluable to avoid difficulties with analysis at the end of the trial. In the context of integrated pest management, the efficacy of an insecticide has to be considered in relation to many other factors, such as host plant resistance, the impact of natural enemies and interactions with different cropping practices (Reed *et al.*, 1985). Randomized block trials are not always suitable and a factorial design incorporating factors, such as different volume or dosage rates and swath widths, or split plots in which the main plots may be the spray treatment, split for other factors, such as crop cultivar. An integrated control programme may need to study particular strategies involving two or more control tactics, such as the combination of antifeedant with an insect growth regulator (Griffiths *et al.*, 1991) or the use of an aggregating pheromone with an antifeedant (Smart *et al.*, 1994). There is particular interest now in the use of transgenic plants containing genes to express insecticidal activity to avoid application of insecticides, but as with other studies on host plant resistance, care must be taken with small-plot experiments as behavioural preferences may affect the results in comparison with larger plots.

10.5.4. Spraying equipment

As field trials are on relatively small areas, a number of specialized sprayers have been developed as plot sprayers. Many of the manually carried sprayers use an adaptation of a compression sprayer using pressurized gas to force the insecticide spray through one or more hydraulic nozzles (Fig. 10.3), often mounted on an off-set boom so that the operator can walk alongside the plot (Matthews, 1984, 1994). A pressure regulator is used to provide constant pressure at the nozzles, which should be fitted with a diaphragm check valve to prevent any insecticide spray dripping from the lance or boom at the edge of a plot. Some have been adapted to apply quite small amounts of experimental material. Many of these plot sprayers were developed principally for herbicide application with the operating pressure set at 1–2 bar, but when used at a higher pressure, with appropriate nozzles, the equipment can be used for insecticide trials. Some experimenters have used motorized knapsack sprayers (Rutherford, 1985), while Robinson (1985)

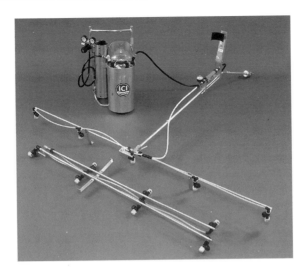

Fig. 10.3. Equipment for small plot trials (Zeneca photo).

used a small electrically-driven air compressor to maintain a constant pressure in the pesticide container. However, in many cases a standard knapsack lever-operated sprayer has been used, but in addition to variation in walking speed, differences in speed of pumping can affect the output of spray. The addition of a spray management valve (Matthews and Thornhill, 1993) to ensure constant pressure at the nozzle removes some of this variability.

Where water supplies are scarce, reduced volume spraying against cotton bollworms, locusts, and armyworms has been evaluated initially using hand-carried spinning disc sprayers (Matthews, 1973; Fisk *et al.*, 1993; Bateman *et al.*, 1994), and vehicle mounted equipment (Symmons *et al.*, 1989; Hewitt and Meganasa, 1993).

10.5.5. Mechanized plot sprayers

Walking speed can vary between operators, so small mechanized plot sprayers are used where possible. These are now designed to minimize operator contact with the pesticide to meet more stringent health and safety requirements. Slater *et al.* (1985) described a single-wheeled motorized sprayer which could also be used as a granule applicator on small plots. An outdoor version of the 'MarDrive' system mounted on a tractor for moving the nozzle across a small plot within a shielded enclosure was described by French (1980), while Skurray (1985) designed a self-propelled gantry which eliminated the need for a passage for the tractor alongside the plots. Crabtree (1993) adapted a hedge cutter arm to mount an off-set boom fitted with a separate array of nozzles, protected within a shield, for each treatment (Fig. 10.4). The spray for each treatment was pre-mixed and kept in

Fig. 10.4. Plot sprayer using adaptation of hedge cutter arm (Crabtree – Zeneca photo).

separate containers, so that several treatments could be applied rapidly in the field. While this equipment has been used primarily for treating cereal crops, such as wheat, it could be adapted for other low row crops. In trials with this equipment, plot size, especially width of the plot, is often determined by the mechanized harvesting equipment being used.

In contrast to the usual volume application rates, extremely high volumes are used in some irrigation systems when pesticides are injected into the irrigation water. Sumner *et al.* (1991) describe field trials to evaluate the effect of nozzle type and other variables on control of *Spodoptera frugiperda* when applying chlorpyrifos insecticide in a sprinkler type centre-pivot irrigation system.

10.5.6. Granule application

Seed treatment and granule application are two other options for more precise placement of insecticides. Precision granule-metering belts were fitted to a precision seed-spacing drill to allow accurate delivery of granules and their incorporation into the soil while sowing (Thompson *et al.*, 1981; Thompson and Wheatley, 1985). Information on techniques of commercial quantities of seed has been published (Jeffs, 1986; Clayton, 1993), but for small samples used in trials, simple mixing of seed with an appropriate formulation of insecticide has usually been carried out in the laboratory. Where solvents have been used, careful volatilization of the solvent is needed under controlled safe conditions.

10.5.7. Sampling and recording data

Adequate sampling of the pests and natural enemies in field trials presents considerable problems in terms of costs, labour needed, interference in the crop and statistical requirements. With increased emphasis on integrated pest management, effects of insecticides on populations of natural enemies and other non-target organisms need to be evaluated as well as the insect pests. Vickerman (1985) discusses sampling plans needed to assess beneficial arthropods in cereal crops using pitfall traps, quadrats and D-vac samples. New equipment for suction sampling has been designed using a vortex to eliminate the need for a net to collect the sample (Arnold, 1994). Hand-held data loggers (Hunt and Raven, 1985) have been used for recording extensive cultivar field trial data that are then transferred to a computer for analysis (Reeves and Law, 1985). Pike and Harris (1985) describe their development of recording yield data, while Hunt and Taylor (1985) developed a management system for handling trials data.

10.6. Discussion

Mortality of pest species was the most immediate concern when assessing the development of neurotoxic insecticides, but with the increasing realization of adverse side-effects in the environment, evaluation techniques have to consider a broader range of effects, especially in relation to the effects on non-target organisms and natural enemies where reduced oviposition could be a critical factor. Therefore, in the development of evaluation techniques, it is important that sufficient observations are made in order that unusual events are not missed. With tests on acylurea insecticides, some larvae apparently survived a moult, but subsequently died of starvation when mandibles were weakened by insufficient muscle attachments (Fisk and Wright, 1992a).

In selecting suitable insecticides for integrated control programmes, some selectivity is clearly important, hence the interest, for example, in specific baculoviruses, but in many crop situations, a farmer has to control a pest complex, or the slower speed of action of a more selective insecticide may be unacceptable. Therefore the tests in the laboratory need to examine as far as possible the range of activity that is likely to be of concern in future field evaluations. Thus, the more selective action of many acylureas is of particular interest in IPM programmes, but possible adverse effects on the immature stages of some natural enemies and on non-target organisms, such as crustacea, must also be considered. The latter is one of many environmental concerns that require specific evaluation using, in some cases, specialized tests. However, under field conditions, such adverse effects can be mitigated by observance of untreated barrier zones around sensitive areas (Riley *et al.*, 1989).

Continued evaluation of insecticides is a key part of the monitoring of resistance in field populations to provide data on which sound management decisions can be made. Much emphasis has been put on the need for a simplified test to evaluate the level of resistance in samples of pest populations collected in the field. Immunoassays have been devised to measure insecticide-degrading esterases of individual insects, such as aphids (Devonshire *et al.*, 1986) and these allow more extensive surveys of the incidence of resistance, for example using aphids collected in suction traps (Tatchell *et al.*, 1988). However, more detailed testing to understand the mechanisms of resistance is essential. There is also a need to look at the wider context of resistance to insecticides by examining whether some other factor such as host plant resistance can enhance parasitism of the pest and thus reduce the number of sprays required and minimize selection for resistance (Verkerk and Wright, 1994).

This chapter can only refer to some of the main techniques used, among the many different approaches to the evaluation of insecticide efficacy. The selection of appropriate methods of laboratory, glasshouse and field trials depends very much on the ultimate aim of the studies and selection criteria used (Krahmer and Russell, 1994). With the escalation of costs, there is a risk of simplified standard tests, whereas investment may be necessary in more detailed evaluations to understand the more complex interactions that occur, for example, between pests and their natural enemies and host plant resistance, especially with the introduction of transgenic crops.

10.7. References

Abbott, W.S. (1925) A method of computing the effectiveness of an insecticide. *Journal of Economic Entomology* 18, 265–267.

Abro, G.H. and Wright, D.J. (1989) Host plant preference and the influence of different cabbage cultivars on the toxicity of abamectin and cypermethrin against *Plutella xylostella* Lepidoptera: Plutellidae. *Annals of Applied Biology* 115, 481–487.

Adams, A.J., Abdalla, M.R., Wyatt, I.J. and Palmer, A. (1987) The relative influence of the factors which determine the spray droplet density required to control the glasshouse whitefly, *Trialeurodes vaporariorum*. *Aspects of Applied Biology* 14, 257–266.

Aikins, J.A. and Wright, D.J. (1985) Toxicity of DDT and malathion to various larval stages of *Mamestra brassicae* L. *Pesticide Science* 16, 73–80.

Arnold, A.J. (1965) A high-speed automatic micrometer syringe. *Journal of Scientific Instruments* 42, 350–351.

Arnold, A.J. (1967) Hand-operated micro-applicator to deliver drops of five sizes. *Laboratory Practice* 16, 56–57.

Arnold, A.J. (1994) Insect suction sampling without nets, bags or filters. *Crop Protection* 13, 73–76.

Ball, B.V., Pye, B.J., Carreck, N.L., Moore, D. and Bateman, R.P. (1994) Laboratory testing of a mycopesticide on non-target organisms: the effects of an oil formulation of *Metarhizium flavoviride* applied to *Apis mellifera*. *Biocontrol Science and Technology* 4, 289–296.

Bateman, R. (1994) Performance of myco-insecticides: Importance of formulation and controlled droplet application. *British Crop Protection Council Monograph* 59, 275–284.

Bateman, R.P., Godonou, I., Kpindu, D., Lomer, C.J. and Paraiso, A. (1992) Development of a novel field bioassay technique for assessing mycoinsecticide ULV formulations. In: Lomer, C.J. and Prior, C. (eds) *Biological Control of Locusts and Grasshoppers*. CAB International, Wallingford.

Bateman, R.P., Price, R.E., Muller, E.J. and Brown, H.D. (1994) Controlling brown locust hopper bands in South Africa with a myco-insecticide spray. *Brighton Crop Protection Conference – Pests and Diseases*, pp. 609–616.

Bell, M.R. and Hayes, J.L. (1994) Areawide management of cotton bollworm and tobacco budworm (Lepidoptera: Noctuidae) through application of a nuclear polyhedrosis virus on early-season alternate hosts. *Journal of Economic Entomology* 87, 53–57.

Berglund, R.N. and Liu, B.Y.H. (1973) Generation of monodisperse aerosol standards. *Environmental Science and Technology* 7, 147–153.

Bishop, B.A. and Grafius, E. (1991) An on-farm insecticide resistance test kit for Colorado potato beetle. *American Potato Journal* 68, 53–54.

Bliss, C.I. (1935) The calculation of the dosage–mortality curve. *Annals of Applied Biology* 22, 134–167.

Brodsgaard, H.F. (1994) Insecticide resistance in European and African Western Flower Thrips (Thysanoptera: Thripidae) tested in a new residue on glass test. *Journal of Economic Entomology* 87, 1141–1146.

Busvine, J.R. (1971) *A Critical Review of the Techniques for Testing Insecticides.* Commonwealth Institute of Entomology, London.

Carter, N. (1992) A European guideline for testing the effects of plant protection products on arthropod natural enemies. In: Brown, R.A., Jepson, P.C. and Sotherton, N.W. (eds) *Interpretation of Pesticide Side Effects on Beneficial Arthropods. Aspects of Applied Biology* 31, 157–163.

Clayton, P.B. (1993) Seed treatment. In: Matthews, G.A. and Hislop, E.C. (eds) *Application Technology for Crop Protection*. CAB International, Wallingford, pp. 329–349.

Coggins, S. and Baker, E.A. (1983) Microsprayers for the laboratory application of pesticides. *Annals of Applied Biology* 102, 149–154.

Cooke, A.S. (ed.) (1993) *The Environmental Effects of Pesticide Drift*. English Nature.

Cooke, B.K. and Hislop, E.C. (1993) Spray tracing techniques. In: Matthews, G.A. and Hislop, E.C. (eds) *Application Technology for Crop Protection*. CAB International, Wallingford, pp. 85–100.

Crabtree, J.H. (1993) The development of a tractor mounted field trials sprayer. *Proc. International Symposium on Pesticide Application*. ANPP/British Crop Protection Council, Strasbourg, pp. 661–668.

Croft, B.A., Briozzo, J. and Carbonell, J.B. (1976) Resistance to organophosphate insecticide in a predaceous mite *Ambelyseius chilenensis*. *Journal of Economic Entomology* 69, 563–565.

Daly, J.C., Fisk, J.H. and Forrester, N.W. (1988) Selective mortality in field trials between strains of *Heliothis armigera* (Lepidoptera: Noctuidae) resistant and susceptible to pyrethroids: Functional dominance of resistance and age class. *Journal of Economic Entomology* 81, 1000–1017.

De Souza, K.R., McVeigh, L.T. and Wright, D.J. (1992) Selection of insecticides for lure and kill studies against *Spodoptera littoralis* (Lepidoptera: Noctuidae). *Journal of Economic Entomology* 85, 2100–2106.

Devonshire, A.L., Moores, G.D. and ffrench-Constant, R.H. (1986) Detection of insecticide resistance by immunological estimation of carboxylesterase activity in *Myzus persicae* (Sulzer) and cross reaction to antiserum with *Phorodon humili* (Schrank) (Hemiptera: Aphididae). *Bulletin of Entomological Research* 76, 97–107.

Dittrich, V. (1962) A comparative study of toxicological test methods on a population of the two-spotted spider mite (*Tetranychus telarius*) *Journal of Economic Entomology* 55, 644-651.

Dohman, G.P. (1994) The effect of pesticides on beneficial organisms in the laboratory and in the field. *British Crop Protection Council Monograph* 59, 201–210.

Drinkwater, T.W. (1994) Comparison of imidacloprid with carbamate insecticides, and the role of planting depth in the control of false wireworms, *Somaticus* species, in maize. *Crop Protection* 13, 341–345.

Edwards, M.H., Kolmes, S.A. and Dennehy, T.J. (1994) Can pesticide formulations significantly influence pesticide behaviour? The case of *Tetranychus urticae* and dicofol. *Entomologia Experimentalis et Applicata* 72, 245–253.

Elzan, G.W., Leonard, B.R., Graves, J.B., Burris, E. and Micinski, S. (1992) Resistance to pyrethroid, carbamate and organophosphate insecticides in field populations of tobacco budworm (Lepidoptera: Noctuidae) in 1990. *Journal of Economic Entomology* 85, 2064–2072.

Elzan, G.W., Martin, S.H., Leonard, B.R. and Graves, J.B. (1994) Inheritance, stability and reversion of insecticide resistance in tobacco budworm (Lepidoptera: Noctuidae) field populations. *Journal of Economic Entomology* 87, 551–558.

Fauziah, I. (1990) Studies on the resistance to acylurea compounds in *Plutella xylostella* L. (Lepidoptera: Yponeumeutidae). PhD thesis, University of London. 259pp.

Finney, D.J. (1971) *Probit Analysis*. Cambridge University Press, Cambridge.

Fisk, T. and Wright, D.J. (1992a) Comparative studies on acylurea insect growth regulators and neuroactive insecticides for the control of the armyworm *Spodoptera exempta* Walk. *Pesticide Science* 35, 175–182.

Fisk, T. and Wright, D.J. (1992b) Response of *Spodoptera exempta* (Walk.) larvae to simulated field spray applications of acylurea insect growth regulators with observations on cuticular uptake of acylureas. *Pesticide Science* 35, 321–330.

Fisk, T., Cooper, J. and Wright, D.J. (1993) Control of *Spodoptera* spp. using ULV formulations of the acylurea insect growth regulator, flufenoxuron: Field studies with *Spodoptera exempta* and effect of toxicant concentration on contact activity. *Pesticide Science* 39, 79–83.

Ford, M.G. and Salt, D.W. (1987) Behaviour of insecticide deposits and their transfer from plant to insect surfaces. In: Cottrell, H.J. (ed.) *Pesticides on Plant Surfaces* Wiley, London pp. 26–81.

Forrester, N.W., Cahill, M., Bird, L.J. and Layland, J.K. (1993) Management of pyrethroid and endosulfan resistance in *Helicoverpa armigera* (Lepidoptera: Noctuidae) in Australia. *Bulletin of Entomological Research*. Supplement 1.

French, P. (1980) A mechanized field sprayer for small plot pesticide trials. *Proc 5th International Conference on Mechanization of Field Experiments*, pp. 135–142.

Furlong, M.J., Verkerk, R.H.J. and Wright, D.J. (1994) Differential effects of the acylurea insect growth regulator teflubenzuron on the adults of two endolarval parasitoids of *Plutella xylostella, Cotesia plutellae* and *Diadegma semiclausum*. *Pesticide Science* 41, 359–364.

Gart, J.J. and Nam, J. (1988) Approximate interval estimation of the ratio of binomial parameters: A review and corrections for skewness. *Biometrics* 44, 323–338.

Giles, D.P. (1989) Principles in the design of screens in the process of agrochemical discovery. *Aspects of Applied Biology* 21, 39–50.

Gomez, K.A. and Gomez, A. A. (1976) *Statistical Procedures for Agricultural Research, with Emphasis on Rice.* IRRI, Philippines.

Goodwin-Bailey, K.A., Holborn, J.M. and Davies, M. (1957) A technique for the biological evaluation of insecticide aerosols. *Annals of Applied Biology* 45, 347–360.

Granett, J. and Retnakaran, A. (1977) Stadial susceptibility of eastern spruce budworm *Choristoneura fumiferana* (Lepidoptera: Tortricidae), to the insect growth regulator Dimilin. *Canadian Entomologist* 109, 893–894.

Griffiths, D.C., Maniar, S.P., Merritt, L.A., Mudd, A., Pickett, J.A., Pye, B.J., Smart, L.E. and Wadhams, L.J. (1991) Laboratory evaluation of pest management strategies combining antifeedants with insect growth regulator insecticides. *Crop Protection* 10, 145–151.

Gunning, R.V. (1993) Comparison of two bioassay techniques for larvae of *Helicoverpa* spp. (Lepidoptera: Noctuidae). *Journal of Economic Entomology* 86, 234–238.

Hall, F.R. and Thacker, J.R.M. (1994) Effects of droplet size on the topical toxicity of two pyrethroids to the cabbage looper *Trichoplusia ni* (Hubner). *Crop Protection* 13, 225–229.

Hassan, S.A. (1977) Standardized technique for testing side-effects of pesticides on beneficial arthropods in the laboratory. *Z. Planzenkrankh. Pflanzenensch* 84, 158–163.

Hassan, S.A. (1985) Standard methods to test the side-effects of pesticides on natural enemies of insects and mites developed by the IOBC/WPRS working group 'Pesticides and Beneficial Organisms'. *OEPP/EPPO Bulletin* 15, 214–255.

Hewitt, A.J. and Meganasa, T. (1993) Droplet distribution densities of a pyrethroid insecticide within grass and maize canopies for the control of *Spodoptera exempta* larvae. *Crop Protection* 12, 59–62.

Hewlett, P.S. (1954) A micro-drop applicator and its use for the treatment of certain small insects with liquid insecticides. *Annals of Applied Biology* 41, 45–64.

Hewlett, P.S. (1962) Toxicological studies on a beetle, *Alphitobius laevigatus* (F.) 1. Dose–response relations for topically applied solutions of four toxicants in a non-volatile oil. *Annals of Applied Biology* 50, 335–349.

Hewlett, P.S. and Plackett, R.L. (1979) *The Interpretation of Quantal Responses in Biology.* Arnold, London.

Hislop, E.C. (1989) Crop spraying under controlled conditions. *Aspects of Applied Biology* 21, 119–120.

Hunt, P.F. and Raven, C.A. (1985) An investigation and development of the use of hand held data loggers in the field. *Aspects of Applied Biology* 10, 313–320.

Hunt, P.F. and Taylor, E.F. (1985) The development of a field trials data management, analysis and reporting system for the microcomputer. *Aspects of Applied Biology* 10, 355–362.

Ismail, F. and Wright, D.J. (1991) Cross-resistance between acylurea insect growth regulators in a strain of *Plutella xylostella* L. (Lepidoptera: Yponomeutidae) from Malaysia. *Pesticide Science* 33, 359–370.

Jeffs, K.A. (1986) *Seed Treatment,* 2nd. edn. British Crop Protection Publications, 332pp.

Jenkins, J.N., Parrott, W.L., McCarty, J.C., Callahan, F.E., Berberich, J.A. and Deatin, W.R. (1993) Growth and survival of *Heliothis virescens* (Lepidoptera: Noctuidae) on transgenic cotton containing a truncated form of the delta endotoxin gene from *Bacillus thuringiensis. Journal of Economic Entomology* 86, 181–185.

Johnstone, D.R., Cooper, J.F., Dobson, H.M. and Turner, C.R. (1989a) The collection of aerosol droplets by resting tsetse flies, *Glossina morsitans* Westwood (Diptera: Glossinidae). *Bulletin of Entomological Research* 79, 613–624.

Johnstone, D.R., Cooper, J.F., Flower, L.S., Harris, E.G., Smith, S.C. and Turner, C.R. (1989b) A means of applying mature aerosol drops to insects for screening biocidal activity. *Tropical Pest Management* 35, 65–66.

Joyce, R.J.V. and Roberts, P. (1959) The determination of the size of plot suitable for cotton spraying experiments in the Sudan El Gezira. *Annals of Applied Biology* 47, 287–305.

Kerns, D.L. and Gaylor, M.J. (1992) Insecticide resistance in field populations of the cotton aphid (Homoptera: Aphididae). *Journal of Economic Entomology* 85, 1–8.

Koopman, P.A.R. (1994) Confidence intervals for the Abbott's formula correction of bioassay data for control response. *Journal of Economic Entomology* 87, 833.

Krahmer, H. and Russell, P.E. (1994) General problems in glasshouse to field transfer of pesticide performance. *British Crop Protection Council Monograph* 59, 3–16.

Learmount, J. (1994) Selection of houseflies (Diptera: Muscidae) with a pyrethroid space spray using a large-scale laboratory method. *Journal of Economic Entomology* 87, 894–898.

Leonard, P.K., Hertlein, M.B., Thompson, G.D. and Paroonagian, D.L. (1994) Prediction of the field efficacy of a novel pyrazoline insecticide (809580) against tobacco budworm (*Heliothis virescens*). *British Crop Protection Council Monograph* 59, 67–80.

MacCuaig, R.D. (1983) *Insecticide Index*, 2nd edn. FAO, Rome.

MacCuaig, R.D. and Watts, W.S. (1968) A simple technique for applying small measured quantities of insecticides to insects. *Bulletin of Entomological Research* 57, 549–552.

Matthews, G.A. (1966) Investigations of the chemical control of insect pests of cotton in Central Africa. I Laboratory rearing methods and tests of insecticides by application to bollworm eggs, II Tests of insecticides with larvae and adults. *Bulletin of Entomological Research* 57, 69–91.

Matthews, G.A. (1973) Ultra-low volume spraying of cotton in Malawi. *Cotton Growers Review* 50, 242–267.

Matthews, G.A. (1984) *Pest Management*. Longman, Harlow.

Matthews, G.A. (1994) A comparison of laboratory and field spray systems. *British Crop Protection Council Monograph* No. 59, 161–171.

Matthews, G.A. and Thornhill, E.W. (1993) Herbicide applications: Equipment design for small scale farmers. *Brighton Crop Protection Conference – Weeds*, pp. 1171–1176.

Miller, P. C. H., Hislop, E.C., Parkin, C.S., Matthews, G.A. and Gilbert, A.J. (1993) The classification of spray generator performance based on wind tunnel assessments of spray drift. *ANPP-British Crop Protection Council Second Symposium on Pesticide Application Techniques*, Strasbourg, pp. 109–115.

Miller, T.A. and Salgado, V.L. (1985) The mode of action of pyrethroids on insects. In Leahy, J.P. (ed.) *The Pyrethroid Insecticides*. Taylor and Francis, London, pp. 43–97.

Morgan, C.P. and Pinniger, D. B. (1987) A sprayer for small scale application of insecticides to test surfaces. *Laboratory Practice* 36, 68–70.

Munthali, D. C. and Scopes, N.E.A. (1982) A technique for studying the biological efficiency of small droplets of pesticide solutions and a consideration of its implications. *Pesticide Science* 13, 60–63.

Needham, P.H. and Devonshire, A.L. (1973) A technique for applying small drops of insecticide solution to *Myzus persicae* (Sulz). *Pesticide Science* 4, 107–111.

Pearce, S.C. (1976) *Field Experimentation with Fruit Trees and other Perennial Plants.* CAB Technical Communication No. 23 Rev.

Perrin, R.M. and Löwer, C. (1994) A comparison of three insecticide resistance monitoring methods for Lepidoptera. *Proc. 1994 Beltwide Cotton Conference,* pp. 1183–1184.

Pike, D.J. and Harris, P.M. (1985) Computer-aided recording of crop yield data. *Aspects of Applied Biology* 10, 363–371.

Plapp, F.W. and Vinson, S.B. (1977) Comparative toxicities of some insecticides to the tobacco budworm and its ichneumonid parasite, *Campoletis sonorensis. Environmental Entomology* 6, 381–384.

Plapp, F.W., McWhorter, G.M. and Vance, G.L. (1987) Monitoring pyrethroid resistance in the tobacco budworm in Texas – 1986. In: *Proc. Beltwide Cotton Production Research Conference 1986.* NCC. Memphis, Tennessee.

Potter, C. (1941) A laboratory spraying apparatus and technique for investigating the action of contact insecticides with some notes on suitable test insects. *Annals of Applied Biology* 28, 142–169.

Potter, C. (1952) An improved laboratory apparatus for applying direct sprays and surface films, with data on the electrostatic charge on atomized sprays. *Annals of Applied Biology* 39, 1–28.

Prior, C., Carey, M., Abraham, Y.J., Moore, D. and Bateman, R.P. (1996) Development of a bioassay method for the selection of entomophagenic fungi virulent to the desert locust *Schistocerca gregaria* (Forskal). *Journal of Applied Entomology* 119, 567–573.

Rayner, A.C. and Halliburton, W. (1955) Rotary device for producing a stream of uniform drops. *Review of Scientific Instruments* 26, 1124–1127.

Reed, W. (1972) Uses and abuses of unsprayed controls in spraying trials. *Cotton Growing Review* 49, 67–72.

Reed, W., Davies, J.C. and Green, S. (1985) Field experimentation. In Haskell, P.T. (ed.) *Pesticide Application: Principles and Practice.* Oxford University Press, pp. 153–174.

Reeves, J.C. and Law, J.R. (1985) A micro-computer based system of large-scale data capture used at the National Institute of Agricultural Botany. *Aspects of Applied Biology* 10, 327–335.

Reichard, D.L. (1990) A system for producing various sizes, numbers and frequencies of uniform-size drops. *Transactions of the American Society of Agricultural Engineers* 33, 1767–1770.

Reichard, D.L., Alm, S.R. and Hall, F.R. (1987) Equipment for studying effects of spray drop size, distribution and dosage on pest control. *Journal of Economic Entomology* 80, 540–543.

Riley, C.M., Wiesner, C.J. and Ernst, W.R. (1989) Off-target deposition and drift of aerially applied agricultural sprays. *Pesticide Science* 26, 159–166.

Robertson, J.L. and Preisler, H.K. (1992) *Pesticide Bioassay with Arthropods.* CRC Press, Boca Raton, Florida.

Robinson, T.H. (1985) A novel sprayer for treatment of small plots. *Aspects of Applied Biology* 10, 523–528.

Rosenheim, J.A. and Hoy, M.A. (1989) Confidence intervals for the Abbott's formula correction of bioassay data for control response. *Journal of Economic Entomology* 82, 331–335.

Rowland, M., Pye, B., Stribley, M., Hackett, B., Denholm, I. and Sawicki, R.M. (1990)

Laboratory apparatus and techniques for the rearing and insecticidal treatment of whitefly *Bemisia tabaci* (Homoptera: Aleyrodidae) under simulated field conditions. *Bulletin of Entomological Research* 80, 209–216.

Rutherford, S. J. (1985) Development of equipment and techniques to enable precise and safe application of pesticides in small and large plot, problem trial situations. *Aspects of Applied Biology* 10, 487–497.

Samu, F., Matthews, G.A., Lake, D. and Vollrath, F. (1992) Spider webs are efficient collectors of agrochemical spray. *Pesticide Science* 36, 47–51.

Sawicki, R.M., Rowland, M.W., Byrne, F.J., Pye, B.J., Devonshire, A.L., Denholm, I., Hackett, B.S., Stribley, M.F. and Dittrich, V. (1989) The tobacco whitefly field control simulator – a bridge between laboratory assays and field evaluation. *Aspects of Applied Biology* 21, 121–122.

Skurray, S.J. (1985) Cereal trials and development of plot machinery as seen by a machinery designer. *Aspects of Applied Biology* 10, 65–73.

Shepherd, H.H. (1958) *Methods of Testing Insects on Insects*, vol. I. Burgess, Minneapolis,

Slater, A.E., Hardisty, J.A. and Yong, K. (1985) Small plot hydraulic sprayer and granule applicator. *Aspects of Applied Biology* 10, 477–486.

Smart, L.E., Blight, M.M., Pickett, J.A. and Pye, B. J. (1994) Development of field strategies incorporating semiochemicals for the control of the pea and bean weevil, *Sitona lineatus* L. *Crop Protection* 13, 127–135.

Stokes, R.H. and Robertson, R.A. (1949) Standard solutions for humidity control at 25°C. *Industry Engineering Chemistry* 41, 2013.

Sumner, H.R., Chalfant, R.B. and Cochran, D. (1991) Influence of chemigation parameters on fall armyworm control in field corn. *Florida Entomologist* 74, 287.

Sundaram, A., Sundaram, K.M.S. and Leung, J.W. (1991) Droplet spreading and penetration of non-aqueous pesticide formulations and spray diluents in Kromekote cards. *Transactions of the American Society Agriculture Engineers* 34, 1941–1951.

Sundaram, K.M.S. (1991) Spray deposit patterns and persistence of diflubenzuron in some terrestrial components of a forest ecosystem after application at three volume rates under field and laboratory conditions. *Pesticide Science* 32, 275–293.

Symmons, P.M., Boase, C.J., Clayton, J.S. and Gorta, M. (1989) Controlling desert locust nymphs with bendiocarb applied by a vehicle-mounted spinning-disc sprayer. *Crop Protection* 8, 324–331.

Tabashnik, B.E. and Cushing, N.L. (1987) Leaf residue vs. topical bioassays for assessing insecticide resistance in the diamond-back moth, *Plutella xylostella*. *FAO Plant Protection Bulletin* 35, 11–14.

Tatchell, G.M., Thorn, M., Loxdale, H.D. and Devonshire, A.L. (1988) Monitoring for insecticide resistance in immigrant populations of *Myzus persicae*. *Brighton Crop Protection Conference – Pests and Diseases* 1, 439–444.

Taylor, N. and Matthews, G.A. (1986) Effect of different adjuvants on the rainfastness of bendiocarb applied to Brussels sprout plants. *Crop Protection* 5, 250–253.

Thacker, J.R.M., Hall, F.R. and Downer, R.A. (1992) The interactions between routes of exposure and physicochemical properties of four water-dilutable permethrin formulations in relation to their activities against *Trichoplusia ni* (Hubner). *Pesticide Science* 36, 239–246.

Thacker, J.R.M., Young, R.D.F., Stevenson, S. and Curtis, D.J. (1995) Microdroplet application to determine the effects of a change in pesticide droplet size on the topical toxicity of chlorpyrifos and deltamethrin to the aphid *Myzus persicae*

(Hemiptera: Aphididae) and the ground beetle *Nebria brevicollis* (Coleoptera: Carabidae). *Journal of Economic Entomology* 80, 1560–1565.

Thompson, A.R. and Wheatley, G.A. (1985) Seeder- and planter-mounted attachments for precision evaluation of granule treatments on small plots. *Aspects of Applied Biology* 10, 465–476.

Thompson, A.R., Suett, D.L., Percivall, A.L., Pradbury, C.E., Edmonds, G.H. and Farmer, C.J. (1981) *Precision Equipment for Sowing and Treating Small Plots with Granular Insecticide.* Annual Report, National Vegetable Research Station for 1980, Wellesbourne, pp. 34–35.

Tunstall, J.P. and Matthews, G.A. (1966) Large-scale spraying trials for the control of cotton insect pests in Central Africa. *Cotton Growing Review* 43, 121–139.

Uk, S. (1978) Portable microtip nozzle assembly for producing controlled monosized droplets for use in field microplot experiments. *British Crop Protection Council Monograph* 22, 121–127.

Verkerk, R.H.J. and Wright, D.J. (1994) The potential for induced extrinsic host plant resistance in IRM strategies targeting the diamondback moth. *Brighton Crop Protection Conference – Pests and Diseases,* pp. 457–462.

Vickerman, G.P. (1985) Sampling plans for beneficial arthropods in cereals. *Aspects of Applied Biology* 10, 191–198.

Voss, G. (1961) A new acaricide test-method for spider mites. *Anzeiger für Schädlinskunde,* 34, 76–77.

Walker, P.T. (1960) Insecticide studies on the maize stalk borer *Busseola fusca* (Fuller) in East Africa. *Bulletin of Entomological Research* 51, 321–351.

Womac, A.R., Williford, J.R., Weber, B.J., Pearce, K.T. and Reichard, D.L. (1992) Influence of pulse spike and liquid characteristics on the performance of uniform droplet generator. *Transactions of the American Society of American Engineers* 35, 71–79.

Young, B.W. (1986) A device for the controlled production and placement of individual droplets. *Pesticide Formulations and Application Systems.* Fifth Symposium, pp. 13–22.

Zhao, J.Z. and Grafius, E. (1993) Assessment of different bioassay techniques for resistance monitoring in the diamondback moth (Lepidoptera: Plutellidae). *Journal of Economic Entomology* 86, 995–1000.

11

Techniques to Evaluate the Efficacy of Natural Enemies

N. Mills

Center for Biological Control, Department of Environmental Science, Policy and Management, University of California at Berkeley, 201 Wellman Hall, Berkeley, CA 94720–3112, USA

11.1. Introduction

Spectacular reductions in the abundance of an invading pest following the successful introduction of a natural enemy from another geographic region, its region of origin, provide the most obvious examples of the efficacy and impact of natural enemies. Permanent reductions of more than a 1000-fold in the equilibrium abundance of a pest are not uncommon in classical biological control and these changes are so clear as to be apparent even to the most casual observer. However, not all such programmes provide such dramatic results and, in addition, we are often interested in the action of indigenous natural enemies and the dynamics of non-pest hosts, and so need to make careful quantitative measurements of the mortality caused by natural enemies.

In this chapter I consider the methods that can be used to measure the impact of natural enemies on arthropod populations during a single host generation, or some other specified period of time. There are two basic approaches. The first, and most extensively used approach, focuses on the natural enemy population and aims to estimate directly the extent of mortality imposed by natural enemies on the host population. In contrast, the second approach focuses on the host population, and aims to estimate indirectly the impact of natural enemies through a comparative analysis of the abundance of the host population in the presence and absence of natural enemies. Both approaches can provide estimates of the host mortality induced by specific natural enemy species, but these estimates must then be included in a lifetable analysis to reveal the role of the natural enemy in the dynamics of its host population.

The majority of examples used are derived from the literature on parasitoids and predators, but many of the methods developed for the measurement of parasitism can also be used to estimate the impact of microbial pathogens and nematodes. Similarly, most examples have been drawn from the literature on arthropod pests, but the same methods can readily be applied to natural habitats and non-pest species.

11.2. Direct Estimation of Natural Enemy Efficacy

11.2.1. *Per cent parasitism or infection*

The most conventional estimate of the impact of parasites (parasitoids, nematodes and microbial pathogens) is the percentage of the host population that is parasitized or infected by the natural enemy. For simplicity, I will use the terms parasitism, parasitized and per cent parasitism to represent the attack of a host by each of these three groups of natural enemies. Per cent parasitism is commonly estimated from a sample of hosts by determining, through rearing or incubation, the number parasitized and calculating per cent parasitism as the ratio of the number parasitized to the total number of hosts in the sample (Van Driesche, 1983).

SAMPLING OF THE HOST POPULATION

Per cent parasitism can be determined by collection of host samples alone, but it is essential that the host individuals sampled are representative of the condition of the whole host population. However, it is known that parasitism can influence the behaviour of attacked hosts and alter the relative distribution of healthy and parasitized hosts (Moore, 1995). For example, Ryan (1985) found that larvae of the larch casebearer sampled from tree foliage in spring were not as heavily parasitized by *Agathis pumila* (Ratz.) as those collected the previous fall, due to parasitized hosts moving toward the base of the tree. Similarly, Newman and Carner (1975) found soybean loopers to be more heavily infected by the fungus *Entomophthora gammae* in samples collected by sweep net from the tops of the plants than in samples collected by beating plants over a ground sheet. Thus the possibility of differential distribution of parasitized and healthy hosts must be examined before a sampling plan is devised for the estimation of the impact of natural enemies.

Considerable care must also be taken to ensure that cross-contamination does not occur between individuals in the host sample after collection from the field. This concern is of greatest importance when monitoring microbial pathogens and entomopathogenic nematodes, but similar effects can also occur with parasitoids. Cross-contamination can be reduced by either restricting mortality data to that which occurs within a specific period of time, which corresponds to the known rate of development of the natural enemy, after collection of the sample, or by the separation of hosts into individual containers as they are collected from the field (Carner, 1980). If the field-collected hosts are maintained on natural food to await the expression of parasitism, then the food source must be washed in 0.05% sodium hypochlorite and rinsed in distilled water to remove any potential contaminants. Contaminated foliage is the key mode of transmission of microbial pathogens and some parasitoids (e.g. microtype eggs of tachinids) and may have a considerable influence on the estimation of parasitism (e.g. Ignoffo *et al.*, 1975).

MARGINAL ATTACK RATE AND APPARENT PARASITISM

Once a sample of hosts has been collected from the field they can either be reared to await the manifestation of parasitism, or they can be dissected to determine the level of parasitism. There is an important distinction here, as the two measurements may provide very different estimates of parasitism from the same sample. Some hosts inevitably die during rearing without showing signs of parasitism and are either excluded from the sample or considered unattacked. However, any differential mortality of parasitized and healthy hosts, during rearing, will clearly bias the estimate of per cent parasitism (Waage and Mills, 1992). This illustrates the distinction between apparent parasitism, the observed percentage of hosts *killed* by a natural enemy in a reared sample of hosts, and the marginal attack rate, the percentage of hosts *attacked* by the natural enemy determined directly from dissection of the sample of hosts (Royama, 1981). The marginal attack rate (m_i) represents the percentage of hosts that would be killed by a natural enemy in the absence of other contemporaneous mortality factors. For the case of two contemporaneous natural enemies (A and B), their marginal attack rates can be calculated from the apparent mortality (d_i) by:

$$m_A = [b - (b^2 - 4cd_A)^{\frac{1}{2}}]/2c$$

$$m_B = d_B/(1 - cm_A)$$

where $b = c(d_A + d_B) + 1 - d_B$ and c is the proportion of hosts dying from natural enemy A when attacked by both A and B. These formulae assume that the death of the host is due to a single natural enemy, A or B, but not to both. In the absence of data on the competitive outcome of the interaction between the natural enemies, c is set at 0.5. Buonaccorsi and Elkinton (1990) discuss the estimation of marginal attack rates when three or more mortality factors interact. The product of the survival rates (1 − marginal attack rate) from each of the contemporaneous factors provides the overall survival rate (1 − total apparent parasitism).

Dissection of the hosts to measure the marginal attack rate is both time-consuming and dependent upon whether the natural enemies can be accurately detected. More recently, molecular techniques such as electrophoresis (Holler and Braune, 1988; Walton *et al.*, 1990), serological assays (Allen *et al.*, 1992) and DNA probes (Stuart and Greenstone, 1996) have been used to detect parasitism of hosts by parasitoids, and both serological assays (Webb and Shelton, 1990) and DNA probes (Keating *et al.*, 1989; Hegedus and Khachatourians, 1993) have been developed for the detection of fungal and viral pathogens. However, these techniques are not yet sufficiently sensitive to detect the egg and young larval stages of parasitoids or early-stage infections by microbial pathogens.

GENERATIONAL MORTALITY

An estimate of per cent parasitism from a single sample of hosts merely represents a passing snapshot of parasitism, and it is well known that such estimates vary widely with time and location. The true impact of a natural enemy is determined from estimates of the generational mortality of the host population that is due to parasitism. Van Driesche (1983) pointed out the gross inaccuracies in the common practice of using either the peak or the mean level of parasitism from a series of point samples to estimate generational mortality. The solution to the accurate estimation of generational mortality from parasitism lies in the selection of correct values for the ratio estimate of per cent parasitism. The numerator of the ratio must include all of the host individuals that are parasitized by a specific natural enemy within a host generation and the denominator must include all of the host individuals recruited to the susceptible stage in the host population. The susceptible stage of the host is those life stages or instars that are susceptible to attack by the natural enemy, which for parasitoids may be a very specific and relatively short period of the host's life cycle, but for microbial pathogens and nematodes may include almost all stages of the life cycle. Host samples that include non-susceptible host stages or that are taken before the natural enemy has completed its attack of the host population will underestimate generational mortality. In contrast, if parasitism delays the development of parasitized individuals relative to healthy individuals, host samples taken after healthy individuals have begun to leave the susceptible stage will tend to overestimate generational mortality.

These effects can best be illustrated using a hypothetical example (Table 11.1). Consider a host population of 100 eggs that is attacked by either a parasitoid or an entomogenous nematode during its larval stage. Parasitism accumulates through the host larval stage, parasitized hosts show delayed development (4.5 days) but are eventually killed as mature larvae, and healthy hosts develop more rapidly (3 days) to move through to the pupal stage. Of the seven sample dates in which host larvae are sampled, the first two underestimate the actual rate of 50% parasitism, while the last four sample dates overestimate parasitism. In addition, the peak of parasitism at 100% (on sample dates 7 and 8) obviously does not reflect the generational parasitism of the natural enemy and even the sum of the parasitized and total host larvae, over the seven samples that include the susceptible larval stage, provide a pooled estimate of 59.8% parasitism, an overestimate of the true impact.

DIFFERENTIAL RATES OF DEVELOPMENT

The simplest correction factor that can be used to improve the estimate of generational parasitism concerns the differential development rate of parasitized and healthy hosts. Parasitized hosts often develop more slowly than healthy hosts, as in the above example, and as a result become over-represented in samples. Russell (1987) pointed out that this can be corrected by reducing the number of parasitized hosts in the calculation of per cent para-

Table 11.1. A hypothetical example of the impact of a natural enemy on a host population.

Sample date	Host eggs	Healthy larvae	Parasitized larvae	Host pupae	% parasitized larvae
1	100	0	0	0	0
2	50	38	12	0	24.0
3	20	45	35	0	43.8
4	0	50	50	0	50.0
5	0	25	50	25	66.7
6	0	10	50	40	83 3
7	0	0	38	50	100
8	0	0	15	50	100
9	0	0	0	50	0

sitism by a factor determined by the ratio of the duration of development of healthy (H) and parasitized (P) hosts:

$$\% \text{ parasitism} = \frac{\text{No. of parasitized hosts} \times (H/P) \times 100}{\text{No. of healthy hosts} + [\text{No. of parasitized hosts} \times (H/P)]}$$

For the hypothetical example used in Table 11.1 the ratio of development times is 0.67. Whereas the sum of the healthy and parasitized individuals from Table 11.1 provides a pooled estimate of 59.8% parasitism, use of the 0.67 correction factor for the number of parasitized hosts provides an accurate estimate of 50% parasitism. Hosts that have been parasitized by other groups of nematodes or by microbial pathogens are often killed and removed from the host population very rapidly. In these cases, the correction can be used to magnify the representation of infected hosts that will tend to be under-represented in samples. One drawback is that the correction is accurate only when the rates of recruitment of hosts to the susceptible stage and of attack by the natural enemy are constant. Although constant recruitment is unusual in field populations the correction can still be useful in improving the estimation of generational parasitism.

FORTUITOUS LIFE HISTORY CHARACTERISTICS

It is interesting to note that the fourth sample from the hypothetical host population (Table 11.1) provides a correct estimate of generational parasitism. This is due to the coincidence, on this particular sample date, of four interacting factors: (i) all attacks by the natural enemy are complete; (ii) none of the attacked hosts has yet been killed and removed from the host population; (iii) there is no further recruitment to the host population from the egg stage; and (iv) none of the healthy host larvae has been lost from the susceptible stage by recruitment to the pupal stage (see Van Driesche, 1983, for further discussion of these factors). Several guilds of endoparasitoids (a classification that defines the pattern of host utilization by parasitoids) and even one guild of ectoparasitoids are known to attack a particular and restricted stage of host development and yet exhibit protracted development such that they complete their development within, and kill, a distinctly different stage of host development (Mills, 1994). An example is the egg–prepupal parasitoid guild, that is characterized by attack of host eggs but completion of development is in host prepupae. For these parasitoids there is a greater intervening period when generational parasitism can be correctly estimated from a single point sample of hosts due to the coincidence of the same four factors noted above. In the case of the egg–prepupal parasitoid guild, this intervening period spans the whole of the host larval stage. A similar situation arises for endoparasitoids when diapause arrests the development of the host at a stage that follows parasitoid attack but before completion of parasitoid development. Both Miller (1954) and Hill (1988) have estimated generational parasitism by larval parasitoids using this approach. Of course, such estimates will only be correct if there is no differential mortality of parasitized and healthy hosts between the time of parasite attack and the collection of the host sample.

There are many examples of parasitism, however, in which biology and life history characteristics do not provide good opportunities for the accurate estimation of generational parasitism from a single host sample. The majority of ectoparasitoids, all non-larval endoparasitoids, entomopathogenic nematodes, and microbial pathogens both attack and complete their development at the same host stage. In these cases, hosts may either die from other causes or develop through to the next stage before all host individuals have been recruited to the susceptible stage. There is no point in time at which all host individuals are present in the susceptible stage and so no single sample point can be used to estimate generational parasitism. This situation is further complicated by the fact that unless the rate of parasitism is constant throughout the host recruitment period, the interaction between host and parasite recruitment will have an overriding influence on the per cent parasitism at any particular point in time. Under such circumstances per cent parasitism must be determined through either recruitment analysis or death rate analysis.

RECRUITMENT ANALYSIS

Unbiased estimation of generational parasitism requires the total number of parasitized hosts to be divided by the total number of hosts entering the susceptible stage. These quantities can be obtained from the direct measurement of the recruitment of host individuals to the susceptible stage and of attacked hosts to the parasitized host category. Van Driesche and Bellows (1988) successfully developed this approach to estimate the generational parasitism of *Pieris rapae* (L.) by the larval parasitoid *Cotesia glomerata* (L.). Recruitment of *P. rapae* to the first larval instar was measured by stripping all larvae from sets of randomly selected collard plants and counting the number of first and second instar larvae 3–4 days later, representing those individuals that have been recruited to the susceptible stage during the sampling interval. Recruitment of *C. glomerata* was then measured by dissection of the host larvae stripped from the collard plants and counting the number of host larvae that contained parasitoid eggs of less than a particular length (age), again representing only those parasitoids that could have been recruited during the sampling interval. These measurements were repeated over the entire recruitment period and the counts summed to provide the total number of parasitized hosts and the total number of hosts recruited to the susceptible stage. An alternative for measurement of the recruitment of parasitoids is to make use of trap (or sentinel) hosts. Hosts from laboratory cultures, at the stage susceptible to parasitism, can be placed out in the field and exposed to parasitism for the same short interval of time before being recovered and reared through or dissected to estimate parasitism (Van Driesche, 1988; Gould *et al.*, 1992a). The trap host method works best when the susceptible host stage is sessile (e.g. Van Driesche *et al.*, 1991), as when mobile, the behaviour and distribution of the trap hosts can have a major influence on the level of parasitism that they experience relative to the wild population (e.g. Gould *et al.*, 1992a).

Recruitment analysis can provide an unbiased estimate of generational parasitism, but poses certain constraints. Accurate measurement of host recruitment is dependent on the recruits not moving from tagged plants or dying from other causes. Similarly recruitment of parasitized hosts is dependent on accurate detection of early stages of the natural enemy and, if sentinel hosts are used, that these latter hosts are no more or less susceptible to attack than natural hosts. In addition, the sample interval must be sufficiently short that the natural enemy does not develop past the recognizable marker stage (e.g. the specified egg length in the *C. glomerata* example) nor that trap hosts develop beyond the susceptible stage. This interval may be particularly short for estimation of attack by microbial pathogens or entomopathogenic nematodes.

DEATH RATE ANALYSIS

As an alternative to the direct measurement of recruitment of susceptible and parasitized hosts, the death rate of hosts from natural enemies over a specific time interval can be measured (Gould *et al.*, 1989). This method consists of collecting field samples of the host at regular and frequent intervals and measuring the apparent death rate of the host sample, the observed proportion of host individuals that die from parasitism, during the interval between one sample and the next. This method differs from the more traditional measurement of apparent parasitism in that the death rate is measured for the interval between samples only rather than retaining the sample until all parasitoids have emerged from the hosts. The generational per cent parasitism is obtained from $(1 - \text{proportion surviving}) \times 100$, where the proportion surviving is the product of the proportion of host individuals that survived the control agent in each of the individual samples (Gould *et al.*, 1992a).

This particular method was developed to measure the marginal death rate from parasitism in gypsy moth populations that suffer contemporaneous mortality from several natural enemies (Gould *et al.*, 1990, 1992a). Death rate analysis is intuitively appealing as it is far less demanding than recruitment analysis, and requires even less effort than the traditional estimation of per cent parasitism due to a shortening of the rearing period for each sample. It can also be applied under a broader range of conditions, such as situations when host recruitment or the initial stages of parasitism cannot readily be determined. The main source of error with this method results from a sample interval that is too long relative to the development rate of the natural enemies. If host mortality during the inter-sample interval differs between the laboratory-reared sample and the wild population in the field, the apparent death rate from parasitism may be biased. Similarly, if an additional natural enemy could both attack and kill the host during the inter-sample interval, this would influence the apparent parasitism by both agents.

11.2.2. Predation

Predation is rather more difficult to measure than parasitism for two reasons: (i) the interaction between a predator and its prey is very brief, a matter of minutes in most cases, which severely reduces the chances of detection by an observer; and (ii) very often there are few, if any, remains of the prey that can be detected after predation has taken place. Nonetheless, there are several methods that can be used to directly estimate predation in the field.

LABORATORY ESTIMATION OF PREDATOR POWER

Perhaps the most obvious way to estimate predation is through observations of the feeding potential of predators in the laboratory. Simple laboratory assays in artificial arenas are used to measure the 'power' of a predator or the number of prey consumed per predator per day, at successive stages of development and in relation to a series of constant temperatures, in the presence of excess prey. These estimates represent the maximum feeding potential of the predator under ideal conditions, without the need to search for and subdue prey and where metabolic costs are minimized. The level of predation in a field population is then estimated from the 'power' of an individual predator, at the average field temperatures experienced, multiplied by the density of predators in the field. Bombosch (1963) and Tamaki *et al.* (1974) developed the concept of predator power and efficiency in a study of heteropteran predation of aphids in peach orchards. The same approach has been extended by Chambers and Adams (1986) to quantify the impact of syrphid predators on populations of cereal aphids in winter wheat.

DIRECT OBSERVATION

Direct observation is not only the best way to make an initial assessment of the diet breadth of a predator, but detailed behavioural observations can also provide a measure of the rate of predation. Edgar (1970) estimated the predation rate of lycosid spiders from observations of the number of hours a day that the spiders are active (t_s), the mean proportion of spiders actively feeding at any one point in time (p) and the mean time in hours that a spider takes to fully consume a single prey item (t_f). The predation rate (r), or the number of prey consumed by an individual spider per day, is given by:

$$r = pt_s/t_f$$

The mean proportion of spiders actively feeding at any one time (p) is used to represent the mean proportion of time that an average spider spends feeding during the day. It is best estimated from a series of point samples of the feeding activity of a population of spiders throughout the activity period. This approach has subsequently been used to estimate spider predation in a variety of crops (Kiritani *et al.*, 1972; Sunderland *et al.*, 1986; Nyffeler *et al.*, 1987). It could readily be applied to many other types of predator and has been used to estimate rates of parasitism by the parasitoids of bark beetles (Mills, 1991). The only drawback to this approach is that the observer must take extra care not to influence the behaviour of the predator during the observation periods.

SENTINEL PREY

As in the estimation of parasitoid recruitment, we can similarly make use of sentinel prey for the estimation of predation. Non-mobile prey stages, such as eggs or pupae, can very readily be placed out in the field at natural densities to monitor losses from predation, but it is important that the sentinel prey are no more or less susceptible to predation than the wild population. For example, Smith (1985) attached freeze-dried female pupae of the gypsy moth to small squares of burlap with melted beeswax and fixed these at different tree heights to measure the extent of predation by small mammals and invertebrate predators. This method has subsequently been used by Cook *et al.* (1994, 1995) to examine gypsy moth predation under a variety of conditions. Similarly, Andow (1992) exposed egg masses of the European corn borer, obtained from laboratory cultures, for 3 days in the field to measure the extent of predation by *Coleomegilla maculata* DeGeer, *Chrysopa* sp. and *Orius insidiosus* (Say). For mobile stages of the prey, such as larvae or adults, individual insects can be tethered using strong sewing thread and exposed to predation for limited time periods. Tethering obviously affects the behaviour of the prey and this influence must be taken into account. Weseloh (1990) estimated that predation of large gypsy moth larvae by ants was almost doubled by tethering, but that a correction factor could be used to measure rates of predation in the field.

TRACES OF PREY REMAINS

Not all examples of predation result in the total loss of prey remains. For example, eggs, pupae and the coverings of Homoptera often remain in place for a sufficient length of time, after predation has occurred, to be used to directly quantify the number of individuals predated. In some cases these more persistent structural components of the prey will also provide distinctive evidence of the type of predator, for example a pair of small holes result from predation by Neuroptera, a single hole from Heteroptera and a peppering of small holes from ants. Andow (1990) was able to characterize the attack of several different insect predators on eggs of the European corn borer, and Smith and Lautenschlager (1981) distinguish the effects of invertebrate and vertebrate predators on gypsy moth pupae. Examples of the successful measurement of predation from prey remains include predation of whitefly nymphs by the coccinellid *Delphastus pusillus* LeConte (Heinz and Parrella, 1994), predation of European corn borer eggs (Andow, 1992) and predation of gypsy moth pupae (Cook *et al.*, 1994).

LABELLING PREY

If prey can be labelled with markers, they can then be released into the field and predators sampled to determine the presence of the label. This approach has been used primarily for qualitative rather than quantitative studies of predation and a wide range of both internal or external markers have been used. Hawkes (1972) successfully coated eggs of the cinnabar moth, *Tyria jacobaeae* (L.), with fluorescent dye and found fluorescent particles in the guts of 90% of the earwigs collected in the vicinity of the marked eggs. Hagler and Durand (1994) were similarly able to externally label pink bollworm eggs and whitefly adults by topical application of a readily available antigen, rabbit immunoglobulin G. This antigen marker was readily detected by ELISA in the gut of 98% of chewing predators but only 30% of predators with sucking mouthparts.

Prey can also be labelled internally by feeding them on a diet that incorporates the label before exposing them to predation in the field. McCarty *et al.* (1980) were able to identify the predators of three lepidopteran prey in soybeans using the radioactive isotope ^{32}P, although Southwood (1978) provides a good discussion of the problems associated with using radioactive markers in predation studies. Rubidium is an internal biological marker that occurs in low concentration in most habitats and can be used without any deleterious effects to study trophic interactions of both predators and parasitoids (Akey *et al.*, 1991). Johnson and Reeves (1995) added 3 g l^{-1} RbCl to the diet of gypsy moth larvae and found a positive relationship between the concentration of rubidium in the carabid predator *Carabus nemoralis* Müller and the number of labelled larvae consumed, suggesting that rubidium could be used to quantify rates of predation in the field.

DETECTION OF PREY REMAINS IN THE GUT OF A PREDATOR

The difficulty of directly observing predation prompted the development of techniques to identify the presence of prey within the gut of predators. The most important factor associated with all of these techniques is that although prey remains may be detected in the gut of a potential predator this does not confirm that the predator killed the prey, as any feeding on dead prey will result in the same gut contents of the predator.

The first developments relied upon the correct identification of fragments of the cuticle of prey dissected from the guts or the faecal pellets of a predator. Sunderland (1975) was able to identify prey types in the guts of carabids, a few staphylinids and earwigs, but many fluid-feeding staphylinids contained no detectable remains. Similarly, Folsom and Collins (1984) were able to determine the prey of dragonfly larvae by detailed examination of faecal pellets. Although Hildrew and Townsend (1982) successfully used gut dissection to quantify predation by caddis larvae, they noted that some invertebrate predators may lose part of their gut content when placed in preservative. The visual detection of the presence of prey has obvious drawbacks, as soft-bodied prey may not provide any detectable fragments and fluid-feeding predators do not ingest recognizable fragments of their prey.

Electrophoresis has been used to detect characteristic esterase banding patterns of prey in the gut of a predator (Murray and Solomon, 1978; Powell *et al.*, 1996). Prey enzymes from the dissected gut of a larger predator, or from the whole body of a small predator, are separated on gels by electrophoresis and revealed by staining. (Symondson and Hemingway provide details of the protocols for polyacrylamide gel electrophoresis in Chapter 12 of this volume.) The technique is generally not very sensitive as the same esterase bands may be present in a specific prey species, the alternative foods of the predator, the prey's food, and even the gut wall of the predator, and so the method is really only applicable to prey that exhibit high esterase activity (Solomon *et al.*, 1985). However, Giller (1986) used electrophoresis to successfully identify 18 types of prey in the gut of *Notonecta* waterbugs and several studies have used the technique to examine predation by mite predators (e.g. Lister *et al.*, 1987; Jones and Morse, 1995).

Serological techniques have also been used to examine the gut contents of predators. These techniques are based upon the inoculation of rabbits or mice with small quantities of prey to induce the production of antisera that can subsequently be used to react with potential prey antigens in the gut contents of predators. The majority of predation studies have made use of polyclonal antisera that are generated from whole insect extracts. Polyclonal antisera are readily developed, can be fine tuned for greater specificity and can detect prey remains in the gut of a predator for a greater period of time after feeding, but have the disadvantage that different batches of antisera are not necessarily consistent in their range of reactivity (Sunderland, 1988; this volume, Chapter 12). More recently monoclonal antibodies have been developed from prey extracts, through advances in hybridoma technology. Monoclonal antibodies are very specific, can be characterized and have consistent and reproducible activity, but have the disadvantages that they are difficult and costly to produce and can only detect prey remains for a relatively short period after predator feeding (Hagler *et al.*, 1994; Symondson and Hemingway, this volume, Chapter 12). The serological reaction between samples and antisera can be assayed using ELISA or immunodot techniques to determine positive responses or even to provide quantitative estimates of the amount of prey in the gut of the predator (Sopp *et al.*, 1992; Hagler *et al.*, 1995). Symondson and Hemingway, in Chapter 12, provides a detailed account of the protocols used for the development of antisera and assays to monitor the gut content of predators.

QUANTIFYING THE IMPACT OF PREDATION

Once the remains of prey have been detected in the gut of a predator, whether by dissection, electrophoresis or serology, the application of these data to the estimation of predation in the field must be considered. Dempster (1960) suggested that the rate of predation could be simply estimated from

$$r = pP/t_{DP}$$

where r is the number of prey eaten by the predator population per unit area per unit time, p is the proportion of sampled predators that contain detectable prey remains, P is the predator density and t_{DP} is the detection period, the length of time that prey remain detectable in the gut of the predator. Dempster's equation assumes that a predator that scores positive for prey remains has consumed only a single prey item during a period of time equal to the detection period. As a result it provides a rather conservative estimate of the predation rate, unless the detection period is very short, as many predators will consume a second prey item before the first has been fully digested and removed from the gut. Rothschild (1966) considered that a positive score for a predator should represent the number of prey consumed by the predator (N_a) over a period of time equal to the detection period, giving

$$r = pN_a P/t_{DP}$$

where the ratio N_a/t_{DP} represents the daily feeding rate of the predator in the insectary. In contrast to Dempster's equation, Rothschild's equation will provide an upper bound to the actual level of predation, since it assumes that predators in the field will feed at the same rate as in the insectary. Rothschild's equation has been used to estimate predation of the sunn pest, *Eurygaster integriceps* Puton (Kuperstein, 1979), the imported cabbageworm, *Pieris rapae* L. (Ashby, 1974), and the Australian soldier fly, *Inopus rubriceps* (Macquart) (Doane *et al.*, 1985).

Nakamura and Nakamura (1977) use a rather different approach to quantify predation rates. They assume that the proportion of predators that score negative in the assay $(1 - p)$ represents the zero term of a random (Poisson) distribution of meal sizes for individuals within the predator population. The mean meal size of a predator (x) from this assumption is often misquoted in the literature and should be $-\ln(1 - p)$, from the zero term of the Poisson distribution (e^{-x}), giving a predation rate of

$$r = -\ln(1 - p)P/t_{DP}$$

The assumption of a random distribution of predator meal size generates an exponential relationship between the mean meal size and the proportion of predators scoring positive, and so predicts less than a single prey consumed per predator until 63% of the predator population scores positive. Like Dempster's equation, this provides a very conservative estimate of the rate of predation and one that is clearly inadequate for a predator that is able to consume many prey individuals to complete a meal. A very similar approach has been used by Lister *et al.* (1987) to estimate predation by an Antarctic mite, *Gamasellus racovitzai* (Trouessart).

Sopp *et al.* (1992) consider it best to estimate the quantity of prey present in the gut of the predator, rather than relying on the proportion of predator individuals that score positive. If a sample of predators taken from the field represent a random sample of stages through the feeding cycle of the predators, then the quantity of prey in the gut of a sample predator $Q_0 = fQ_i$, where Q_i is the mean quantity of prey consumed initially by a predator and f is the mean proportion of the meal that remains at the time of sampling. Replacing the proportion of positive reactions, p, with the quantity of prey initially consumed by a predator (Q_i) in Dempster's (1960) equation gives:

$$r = Q_0 p/ft_{DP}$$

Q_0 can be estimated from a calibration curve relating absorbance from an ELISA assay, or the hue of a dot blot from an immunodot assay, to the concentration of prey. f is dependent on the shape of the digestion decay rate of the prey antigen in the gut of the predator, which is typically described by an exponential decay function (Sopp and Sunderland, 1989; Symondson and Liddell, 1995). As an exponential function does not completely decay in finite time, f must be assumed to represent the proportion of the prey remaining in the gut of the predator at the midpoint of the detection period for the assay. This equation replaces Dempster's (1960) assumption that a positive score represents the consumption of a single prey, with the estimate of the true meal size of a predator based on the quantity of prey in its gut at the time of sampling and the known decay rates of prey antigens. It still assumes, however, that a predator takes discrete meals which are digested and voided before another meal is taken, and so may still underestimate the predation rate of a predator that feeds more continuously.

11.3. Indirect Estimation of Natural Enemy Efficacy

In place of making direct estimates of the per cent parasitism of hosts or the predation rate of predators, the efficacy of natural enemies can be indirectly estimated by comparison of the abundance of host populations in the presence and absence of natural enemies. This approach has been extensively used to estimate the impact of predators and parasitoids, but not for nematodes or microbial pathogens. Predators or parasitoids can be excluded from experimental plots through a variety of techniques.

11.3.1. Physical barriers

The type of exclusion barrier used depends very much on the types of natural enemies that attack the target host. For insect parasitoids and flying predators the most commonly used barrier is a field cage of sufficient size to cover a small plot of low growing plants, a whole shrub or just part of a branch of a tree (Luck *et al.*, 1988). For predators that climb plants from the ground, such as carabids, spiders and ants, trenches filled with insecticide soaked straw, guttering sunk flush with the ground and filled with water, vertical barriers sunk into the ground or sticky bands around the base of trees provide a suitable exclusion barrier (Wratten, 1987).

For example, to estimate natural enemy impact on the Russian wheat aphid, *Diuraphis noxia* (Mordv.), Hopper *et al.* (1995) used cages constructed of a polyester organza sleeve over a steel frame (25 cm diameter and 120 cm tall) that was sunk 25 cm into the ground covering 5–15 tillers of wheat. In this study three comparisons were made: open plots with no cage; cages with the sleeve left open at plant canopy height; and cages with the sleeve closed over the tops of the plants. The size of the mesh on a cage can also be altered to selectively exclude different groups of natural enemies (e.g. Kring *et al.*, 1985). In contrast, Winder (1990) estimated the impact of non-flying ground predators on the grain aphid, *Sitobion avenae* (F.), using vertical exclusion barriers. 2×2 m plots of wheat were surrounded by 70 cm high vertical polythene barriers sunk 25 cm into the soil and four pitfall traps were used to remove predators from each exclusion plot. Although barriers are primarily used to exclude natural enemies, they can also be used to increase the abundance of natural enemies in experimental plots. This approach has most frequently been used in association with ground predators, where 'ingress-only' trenches, with an overhanging flange on the outside edge of the plot, allow predators to enter the plot but not to leave the plot. Chiverton (1986) used a combination of exclusion and inclusion plots to find an inverse correlation between the number of bird cherry-oat aphids, *Rhopalosiphum padi* L., per shoot and the number of polyphagous ground predators in experimental plots of spring barley.

The advantages and disadvantages of exclusion barriers have been reviewed by Wratten (1987) and Luck *et al.* (1988). Two major disadvantages of field cages are that the microclimate within a cage is altered (temperature, humidity, light intensity, wind speed) and that emigration of the host in exclusion plots is prevented. The population growth rate of a phytophagous host may be either directly affected by a change of microclimate or indirectly affected through a change in the physiology of the plant and hence its suitability as a food plant. The prevention of emigration can also affect the population growth rate of a phytophagous host and is particularly important for hosts, such as aphid or mites, that have a high rate of reproduction. Although physical barriers, used for the exclusion of ground predators, are far less likely to influence host population growth rates, these barriers are not always successful in eliminating natural enemies from exclusion plots and so it is essential to monitor predator densities, as well as prey densities, in the experimental plots.

11.3.2. Chemical barriers

DeBach (1946) developed the insecticide check as a technique to evaluate the impact of both parasitoids and predators. Dilute applications of an insecticide exclude the natural enemies from a series of replicate plots and the host or prey abundance is contrasted between treated and untreated plots. The key requisite for this approach is to find a selective insecticide that has no direct or indirect effects on the host but at the same time is highly toxic to the natural enemy. More recent applications of this technique for estimating the impact of predators include Kenmore *et al.* (1984), Braun *et al.* (1989) and Nagai (1990). Although the insecticidal check can provide a dramatic visual interpretation of the impact of natural enemies, there are a number of problems in quantifying natural enemy impact from insecticide exclusion. For example, insecticides are known to stimulate the reproductive rate of some hosts, such as mites, thrips and delphacids; to affect the sex ratio of others, such as scales, and to affect the physiology of the plant (Luck *et al.*, 1988). The resulting differences between host populations in treated and untreated plots must therefore be treated with some caution.

11.3.3. Hand removal

Hand removal has occasionally been used as a method of predator exclusion (Luck *et al.*, 1988) but appears not to be very practical as it is very time consuming and is applicable only to large, slow-moving predators.

11.4. Lifetable Analysis

Having obtained either a direct or indirect estimate of the extent of host mortality that can be attributed to a particular natural enemy, this is one measure of its efficacy, but it is even more valuable to have a series of estimates that can be incorporated into a lifetable analysis to determine the role of the natural enemy in the dynamics of the host population. It is beyond the scope of this chapter to consider the methods used for the construction and analysis of lifetables, and so readers are referred to Chapter 4 and to reviews by Bellows *et al.* (1992), Van Driesche and Bellows (1996) and Kidd and Jervis (1996). Varley and Gradwell (1960) introduced key-factor analysis as a technique for the analysis of lifetable data and in their classic study of the dynamics of winter moth populations in England (Varley *et al.*, 1973), they were able to show that the disappearance of young larvae was the key factor responsible for the variation in population abundance between generations, that parasitism and disease, although common, had little influence on population changes, and that density-dependent pupal predation by shrews and carabid beetles was responsible for regulating the population about its equilibrium density. In a more recent analysis of 49 population studies, Stiling (1988) found natural enemies to be the most common key factor, particularly for lepidopteran populations. Other more recent examples that illustrate the potential for evaluating the efficacy of natural enemies in lifetable analyses include Dowell *et al.* (1979), Van Driesche and Taub (1983), Ryan (1990), Münster-Swendsen (1991), Gould *et al.* (1992b) and Roland (1994).

11.5. References

Akey, D.H., Hayes, J.L. and Fleischer, S.J. (1991) Use of elemental markers in the study of arthropod movement and trophic interactions. *Southwest Entomologist* Supplement 14.

Allen, W.R., Trimble, R.M. and Vickers, P.M. (1992) ELISA used without host trituration to detect larvae of *Phyllonorycter blancardella* (Lepidoptera: Gracillaridae) parasitized by *Pholetesor ornigis* (Hymenoptera: Braconidae). *Environmental Entomology* 21, 50–56.

Andow, D.A. (1990) Characterization of predation on egg masses of *Ostrinia nubilalis* (Lepidoptera: Pyralidae). *Annals of the Entomological Society of America* 83, 482–486.

Andow, D.A. (1992) Fate of eggs of first generation *Ostrinia nubilalis* (Lepidoptera: Pyralidae) in three conservation tillage systems. *Environmental Entomology* 21, 388–393.

Ashby, J.W. (1974) A study of arthropod predation of *Pieris rapae* L. using serological and exclusion techniques. *Journal of Applied Ecology* 11, 419–425.

Bellows, T.S., Van Driesche, R.G. and Elkinton, J.S. (1992) Life-table construction and analysis in the evaluation of natural enemies. *Annual Review of Entomology* 37, 587–614.

Bombosch, S. (1963) Untersuchungen zur vermehrung von *Aphis fabae* Scop. in Samenrübenbestanden unter besonderer Berücksichtigung der Schwebfliegen (Diptera: Syrphidae). *Zeitschrift für Angewandte Entomologie* 52, 105–141.

Braun, A.R., Bellioti, A.C., Guerrero, J.M. and Wilson, L.T. (1989) Effect of predator exclusion on cassava infested with tetranychid mites (Acari: Tetranychidae). *Environmental Entomology* 18, 711–714.

Buonaccorsi, J.P. and Elkinton, J.S. (1990) Estimation of contemporaneous mortality factors. *Researches on Population Ecology* 32, 151–171.

Carner, G.R. (1980) Sampling pathogens of soybean insect pests. In: Kogan, M. and Herzog, D.C. (eds) *Sampling Methods in Soybean Entomology*. Springer-Verlag, New York, pp. 559–573.

Chambers, R.J. and Adams, T.H.L. (1986) Quantification of the impact of hoverflies (Diptera: Syrphidae) on cereal aphids in winter wheat: an analysis of field populations. *Journal of Applied Ecology* 23, 895–904.

Chiverton, P.A. (1986) Predator density manipulation and its effects on populations of *Rhopalosiphum padi* (Hom.: Aphididae) in spring barley. *Annals of Applied Biology* 109, 49–60.

Cook, S.P., Hain, F.P. and Smith, H.R. (1994) Oviposition and pupal survival of gypsy moth (Lepidoptera: Lymantriidae) in Virginia and North Carolina pine-hardwood forests. *Environmental Entomology* 23, 360–366.

Cook, S.P., Smith, H.R., Hain, F.P. and Hastings, F.L. (1995) Predation by gypsy moth (Lepidoptera: Lymantriidae) pupae by invertebrates at low small mammal population densities. *Environmental Entomology* 24, 1234–1238.

DeBach, P. (1946) An insecticidal check method for measuring the efficacy of entomophagous insects. *Journal of Economic Entomology* 39, 695–697.

Dempster, J.P. (1960) A quantitative study of the predators of the eggs and larvae of the broom beetle, *Phytodecta olivacea* Forster, using the precipitin test. *Journal of Animal Ecology* 29, 149–167.

Doane, J.F., Scotti, P.D., Sutherland, O.R.W. and Pottinger, R.P. (1985) Serological identification of wireworm and staphylinid predators of the Australian soldier fly (*Inopus rubriceps*) and wireworm feeding on plant and animal food. *Entomologia experimentalis et applicata* 38, 65–72.

Dowell, R.V., Fitzpatrick, G.E. and Reinert, J.A. (1979) Biological control of citrus blackfly in southern Florida. *Environmental Entomology* 8, 595–597.

Edgar, W.D. (1970) Prey and prey feeding behaviour of adult females of the wolf spider *Pardosa amentata* (Clerk). *Netherlands Journal of Zoology* 20, 487–491.

Folsom, T.C. and Collins, N.C. (1984) The diet and foraging behavior of the larval dragonfly *Anax junius* (Aeschnidae), with an assessment of the role of refuges and prey activity. *Oikos* 42, 105–113.

Giller, P.S. (1986) The natural diet of the Notonectidae: field trials with electrophoresis. *Ecological Entomology* 11, 163–172.

Gould, J.R., Van Driesche, R.G., Elkinton, J.S. and Odell, T.M. (1989) A review of techniques for measuring the impact of parasitoids of lymantriids. In: Wallner, W.E. (ed.) *The Lymantriidae: a Comparison of Features of New and Old World Tussock Moths*. USDA Forest Service General Technical Report NE-123, pp. 517–531.

Gould, J.R., Elkinton, J.S. and Wallner, W.E. (1990) Density-dependent suppression of experimentally created gypsy moth, *Lymantria dispar* (Lepidoptera: Lymantriidae), populations by natural enemies. *Journal of Animal Ecology* 59, 213–233.

Gould, J.R., Elkinton, J.S. and Van Driesche, R.G. (1992a) Suitability of approaches for measuring parasitoid impact on *Lymantria dispar* (Lepidoptera: Lymantriidae) populations. *Environmental Entomology* 21, 1035–1045.

Gould, J.R., Bellows, T.S. and Paine, T.D. (1992b) Evaluation of biological control of *Siphoninus phillyreae* (Haliday) by the parasitoid *Encarsia partenopea* (Walker), using life table analysis. *Biological Control* 2, 127–134.

Hagler, J.R. and Durand, C.M. (1994) A new method for immunologically marking prey and its use in predation studies. *Entomophaga* 39, 257–265.

Hagler, J.R., Naranjo, S.E., Bradley-Dunlop, D., Enriquez, F.J. and Henneberry, T.J. (1994) A monoclonal antibody to pink bollworm (Lepidoptera: Gelechiidae) egg antigen: a tool for predator gut analysis. *Annals of the Entomological Society of America* 87, 85–90.

Hagler, J.R., Buchmann, S.L. and Hagler, D.A. (1995) A simple method to quantify dot blots for predator gut content analyses. *Journal of Entomological Science* 30, 95–98.

Hawkes, R.B. (1972) A flourescent dye technique for marking insect eggs in predation studies. *Journal of Economic Entomology* 65, 1477–1478.

Hegedus, D.D. and Khachatourians, G.G. (1993) Construction of cloned DNA probes for the specific detection of the entomopathogenic fungus *Beauveria bassiana* in grasshoppers. *Journal of Invertebrate Pathology* 62, 233–240.

Heinz, K.M. and Parrella, M.P. (1994) Biological control of *Bemisia argentifolii* (Homoptera: Aleyrodidae) infesting *Euphorbia pulcherrima*: evaluations of releases of *Encarsia luteola* (Hymenoptera: Aphelinidae) and *Delphastus pusillus* (Coleoptera: Coccinellidae). *Environmental Entomology* 23, 1346–1353.

Hildrew, A.G. and Townsend, C.R. (1982) Predators and prey in a patchy environment: a freshwater study. *Journal of Animal Ecology* 51, 797–815.

Hill, M.G. (1988) Analysis of the biological control of *Mythimna separata* (Lepidoptera: Noctuidae) by *Apanteles ruficrus* (Hymenoptera: Braconidae) in New Zealand. *Journal of Applied Ecology* 25, 197–208.

Holler, C. and Braune, H.J. (1988) The use of isoelectric focusing to assess percentage hymenopterous parasitism in aphid populations. *Entomologia Experimentalis et Applicata* 47, 105–114.

Hopper, K.R., Aidara, S., Agret, S., Cabal, J., Coutinot, D., Dabire, R., Lesieux, C., Kirk, G., Reichert, S., Tronchetti, F. and Vidal, J. (1995) Natural enemy impact on the abundance of *Diuraphis noxia* (Homoptera: Aphididae) in wheat in southern France. *Environmental Entomology* 24, 402–408.

Ignoffo, C.M., Puttler, B., Marston, N.L., Hostetter, D.L. and Dickerson, W.A. (1975) Seasonal incidence of the entomopathogenic fungus, *Spicaria rileyi* associated with noctuid pests of soybeans. *Journal of Invertebrate Pathology* 25, 135–137.

Johnson, P.C. and Reeves, R.M. (1995) Incorporation of the biological marker Rubidium in gypsy moth (Lepidoptera: Lymantriidae) and its transfer to the predator *Carabus nemoralis* (Coleoptera: Carabidae). *Environmental Entomology* 24, 46–51.

Jones, S.A. and Morse, J.G. (1995) Use of isoelectric focusing electrophoresis to evaluate citrus thrips (Thysanoptera: Thripidae) predation by *Euseius tularensis* (Acari: Phytoseiidae). *Environmental Entomology* 24, 1040–1051.

Keating, S.T., Burand, J.B. and Elkinton, J.S. (1989) DNA hybridization assay for detection of gypsy moth nuclear polyhedrosis virus in infected gypsy moth (*Lymantria dispar* L.) larvae. *Applied Environmental Microbiology* 55, 2749–2754.

Kenmore, P.E., Carino, F.O., Perez, C.A., Dyck, V.A. and Gutierrez, A.P. (1984) Population regulation of the rice brown planthopper (*Nilaparvata lugens* Stal) within rice fields in the Philippines. *Journal of Plant Protection in the Tropics* 1, 19–37.

Kidd, N.A.C. and Jervis, M.A. (1996) Population dynamics. In: Jervis, M. and Kidd, N. (eds) *Insect Natural Enemies. Practical Approaches to their Study and Evaluation.* Chapman & Hall, London, pp. 293–374.

Kiritani, K., Kawahara, S., Sasaba, T. and Nakasuji, F. (1972) Quantitative evaluation of predation by spiders on the green rice leafhopper, *Nephotettix cincticeps*, by a sight-count method. *Researches on Population Ecology* 13, 187–200.

Kring, T.J., Gilstrap, F.E. and Michels, G.J. (1985) Role of indigenous coccinellids in regulating greenbugs (Homoptera: Aphididae) on Texas grain sorghum. *Journal of Economic Entomology* 78, 269–273.

Kuperstein, M.L. (1979) Estimating carabid effectiveness in reducing the Sunn pest, *Eurygaster integriceps* Puton (Heteroptera: Scutellaridae) in the USSR. *Entomological Society of America Miscellaneous Publication* 11, 80–84.

Lister, A., Usher, M.B. and Block, W. (1987) Description and quantification of field attack rates by predatory mites: an example using an electrophoresis method with a species of Antarctic mite. *Oecologia* 72, 185–191.

Luck, R.F., Shepard, B.M. and Kenmore, P.E. (1988) Experimental methods for evaluating arthropod natural enemies. *Annual Review of Entomology* 33, 367–391.

McCarty, M.T., Shephard, M. and Turnipseed, G. (1980) Identification of predaceous arthropods in soybeans by using autoradiography. *Environmental Entomology* 9, 199–203.

Miller, C.A. (1954) A technique for assessing spruce budworm larval mortality caused by parasites. *Canadian Journal of Zoology* 33, 5–17.

Mills, N.J. (1991) Searching strategies and attack rates of parasitoids of the ash bark beetle (*Leperisinus varius*) and its relevance to biological control. *Ecological Entomology* 16, 461–470.

Mills, N.J. (1994) Parasitoid guilds: defining the structure of the parasitoid communities of endopterygote insect hosts. *Environmental Entomology* 23, 1066–1083.

Moore, J. (1995) The behavior of parasitized anaimals. *Bioscience* 45, 89–96.

Munster-Swendsen, M. (1991) The effect of sublethal neogregarine infections in the spruce needleminer, *Epinotia tedella* (Cl.) (Lepidoptera: Tortricidae). *Ecological Entomology* 16, 211–219.

Murray, R.A. and Solomon, M.G. (1978) A rapid technique for analysing diets of invertebrate predators by electrophoresis. *Annals of Applied Biology* 90, 7–10.

Nagai, K. (1990) Suppressive effect of *Orius* sp. (Hemiptera: Anthocoridae) on the population density of *Thrips palmi* Karny (Thysanoptera: Thripidae) in eggplant in an open field. *Japanese Journal of Applied Entomology and Zoology* 34, 109–114.

Nakamura, M. and Nakamura, K. (1977) Population dynamics of the chestnut gall wasp (Cynipidae, Hym.). 5. Estimation of the effect of predation by spiders on the mortality of imaginal wasps based on the precipitin test. *Oecologia* 27, 97–116.

Newman, G.G. and Carner, G.R. (1975) Disease incidence in soybean loopers collected by two sampling methods. *Environmental Entomology* 4, 231–232.

Nyffeler, M., Dean, D.A. and Sterling, W.L. (1987) Evaluation of the importance of the striped lynx spider, *Oxyopes salticus* (Araneae: Oxyopidae), as a predator in Texas cotton. *Environmental Entomology* 16, 1114–1123.

Powell, W., Walton, M.P. and Jervis, M.A. (1996) Populations and communities. In: Jervis, M. and Kidd, N. (eds) *Insect Natural Enemies. Practical Approaches to Their Study and Evaluation.* Chapman & Hall, London, pp. 223–292.

Roland, J. (1994) After the decline: what maintains low winter moth density after successful biological control? *Journal of Animal Ecology* 63, 392–398.

Rothschild, G.H.L. (1966) A study of a natural population of *Conomelus anceps* Germar (Homoptera: Delphacidae) including observations on predation using the precipitin test. *Journal of Animal Ecology* 35, 413–434.

Royama, T. (1981) Evaluation of mortality factors in insect life table analysis. *Ecological Monographs* 5, 495–505.

Russell, D.A. (1987) A simple method for improving estimates of percentage parasitism by insect parasitoids from field sampling of hosts. *New Zealand Entomologist* 10, 38–40.

Ryan, R.B. (1985) A hypothesis for decreasing parasitization of larch casebearer (Lepidoptera: Coleophoridae) on larch foliage by *Agathis pumila*. *Canadian Entomologist* 117, 1573–1574.

Ryan, R.B. (1990) Evaluation of biological control: introduced parasites of larch casebearer (Lepidoptera: Coleophoridae) in Oregon. *Environmental Entomology* 19, 1873–1881.

Smith, H.R. (1985) Wildlife and gypsy moth. *Wildlife Society Bulletin* 13, 166–174.

Smith, H.R. and Lautenschlager, R.A. (1981) Gypsy moth predators. In: Doane, C.C. and McManus, M.L. (eds) *The Gypsy Moth: Research Toward Integrated Pest Management*. USDA Forest Service Technical Bulletin 1584, pp. 96–125.

Solomon, M.G., Murray, R.A. and Van Der Geest, L.P.S. (1985) Analysis of prey by means of electrophoresis. In: Helle, W. and Sabelis, M.B. (eds) *Spider Mites. Their Biology, Natural Enemies and Control. Volume IB*. Elsevier Science Publishers, Amsterdam, pp. 171–173.

Sopp, P.I. and Sunderland, K.D. (1989) Some factors affecting detection period of aphid remains in predators using ELISA. *Entomologia Experimentalis et Applicata* 51, 11–20.

Sopp, P.I., Sunderland, K.D., Fenlon, J.S. and Wratten, S.D. (1992) An improved quantitative method for estimating invertebrate predation in the field using an enzyme-linked immunoabsorbent assay. *Journal of Applied Ecology* 29, 295–302.

Southwood, T.R.E. (1978) *Ecological Methods*. Chapman & Hall, London, 524pp.

Stiling, P. (1988) Density-dependent processes and key factors in insect populations. *Journal of Animal Ecology* 57, 581–594.

Stuart, M.K. and Greenstone, M.H. (1996) Serological diagnosis of parasitism: a monoclonal anitibody-based immunodot assay for *Microplitis croceipes* (Hymenoptera: Braconidae). In: Symondson, W.O.C. and Liddell, J.E. (eds) *The Ecology of Agricultural Pests: Biochemical Approaches*. Chapman & Hall, London, pp. 300–321.

Sunderland, K.D. (1975) The diet of some predatory arthropods in cereal crops. *Journal of Applied Ecology* 12, 507–515.

Sunderland, K.D. (1988) Quantitative methods for detecting invertebrate predation occurring in the field. *Annals of Applied Biology* 112, 201–224.

Sunderland, K.D., Fraser, A.M. and Dixon, A.F.G. (1986) Field and laboratory studies on money spiders (Linyphiidae) as predators of cereal aphids. *Journal of Applied Ecology* 23, 433–447.

Symondson, W.O.C. and Liddell, J.E. (1995) Decay rates for slug antigens within the carabid predator *Pterostichus melanarius* monitored with a monoclonal antibody. *Entomologia Experimentalis et Applicata* 75, 245–250.

Tamaki, G., McGuire, J.U. and Turner, J.E. (1974) Predator power and efficiency: a model to evaluate their impact. *Environmental Entomology* 3, 625–630.

Van Driesche, R. (1983) Meaning of 'percent parasitism' in studies of insect parasitoids. *Environmental Entomology* 12, 1611–1622.

Van Driesche, R. (1988) Field measurement of population recruitment of *Apanteles glomeratus* (L.) (Hymenoptera: Braconidae), a parasitoid of *Pieris rapae* (L.) (Lepidoptera: Pieridae), and factors influencing adult parasitoid foraging success on kale. *Bulletin of Entomological Research* 78, 199–208.

Van Driesche, R. and Bellows, T. (1988) Use of host and parasitoid recruitment in quantifying losses from parasitism in insect populations. *Ecological Entomology* 13, 215–222.

Van Driesche, R. and Bellows, T. (1996) *Biological Control*. Chapman & Hall, New York, 539pp.

Van Driesche, R. and Taub, G. (1983) Impact of parasitoids on *Phyllonorycter* leafminers infesting apple in Massachusetts, USA. *Protection Ecology* 5, 303–317.

Van Driesche, R., Bellows, T.S., Elkinton, J.S., Gould, J.R. and Ferro, D.N. (1991) The meaning of percent parasitism revisited: solutions to the problem of accurately estimating total losses from parasitism. *Environmental Entomology* 20, 1–7.

Varley, G.C. and Gradwell, G.R. (1960) Key factors in insect population studies. *Journal of Animal Ecology* 29, 399–401.

Varley, G.C., Gradwell, G.R. and Hassell, M.P. (1973) *Insect Population Ecology: an Analytical Approach*. Blackwell, Oxford.

Waage, J.K. and Mills, N.J. (1992) Understanding and measuring the impact of natural enemies on pest populations. In: Markham, R.H., Wodageneh, A. and Agboola, S. (eds) *Biological Control Manual. Volume I. Principles and Practice of Biological Control*. International Institute of Tropical Agriculture, Cotonou, Benin, pp. 84–114.

Walton, M.P., Powell, W., Loxdale, H.D. and Allen-Williams, L. (1990) Electrophoresis as a tool for estimating levels of hymenopterous parasitism in field populations of the cereal aphid, *Sitobion avenae*. *Entomologia Experimentalis et Applicata* 54, 271–279.

Webb, S.E. and Shelton, A.M. (1990) Effect of age structure on the outcome of viral epizootics in field populations of imported cabbageworm (Lepidoptera: Pieridae). *Environmental Entomology* 19, 111–116.

Weseloh, R.M. (1990) Estimation of predation rates of gypsy moth larvae by exposure of tethered caterpillars. *Environmental Entomology* 19, 448–455.

Winder, L. (1990) predation of the cereal aphid *Sitobion avenae* by polyphagous predators on the ground. *Ecological Entomology* 15, 105–110.

Wratten, S.D. (1987) The effectiveness of native natural enemies. In: Burn, A.J., Coaker, T.H. and Jepson, P.C. (eds) *Integrated Pest Management*. Academic Press, London, pp. 89–112.

12 Biochemical and Molecular Techniques

W.O.C. Symondson and J. Hemingway
*School of Pure & Applied Biology, University of Wales, Cardiff,
PO Box 915, Cardiff CF1 3TL, UK*

12.1. Introduction

Biochemical and molecular methods of examining ecological processes, and interactions between organisms, have come to dominate many areas of entomological research in recent years. Although this is largely a result of rapid technological development, generating a number of highly sensitive and precise systems for detection and analysis, the driving force behind these developments is the wish to acquire unequivocal data, whether quantitative, qualitative or both. The main areas of research to benefit from these developments are studies of population genetics (gene flow, local and metapopulation dynamics, polymorphism, speciation); taxonomy and systematics (identification of species, biotypes and phylogenies); and studies of trophic interactions, including predation, parasitism, symbiosis and infection. A major sub-area of population genetics is currently the study of insecticide resistance, and this subject will be treated separately as a case study, illustrating as it does the value of combining a number of different biochemical and molecular approaches. This chapter will concentrate upon techniques relevant to these areas of research. Other biochemical topics, such as host plant selection and resistance, chemical communication systems in insects, and the use of semiochemicals in biological control programmes, will be dealt with elsewhere.

The range of techniques now available is expanding rapidly, with new developments overtaking and frequently replacing earlier methodologies. The choice of the right system for a particular task is not always obvious, and can be determined as much by the availability of resources and expertise as by the appropriateness of the technology. There is also often little point in gathering more detailed information than is required to answer the questions posed. For example, you do not need to analyse the complete genetic codes of two populations of morphologically identical insects if all you wish to do is find a means of separating them, and a simple enzyme electrophoretic analysis might do this for you. The techniques included below have been selected for their proven utility in a number of situations. Important variations on these are mentioned more briefly, and relevant references cited that include full experimental details. New developments in biochemical and molecular entomology, which appear to be of considerable importance for future research, will also be reported. A chapter such as this can give no more than an brief overview of the techniques available, and recourse to specialist texts is therefore strongly advised.

Most of the systems used, and covered here, could be broadly categorized as antibody, electrophoretic and DNA methods. Many biochemical techniques have been developed for very specific purposes, or have to be radically adapted for the species under study. The chapter ends, therefore, with a detailed case study in which many different techniques have been used to study insecticide resistance in insects, and incorporates descriptions of biochemical and molecular techniques not covered elsewhere.

12.2. Antibody Techniques

The history of the use of antibodies in entomology goes back to the turn of the century. Immunization of mammals with serum proteins from other species stimulates a complex series of immune responses (Roitt et al., 1989), one of which is the proliferation of antibodies that bind specifically with sites (antigenic determinants or epitopes) on the antigens injected. These antibodies can be harvested from blood sera, and used in a number of different assay systems to test for the presence of target proteins (or other biological materials). It was discovered at an early stage that antisera raised against one species would react to different degrees with antigens from other species, the strength of the reaction approximating to taxonomic distance (Nuttall, 1904; Benjamin et al., 1984). This property was first exploited for the identification of bloodmeals in haematophagous arthropods such as ticks, mosquitoes, sandflies, midges and tsetse flies (Bull and King, 1923; Weitz, 1956; Washino and Tempelis, 1983). Later it was found that similar techniques could be used to investigate predation (reviewed in Sunderland, 1988; Greenstone, 1996), parasitism (Ragsdale and Kjer, 1989; Stuart and Greenstone, 1996) and infection (Greenstone, 1983; Oien and Ragsdale, 1992).

12.2.1. Polyclonal antisera

Although serum proteins, such as immunoglobulins and serum albumins, have been used as antigens in studies of haematophagous insects, most large proteins and polysaccharides will elicit an antigenic response if injected into a mouse or rabbit (Tijssen, 1985). Such antigenic molecules usually possess many sites to which antibodies can bind, each site on proteins, for example, consisting of as few as eight to ten amino acids in a particular stereochemical configuration. The immune response will stimulate the proliferation of many different antibodies to each of these sites, with different binding properties in terms of affinity and specificity, creating in total a 'polyclonal' mixture in the blood of the host. The intention usually is to create an antiserum that will react with antigens from a target group of hosts or prey and no other. However, the greater the diversity of antibodies in the antiserum, the greater the chance of cross-reactivity with a site on biological material from a non-target organism.

Wherever possible, therefore, the range of antigens should be limited. In practice useful antisera to broad groups of target prey, such as aphids, have been obtained using crude whole-body extracts as antigens. Crook and Sunderland (1984) produced a general aphid-specific antiserum in this way, which reacted strongly with the species used in antigen preparation, but less strongly with other species. Chiverton (1987) improved the specificity of his antiserum by using proteins from a single aphid species (*Rhopalosiphum padi* L.) as antigen, reducing cross-reactivity with other aphids to lower levels. Until recently such whole-body mixtures were the norm in studies of predation on insects (e.g. Dempster, 1960; Sutton, 1970; Service, 1973; Ashby, 1974; Lund and Turpin, 1977; Ragsdale *et al.*, 1981; Allen and Hagley, 1982; Dennison and Hodkinson, 1983; Doane *et al.*, 1985; Cameron and Reeves, 1990). Further limitation of antigen heterogeneity can be obtained by using haemolymph proteins from larger invertebrates such as slugs (Symondson and Liddell, 1993a). This principle has been applied to bloodmeal identification for many years, with progressive development from earlier immunoassays, using whole blood sera (Weitz, 1960; Tempelis, 1975; Boorman *et al.*, 1977), which frequently displayed high levels of cross-reactivity, to more recent assays using antisera raised against a single class of host immunoglobulin (IgG) (Service *et al.*, 1986; Blackwell *et al.*, 1994), with a consequent increase in specificity.

A general protocol from the production of a polyclonal antiserum is given in Appendix A (detailed descriptions in Tijssen, 1985; Hudson and Hay, 1989; Liddell and Cryer, 1991). Where cross-reactivity occurs it is often possible to improve specificity in one of two ways. The simplest solution is to absorb the cross-reacting antibodies by mixing the antiserum with proteins from the non-target species concerned. Cross-reacting antibodies will then precipitate out of solution, and can be discarded after centrifugation (Catty and Raykundalis, 1988; Symondson and Liddell, 1993a). Alternatively, the specificity of the antiserum can also be improved by affinity purification. Antibodies can be passed through a column (e.g. Sepharose 4B, Pharmacia) to which the non-target proteins concerned are bound. The antibodies that pass through the column are specific for the antigen, while cross-reacting antibodies can be discarded with the column (Schoof *et al.*, 1986). Only if cross-reactivity is severe, and the titre of the antiserum is seriously reduced by these processes, is it necessary to make a new antiserum with a narrower range of antigens. Antigens can be purified and

Table 12.1. Recent studies in which enzyme-linked immunosorbent assays and immunodot assays were developed to study predation and blood-feeding by insects, using polyclonal or monoclonal antibodies.

Target	Predator	References
Nezara viridula (L.) (Hemip.: Pentatomidae)	Insecta, Aranae	Ragsdale *et al.*, 1981
Dendroctonous frontalis Zimmerman (Coleop.: Scolytidae)	NR	Miller, 1981
Orgyia pseudotsugata McDunnough (Lepid.: Lymantriidae)	*Podisus maculiventris* (Say) (Hemip.: Pentatomidae)	Fichter and Stephen, 1981
Aphididae	Carabidae	Crook and Sunderland, 1984
	Forficula auricularia L. (Dermap.: Forficulidae)	
Ostrinia nubilalis Hübner (Lepid.: Pyralidae)	*Pterostichus cupreus* L. (Coleop: Carabidae)	Lovei *et al.*, 1985
Culex tarsalis Coquillett (Dipt.: Culicidae)	Notonectidae, Belostomatidae, Coenagrionidae	Schoof *et al.*, 1986
Vertebrate blood	*Anopheles stephensi* Liston, *Aedes aegypti* (L.),	Service *et al.*, 1986
	Culex quinquefasciatus Say (Dipt.: Culicidae)	
Rhopalosiphum padi L. (Hemip.: Aphididae)	Carabidae, Araneae, Opiliones, other polyphagous predators	Chiverton, 1987
Aphis fabae (Scopoli) (Hemiptera: Aphididae)	*Pterostichus melanarius* Illiger (Col.: Carabidae)	Hance and Renier, 1987
Pieris brassicae (L.) (Lepid.:Pieridae)	Coleoptera	Kapuge *et al.*, 1987
Sitobion avenae (F.) (Hemip.: Aphididae)	*Bembidion lampros* Herbst (Coleop.: Carabidae)	Lovei *et al.*, 1987
Aphididae	Insecta, Aranae	Sunderland *et al.*, 1987
Helicoverpa zea (Boddie) (Lepid.: Noctuidae)	*P. maculiventris*	Greenstone and Morgan, 1989
Sitobion avenae (F.) (Hemip.: Aphididae)	Linyphiidae, Carabidae, Staphylinidae	Sopp and Sunderland, 1989
Lymantria dispar (L.) (Lepid.: Lymantriidae)	Carabidae, Arachnida, Chilopoda	DuDevoir and Reeves, 1990
H. zea	*Phidippus audax* (Hentz) (Araneae: Salticidae)	Stuart and Greenstone, 1990
	P. maculiventris	

Lymantria dispar (L.) (Lepid.: Lymantriidae)	Carabidae	Cameron and Reeves, 1990
Lygus hesperus Knight (Hemip.: Miridae)	Hemiptera	Hagler *et al.*, 1992
Vertebrate blood	Phlebotomine sandflies	Ngumbi *et al.*, 1992
S. avenae	Carabidae, Staphylinidae, Araneae	Sopp *et al.*, 1992
Lopidae nigridea Uhler, *Coquillettia insignis* Uhler (Hemip.: Miridae)	Insecta, Aranae	McIver and Tempelis, 1993
Slugs (Mollusca: Pulmonata)	*Abax parallelepipedus* (Piller and Mitterpacher) (Coleop.: Carabidae)	Symondson and Liddell, 1993a
	Phalangium opilio (Opiliones)	
Slugs	*A. parallelepipedus, Pterostichus madidus* (F.) (Coleop.: Carabidae)	Symondson and Liddell, 1993d
Pectinophora gossypiella (Saunders) (Lepid.: Gelechiidae)	*Collops vittatus* (Say) (Coleop.: Melyridae) *Hippodamia convergens* Guérin-Méneville (Coleop.: Coccinellidae)	Hagler and Naranjo, 1994a
Bemisia tabaci (Gennadius) (Homop.: Aleyrodidae)	Heteroptera	Hagler and Naranjo, 1994b
P. gossypiella	*Culicoides impunctatus* Goetghebuer (Dipt.: Ceratopogonidae)	Blackwell *et al.*, 1994, 1995
Mammalian blood	*P. maculiventris*	
H. zea	*Glossina pallidipes* Austin and *G. longipennis* Corti (Dipt.: Glossinidae)	Greenstone and Trowell, 1994
Vertebrate blood		Sasaki *et al.*, 1995
Slugs	*Pterostichus melanarius* (Illiger) (Coleop.: Carabidae)	Symondson and Liddell, 1995, Symondson *et al.*, 1996
Helicoverpa armigera Hübner (Lepid.: Noctuidae)	Aranae, Coleoptera, Heteroptera and Lepidoptera	Sigsgaard, 1996

characterized to some degree by protein electrophoresis on agarose or polyacrylamide gels prior to immunization (Tijssen, 1985).

The main entomological uses of polyclonal antisera have been for predation studies and bloodmeal identification. Table 12.1 provides a list of recent studies involving enzyme-linked immunosorbent assays (ELISA) to detect prey remains or bloodmeals in predator gut samples, while a more extensive list including earlier assay techniques can be found in Greenstone (1996). Other uses of polyclonal antisera include the detection of infection. Crook and Payne (1980) compared the ability of different ELISA formats to detect baculoviruses, with possible biocontrol potential, from *Pieris brassicae* (L.), *Agrotis segetum* Dennis and Schiffermuller and *Cydia pomonella* (L.). Greenstone (1983) produced an antiserum against an entomopathogenic microsporidium (*Amblyospora* sp.) that infects the mosquito *Culex salinarius* Coquillett. Ragsdale and Oien (1996) developed an assay for the detection of an exotic microsporidium, *Nosema furnacalis* (Guenee), within the European Corn Borer, *Ostrinia nubilalis* (Hübner), as part of a biocontrol assessment programme. They were able to obtain a species-specific assay by raising antisera to partially purified proteins from germinated spores of the microsporidium, which did not cross-react with spore coat proteins (or therefore with ungerminated spores), or with other species within the same genus (Oien and Ragsdale, 1992).

Polyclonal antisera are cheap, quick and simple to produce, but have certain disadvantages. Once a batch of antiserum has run out, a new one has to be made which, even if generated in the same individual mammal, will often have different characteristics in terms of sensitivity and specificity. This factor militates against reproducibility and comparability of results over time. Cross-reactivity can present difficulties even where single proteins, such as immunoglobulins, are used as antigens (Blackwell *et al.*, 1994). Where greater precision is required, monoclonal antibodies can provide an answer to such problems (Section 12.2.3).

12.2.2. Assays

A great many different assay systems have been used over the years for immunological detection of target antigens, and excellent reviews of these, as applied in agricultural and ecological entomology, can be found in Boreham and Ohiagu (1978), Washino and Tempelis (1983), Sunderland (1988), Powell and Walton (1995) and Greenstone (1996). Most of the earlier work exploited the fact that antibodies, which are bivalent (two arms), can form matrices with antigen molecules containing more than one binding site, which can be observed directly as a white precipitate. Such precipitin techniques came in a number of inventive forms, all designed to bring antibodies and antigen together within a fluid or gel medium (Crowle, 1980). The most important methods used in studies of predator–prey relationships, and bloodmeal identification, include: the ring test, in which antibodies and antigens are brought together within a capillary tube, and a precipitate forms at the point of contact (Weitz, 1960; Nakamura and Nakamura, 1977; Gardener *et al.*, 1981; Dennison and Hodkinson, 1983); the Ouchterlony double diffusion test, in which antiserum and antigen are allowed to diffuse towards each other from wells cut into a gel on a glass slide or petri dish, lines of precipitate forming where positive results occur (Tod, 1973; Lund and Turpin, 1977; Whallon and Parker, 1978; Kuperstein, 1979; Lesiewicz *et al.*, 1982); a number of immunoelectrophoretic methods, in which component antigens are separated prior to precipitin reactions, allowing prey-specific arcs to be identified (Pettersson, 1972; Hance and Rossignol, 1983), or in which antigens and antiserum are brought together more rapidly (and with increased sensitivity) under an electric field (Sergeeva, 1975; Leslie and Boreham, 1981; Allen and Hagley, 1982; Doane *et al.*, 1985).

Clearly visible matrices can also be formed by coating antigen, or antibodies, on to particles, such as erythrocytes or latex beads, agglutination indicating a positive reaction. A number of different agglutination, and agglutination inhibition, assays are described by Nichols and Nakamura (1980) and Greenstone (1977, 1996).

None of the assay techniques described above, nor others covered by some of the reviews quoted, is much used today. The assay of choice is now the enzyme-linked immunosorbent assay (ELISA), due to its extreme sensitivity and suitability for quantitative analyses (Table 12.1). There are many different assay formats, most operating on 96-well microtitration plates (details in Voller *et al.*, 1979; Tijssen, 1985; Voller and Bidwell, 1986). Two systems dominate: the indirect ELISA and the double antibody sandwich ELISA (Appendix B and Fig. 12.1). Both systems rely upon the direct relationship between the quantity of antigen in a test sample and colour development, caused by the action of the chosen enzyme catalysing the change of the enzyme substrate from a colourless to a coloured product. Such a reaction can be detected spectroscopically, or visually, at levels of antigen concentration very much lower than would be required to form an antibody–antigen precipitate. An extensive table of detection limits achieved over the years in entomological predation studies using various assay techniques can be found in Greenstone (1996), and clearly shows the advantages of ELISA tests which are routinely capable of detecting < 10 ng of prey antigen.

The simplest, and probably the most sensitive, assay format is the indirect ELISA (Fig. 12.1). Test molecules, such as predator gut samples, are bound to the polystyrene solid phase. Antibodies are then added, and where target antigen is present these will be bound by antibody–antigen reactions. Commercially available anti-species antibodies, conjugated with an enzyme such as horseradish peroxidase or alkaline phosphatase, can then be used to detect such antibody–antigen reactions by binding with the first antibody. The quantity of enzyme in the system will then be proportional to the amount of antigen present, and this can be measured by the colour change induced in the enzyme substrate. Concentrations of antigen,

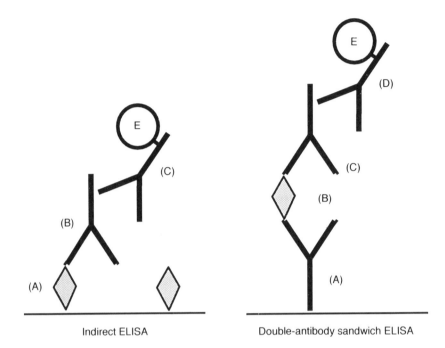

Indirect ELISA   Double-antibody sandwich ELISA

Fig. 12.1. Schematic representation of two common enzyme-linked immunosorbent assay formats. In the indirect assay, (A) represents the specific antigen, bound to the ELISA plate, (B) a rabbit antibody binding with that antigen, and (C) an anti-rabbit antibody, conjugated with an enzyme, bound to the rabbit antibody. In the double-antibody sandwich, (A) represents, for example, polyclonal rabbit antibodies bound to the ELISA plate, (B) captured antigen molecules, (C) a mouse monoclonal antibody also bound to the antigen, and (D) an anti-mouse enzyme conjugate. For explanation of the assays, see Appendix B.

antibody and conjugate can vary considerably, and need to be established at an early stage to optimize the assay in terms of sensitivity, while minimizing background effects. The basic ELISAs are the same whether using polyclonal (Section 12.2.1) or monoclonal (Section 12.2.3) antibodies. Examples of the use of indirect assay in practice, to analyse large numbers of field samples, may be found in Hance and Rossignol (1983), Hagler and Naranjo (1994a,b), Symondson and Liddell (1993d), Sigsgaard (1996) and Symondson *et al.* (1996). Double antibody sandwich ELISAs were used by, for example, Chiverton (1987), Sunderland *et al.* (1987), DuDevoir and Reeves (1990) and Sopp *et al.* (1992). In general, the indirect ELISA is considered to be more sensitive, while the two-site double antibody sandwich is more specific.

An important variant on the ELISA format is the dot-ELISA, or immunodot assay, in which samples are tested on nylon or nitrocellulose membranes instead of ELISA plates. Such systems have advantages in terms of assay time, while retaining high sensitivity, but are less suitable where antigen quantification is important. Dot-ELISAs have been used to analyse the gut contents of predators and to detect parasitism, and are described in Stuart and Greenstone (1990, 1996) and Greenstone and Trowell (1994).

12.2.3. Monoclonal antibodies

Antibodies are produced in mammals by B lymphocytes. As part of the immune response, pre-existing lymphocyte clones generating antibodies, that bind to sites on invading antigens, proliferate, raising the titre (or concentration) of such antibodies in the blood. Unfortunately it is not possible to culture such clones *in vitro* in order to generate useful monoclonal antibodies directly from single clones. However, Köhler and Milstein (1975) found that by fusing lymphocytes with myeloma cells, originally derived from the immune system, hybrid cells, or hybridomas, could be created which could both generate antibodies and be cultured as immortal cell lines *in vitro*. By selecting hybridoma cell lines that produce antibodies to a single targeted site on a protein, a very precise probe can be created for use in immunoassays. Recently this technology has been adopted by entomologists in studies of predation on agricultural pests such as the whitefly *Bemisia tabaci* (Gennadius) (Hagler *et al.*, 1993), the noctuid moth *Helicoverpa zea* (Boddie) (Greenstone and Morgan, 1989), the grain beetles *Trogoderma granarium* Everts (Stuart *et al.*, 1994) and the slug *Deroceras reticulatum* (Müller) (Symondson and Liddell, 1996a). In each instance it was important to distinguish between the remains of individual species of pest, in predator gut samples, and remains from related species found in the same ecosystem and consumed by the same predators. Other uses for monoclonal antibodies in studies involving arthropods have been in phylogenetic studies of the Heliothinae (Lepidoptera: Noctuidae) (Greenstone *et al.*, 1991) and scorpions (Buthidae) (Billiald and Goyffon, 1992), the detection and identification of tracheal mites, *Acarapis woodi* Rennie, in honey bees (Ragsdale and Kjer, 1989), and the detection and identification of parasitoid larvae, *Microplitis croceipes* (Cresson), within their hosts, *H. zea* (Stuart and Greenstone, 1996). Reviews of the work of all these research groups can be found in Symondson and Liddell (1996c). Monoclonal antibodies were also used to detect American foulbrood, *Bacillus larvae* White, in honey bees, *Apis mellifera* L., and could successfully distinguish this bacterium from other species within the same genus (Olsen *et al.*, 1990).

Anyone contemplating making monoclonal antibodies is strongly advised to consult specialist publications such as Goding (1986), Liddell and Cryer (1991) or Liddell and Weeks (1995). Although the techniques involved are beyond the scope of this chapter, all can be undertaken in laboratories with cell culturing facilities, and in particular a laminar air flow cabinet in which most of the work will be carried out. The only other major item of equipment required is an incubator with CO_2 control.

Figure 12.2 presents a flow diagram of the steps in monoclonal antibody

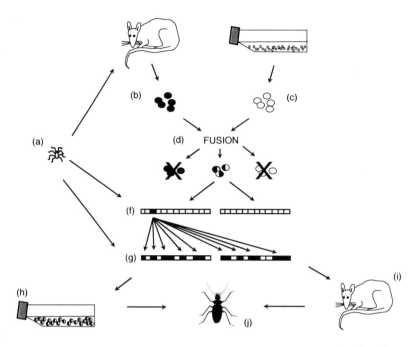

Fig. 12.2. Flow diagram of the steps involved with monoclonal antibody production. For an explanation see text (Section 12.2.3).

production. These antibodies are normally raised in mice, and the immunization protocol is similar to that for rabbits (Appendix A). Each mouse is injected subcutaneously with 10–50 µg of immunogen per 100 µl dose (a), made up as emulsions of 50% complete and incomplete Freund's adjuvant for the first two immunizations respectively, one month apart. This is followed by a final intravenous booster 4 days before fusion.

The mouse is then killed and a single cell suspension made of the spleen cells. The simplest protocol for fusing spleen cells (b) with myeloma cells (c) uses polyethylene glycol (Liddell and Cryer, 1991). Fusion is a crude process (d), which results in a mixture of lymphocytes (unfused spleen cells), unfused myeloma cells and hybridomas (e). The lymphocytes will subsequently die because they cannot be propagated *in vitro*. The main danger is that myeloma cells will subsequently grow up at the expense of the antibody-producing lymphocyte–myeloma hybrids. This is avoided by using a myeloma cell line that lacks the enzyme hypoxanthine guanine phosphoribosyl transferase (HGPRT$^-$), an enzyme necessary for one of two pathways for DNA synthesis. The 'salvage' pathway can be used by cells when the main biosynthetic pathway is blocked by the antibiotic aminopterin. By growing the mixture of myelomas and hybridomas in a medium containing hypoxanthine, aminopterin and thymidine (HAT medium), the HGPRT$^-$

myelomas will die while the HGPRT$^+$ hybridomas, inheriting the necessary genes from the spleen cells, will survive.

Mixtures of hybridoma cells can then be subcultured (f and g) and cell lines producing antibodies that bind specifically with target proteins isolated. Antibodies generated by hybridoma cells can be screened using indirect ELISAs (Appendix B). Bulk supplies of monoclonal antibodies are then produced by *in vitro* (h) or *in vivo* (i) methods (Liddell and Cryer, 1991; Liddell and Weeks, 1995).

The whole process can easily take six months to produce a useful monoclonal antibody. Thereafter, however, limitless supplies of that antibody can be easily and rapidly generated, providing a highly specific and uniform reagent for use in ELISAs and dot-ELISAs to screen predator gut samples (j), or in histology and immuno-diagnostic systems. A major disadvantage of polyclonal antisera, by contrast, is that supplies are limited; no two polyclonal antisera have the same properties (even when generated in the same individual mammal), in terms of sensitivity, specificity and affinity for target antigens, and therefore it is difficult to compare results over time or between laboratories. The fact that monoclonal antibodies bind to a single epitope on a target prey protein, representing as few as eight to ten amino acids, means that cross-reactions are unlikely. A polyclonal antibody may contain thousands of different antibody specificities, and the chances that one of these will cross-react with non-target prey proteins is clearly much higher. The additional precision that monoclonal antibodies provide has allowed antibodies to be generated that can distinguish between very closely related prey, such as noctuid moths within the Heliothinae (Greenstone *et al.*, 1991), different species of whitefly (Hagler *et al.*, 1993) and slugs within the *Deroceras* genus (Symondson and Liddell, 1996a). In several instances, monoclonal antibodies have now been created that are specific to particular stages, such as specific instars of *H. zea* (Greenstone and Morgan, 1989), or the eggs of a number of invertebrates including the bug *Lygus hesperus* (Hagler *et al.*, 1991), the whitefly *B. tabaci* (Hagler *et al.*, 1993), the pink bollworm *Pectinophora gossypiella* (Saunders) (Hagler *et al.*, 1994) and the slugs *D. reticulatum* and *Arion ater* (L.) (Symondson *et al.*, 1995). Monoclonal antibodies usually have shorter detection periods than polyclonal antisera; whereas the former depends upon the rate at which a single protein, or epitope on a protein, becomes denatured, polyclonal antisera will contain antibodies to all the antigenic sites in a mixture of prey proteins, including those that resist digestion for extended periods. For example, prey remains can be detected for up to 13 days with a polyclonal anti-slug haemolymph antiserum (Symondson and Liddell, 1993c), but less than 24 h with an anti-arionid haemocyanin monoclonal antibody (Symondson and Liddell, 1993b). The monoclonal system can therefore define more precisely the time interval within which predation must have occurred.

12.2.4. Recombinant antibodies

Currently work is progressing on the production of recombinant antibodies by genetic engineering for medical applications (Williamson *et al.*, 1993; Winter *et al.*, 1994). The aim is to create antibody gene libraries from sensitized lymphocytes, from which phage display antibodies can be created and propagated in *Escherichia coli*. These techniques have the potential to generate monoclonal antibodies much more rapidly than by hybridoma techniques, with a greater range of specificities. They offer the opportunity of increasing the affinity and specificity of antibodies by genetic manipulation, and have considerable potential in studies of trophic interactions (Liddell and Symondson, 1996).

Figure 12.3 presents a flow diagram of steps required to produce recombinant antibodies although systems other than the one described exist. In this protocol (Liddell and Weeks, 1995) a mouse (b) is immunized with proteins from the target pest (a) and RNA extracted from the lymphocytes (c). This is then transcribed to cDNA (d) and amplified by PCR (e). The antibody heavy and light chains are then isolated by gel electrophoresis (f), cut with appropriate restriction enzymes, purified and ligated into a phagemid expression vector (g). This is then transfected into *E. coli* by electroporation (h). Co-infection with helper phage (i) allows the release of phage into the surrounding medium, expressing antibody binding sites on its surface. This

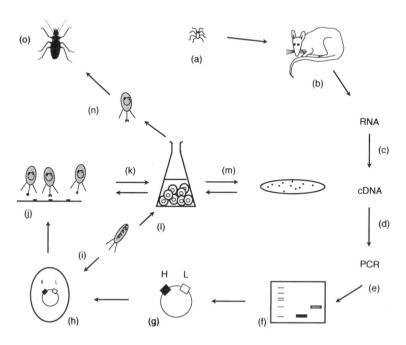

Fig. 12.3. Flow diagram of the steps involved with the production of recombinant antibodies. For a full explanation see text (Section 12.2.4).

phage antibody can then be screened in a process similar to ELISA, in which affinity purification by absorption to antigen can take place on a plate (j). After repeated rounds of purification and growth (k and l), single mono-clonal colonies, generating antibody specific for the target protein, can be isolated on agar (m), cultured and reinfected with helper phage to release monoclonal phage antibody (n) for use in ELISA tests of predator gut samples (o).

Although currently there are no published papers in which useful anti-bodies have been created by these means, the advantages of this approach are such that it is likely to be of increasing importance in the future. Establishing tissue culture facilities for conventional monoclonal antibody production is expensive, and requires both skills and facilities not generally available in entomological and agricultural research establishments. Molecular facilities are, by contrast, becoming universally available. Currently work is proceeding to create recombinant antibodies to the slug *Arion intermedius* Normand and a number of species of whitefly and aphid (W.O.C. Symondson, N. Walkley, T. Gasull and J.E. Liddell, unpublished data).

12.2.5. *Quantifying rates of consumption and predation in the field from ELISA data*

ELISA data can provide information at a number of different levels. At its simplest, it can help to determine which predators are feeding upon a par-ticular target prey. Many studies make no attempt to quantify the mass of prey within individual predators, and simply use a proportional analysis to calculate which may be the most important predators in relation to the per-centage containing prey antigens. This can then be related to predator numbers in the field, although density cannot always be accurately assessed (Sunderland *et al.*, 1995). However, comparative studies between predator species indicate that differential digestion rates can be an important source of error (Sunderland *et al.*, 1987). For example, a study of two carabid beetles, *Pterostichus madidus* (F.) and *Abax parallelepipedus* (Piller and Mitterpacher), showed that the same species of prey could be detected in the former for a period approximately 2.5 times as long as in the latter under identical conditions (Symondson and Liddell, 1993c). In other words, more than 2.5 times as many *P. madidus* may need to test positive before it could be considered a more important slug predator than *A. parallelepipedus*. This of course takes no account of the mean meal size consumed by each species, nor indeed the number of prey killed. Digestion rates can in turn be affected by temperature (Sopp and Sunderland, 1989; Hagler and Cohen, 1990), meal size (McIver, 1981), prey species (Symondson and Liddell, 1993c) and subsequent feeding on alternative prey (Lovei *et al.*, 1987, 1990; Symondson and Liddell, 1995). Certain predators such as spiders are known to change their rate of digestion in response to starvation (Nakamura, 1977; Fichter and Stephen, 1984; Lucas and Graften, 1995), and therefore all these factors must be considered and calibrated for each predator–prey relationship.

Quantities of antigen-recognizable prey remains in the gut of a predator can be calculated from the weight of material in the gut of a predator and ELISA data, where protein standards have been included on ELISA plates (Sopp and Sunderland, 1989; Symondson and Liddell, 1993a). Indeed similar calculations may be made where densitometry is used to quantify predation using electrophoretic systems (e.g. Lister *et al.*, 1987). This does not, however, provide direct information on the quantity originally consumed. ELISA tests cannot distinguish, for example, between a large quantity of well digested target prey and a small quantity of fresh material mixed with non-reactive non-target prey remains. In both situations the crop weights, absorbance reading and calculated quantities of target prey material could be the same, but the biomass of this prey consumed would be very different. It should in theory be possible to estimate the size of the meal consumed, and indeed the time at which feeding occurred, by using combinations of two monoclonal antibodies, or both a monoclonal and polyclonal system in conjunction (Symondson and Liddell, 1993b; Sunderland, 1996). By monitoring the decay of two different antigens, within the same prey, that are digested at different rates, the changing ratio of one antigen to the other could be used to calculate these parameters. It should also be possible to estimate the time at which a prey item was consumed, as a separate parameter, by using a monoclonal antibody that detects a highly labile protein or epitope. Thus by selecting a monoclonal antibody with a short detection period, the time interval within which prey must have been consumed can be minimized, although this approach would necessitate the collection of large numbers of replicates trapped at intervals throughout the activity period of the predator (Sunderland, 1996).

Whether consumption rate or predation rate is the most important measure depends upon the type of prey. Pests such as slugs vary so greatly in size that a neonate individual could easily be one-thousandth of the mass of an adult. The damage they inflict on crops is approximately proportional to their mass, and therefore the number of slugs killed by predators is less important than the biomass taken. In practice it is the biomass killed, rather than the biomass consumed, that needs to be estimated, and there are many potential sources of error (Sunderland, 1996). For example, ELISA data cannot distinguish predation from scavenging. There may be food chain errors where antigen detected may be from secondary prey items, prey material already in the gut of an invertebrate consumed by the predator under study. Larger prey items may be only partially consumed, or partial consumption may result from satiation at high prey density (Lucas and Graften, 1985; Ernsting and van der Werf, 1988). In some instances a predator may take a bite out of a prey individual which subsequently escapes, and recovers, and this material would subsequently be detected in an ELISA. Conversely, the prey may be injured following an attack and subsequently die, with no ingestion of prey occurring. Predation on invertebrates made vulnerable by disease may make no contribution to pest control, and may even limit the spread of natural diseases and microbial biocontrol agents. Conversely, predators feeding on diseased material may help to disseminate the pathogen (Young and Kring, 1991).

In most instances, however, knowledge of the numbers of prey killed (the predation rate) is required, particularly for insects, such as aphids, where the size range in more limited and growth rates are rapid. All of the errors mentioned above can affect attempts to convert consumption rates to predation rates, and additional information is normally required from behavioural studies in order to estimate the importance of each factor and its relevance to predation models. Many of these factors, but particularly prey size choice, partial consumption of prey and scavenging, depend upon prey density, the size range available, predator hunger levels, time of year, predator sex and many other interacting variables. An excellent review of these factors may be found in Sunderland (1996).

Detailed discussions of the merits of the various equations used to estimate predation rates in the field proposed to date are given in Section 11.4 and by Sopp *et al.* (1992).

12.3. Electrophoresis

Electrophoresis can be defined as the movement of ions, or charged molecules, in a fluid under the influence of an electric field, with positively charged particles moving to the cathode and negative particles to the anode. However, this simple definition does little to illustrate the broad array of biological techniques that rely upon this basic process. The areas of greatest significance to invertebrate ecology are probably protein electrophoresis, especially the detection of certain enzymes, immunoelectrophoresis (covered briefly under immunoassays above) and DNA electrophoresis (covered under DNA Techniques in Section 12.4 and Appendix D). In this section we propose to concentrate mainly upon protein electrophoresis, and its use in studies of taxonomy and systematics, population genetics, predation, parasitism and other trophic interactions.

The rate of movement of ions through a supporting medium, such as a gel, at any given electric field strength, depends upon a number of factors, including the net ionic charge, the size and shape of the molecule, and the density of the medium. As the charge, and therefore electrophoretic properties, of many proteins can change at different pH levels (zwitterions), experiments must normally be conducted under standardized, buffered conditions. Thus the aim of the electrophoretic process is to separate biological materials, preferably into clearly resolved bands that are visualized by staining. A major variant on this is isoelectric focusing, in which proteins are separated primarily by charge, rather than by molecular mass. This is achieved by electrophoresis through a pH gradient, with bands migrating to their respective isoelectric points (Hames and Rickwood, 1981). The main types of supporting media used to date have been composed of starch, cellulose acetate, agarose and polyacrylamide, although recently the latter has tended to dominate in enzyme electrophoresis because of its superior resolving power. Different media are useful for different purposes, and each has its advantages. Cellulose acetate is also currently popular because it is cheaper and safer to use than polyacrylamide, while retaining reasonably good band resolution. Standard texts on the wide range of techniques that can be employed include Chrambach and Rodbard (1971), Gordon (1975), Sargent and George (1975), Hames and Rickwood (1981), Richardson *et al.* (1986), Pasteur *et al.* (1988), Hillis and Moritz (1990) and Quicke (1993). Overviews of the use of electrophoresis in studies of agricultural pests may be found in Menken and Ulenberg (1987), Loxdale and den Hollander (1989), Loxdale (1994), Powell *et al.* (1996) and Symondson and Liddell (1996c).

The aim in electrophoresis is usually to produce banding patterns of stained proteins on a gel. Differences in the structure of these proteins will, if great enough, result in changes in mobility within the gel. Where such differences exist, and are characteristic, for example, of a species or population within a species, they can be used as diagnostic markers in ecological studies. At the simplest level, general protein stains can be used to visualize all the proteins in a sample. Ferkovich *et al.* (1983) used Coomassie Brilliant Blue stain to examine changes in haemolymph protein bands within the armyworm, *Spodoptera frugiperda* (J.E. Smith), associated with parasitism by the braconid wasp *Cotesia marginiventris* (Cresson), and comparable methods have been used in the study of scorpions (Vachon and Goyffon, 1978) and slug haemolymph proteins (Symondson and Walton, 1994) for taxonomic purposes. However, although in a similar study by Dahlman and Greene (1981) of the haemolymph proteins of *Manduca sexta* (L.) parasitized by *Apanteles congregatus* (Say), general protein bands were visualized, specific stains for mucoproteins using Bismark Brown Y (Leach, 1947), phospholipids with Nile Blue Sulphate (Menschik, 1953), polysaccharides with the periodic acid-Schiff's reaction (Chippendale and Beck, 1966) and sterol-proteins with phosphomolybdic acid in ethanol (Wang and Patton, 1968) provided further discrimination. Similarly Smilowitz and Smith (1977) used Aniline blue-black as a general protein stain, but specific stains for glycoproteins, lipoproteins, phospholipids and sterols, in their study of the haemolymph proteins of parasitized *Pieris rapae* (L.). However, although these techniques can still be useful in certain instances, the vast majority of electrophoretic studies now use specific enzyme stains, especially for esterase activity.

12.3.1. *Enzyme electrophoresis*

The enzyme groups detected in this form of electrophoresis are mainly those responsible for fundamental processes in living organisms, and therefore failure to detect them is usually a technical problem associated with storage, sample preparation or electrophoretic methodology, rather than the absence of these enzymes from the invertebrates studied (Powell *et al.*, 1996). They have proved to be so widely useful for studies of population genetics, invertebrate systematics and taxonomy, predation and parasitism that they are now the clear method of choice for such electrophoretic investigations. In agricultural entomology, enzyme variability within and between species and populations has provided researchers with a powerful tool for the identification and monitoring of pests, particularly where morphological difference are absent or uncertain. By selecting rapidly evolving enzymes, clear differences can usually be found even between sibling species. Indeed, insects in general display much higher levels of allozyme variability than, for example, vertebrates (Nevo *et al.*, 1984), although among insect groups, variability between Hymenoptera is generally lower (Berkelhamer, 1983).

The terminology applied to enzymes can sometimes be confusing to non-specialists. The term isozyme (or iso-enzyme) refers to multiple forms of an enzyme occurring within a species as a result of the presence of more than one structural gene (Menken and Ulenberg, 1987). Where such genetic variability is the result of multiple alleles at a single locus, these are referred to as allozymes (allelo-enzymes), and as such are inherited in a Mendelian fashion. Allozymes therefore are particularly useful for examining inheritance patterns in population genetics. Co-dominant expression of allozymes results in, for example, single bands for monomeric homozygotes (e.g. FF for a fast band, and SS for a slow band), but two bands for a heterozygote (FS). As enzymes are commonly expressed in polymeric forms (Richardson *et al.*, 1986), complex patterns can arise that can provide diagnostic information. Further variability in banding patterns can arise from transposable elements within enzyme genes (Burkhart *et al.*, 1984), and post-translational modifications of enzymes caused by diet (Schwartz and Sofer, 1976) or the method of enzyme extraction (Ferguson, 1980). Although direct study of differences in DNA sequences provides more detailed and precise information about the same genes (Section 12.4), such detail is rarely required, and electrophoresis can often still provide a simpler, less expensive alternative. A major disadvantage of electrophoresis is that in most cases the insects have to be destroyed, whereas in a molecular study DNA can be extracted from minute samples, such as a leg, from even the smallest insects (Hemingway *et al.*, 1996).

No attempt will be made here to describe the wide range of electrophoretic equipment that is now available, nor to go into its use. This information is widely available, and is usually provided with such equipment in the form of a manual. However, a general protocol is given (Appendix C) for polyacrylamide gel electrophoresis (PAGE). Variants on this can be found in the examples of the use of PAGE and other forms of electrophoresis in the entomological studies reviewed below. A clear description of laboratory techniques may be found in Powell *et al.* (1996).

A surfactant such as sodium dodecyl sulphate (SDS) is frequently used prior to electrophoresis in order to solubilize membrane-bound proteins and to dissolve protein aggregates. However, SDS is a denaturing agent, giving proteins a strongly negative charge. This has the advantage that protein mobilities will be largely a product of molecular mass, but the disadvantage that denatured enzymes will be deactivated, and thus be incapable of turning their respective substrates into visible bands on a gel. Alternative non-ionic surfactants are therefore used in enzyme electrophoresis such as Triton X-100, which preserves enzyme activity. This has the advantage that even single amino acid substitutions are detectable in approximately 20–30% of cases where differences in charge are sufficient (Menken and Ulenberg, 1987). The term 'electromorphs' has been used to refer to the majority of cases where such single (or low numbers of) amino acid substitutions result in molecules that cannot be separated by molecular mass and/or charge under standard conditions (Berlocher, 1984). Successive changes in electrophoretic conditions, such as altering the pH, gel concentration or buffering system can often overcome this 'cryptic variation', as can the use of isoelectric focusing. It should be remembered, however, that there is often a low correlation between numbers of amino acid substitutions and migration distances (Fuerst and Ferrell, 1980).

12.3.2. *Applications for enzyme electrophoresis in agricultural entomology and ecology*

The numbers of relevant applications and case histories for the use of enzyme electrophoresis are now vast, and no attempt can be made here at a comprehensive review. Menken and Ulenberg (1987) provide a table of around 250 references, covering papers dealing with important pest species, for the period 1970–1986, listed by pest. These have been categorized by subject, and cover biochemical systematics (rates of species divergence, intra- and interspecific variation and hybridization), inheritance and non-genetic variation, developmental differences (plus those between social insect caste and between different tissues), parasitism, selection and genetic drift in isolated populations (such as laboratory strains), introduced species, insecticide resistance genes, mating and reproduction. Further references and detailed studies on agricultural pests may be found in Loxdale and den Hollander (1989).

TAXONOMY, SYSTEMATICS AND POPULATION GENETICS

Enzyme electrophoresis can be of crucial importance for identifying sibling species (species that are morphologically indistinguishable), particularly where pest and non-pest species occur in the same habitat. In certain instances, for example, one species may be susceptible to a parasitoid biological control agent but the other not. An understanding of the life history of a pest can be essential for effective pest management, but can vary considerably between closely related species.

Interspecific variation was demonstrated, for example, by Stock *et al.*

(1984), who were able to show that two pine beetles, *Dendroctonus ponderosae* Hopkins and *D. jeffreyi*, previously considered to be a single species and morphologically identical, were in fact sibling species specializing on different hosts but occurring sympatrically. However, a study of Colorado beetle populations in North America found a distinct Mexican population that is now considered a subspecies of the more widespread *Leptinotarsa decemlineata* (Say), found in the USA, Canada and Europe (Hsiao, 1989). Unlike the bark beetles, the two subspecies could interbreed, and in this instance the genetic difference was considerably less than that between *L. decemlineata* and its congener, *L. haldemani*. Six sibling species of treehopper (Homoptera: Membracidae) were identified by Guttman *et al.* (1981) that had formerly been considered host races of *Enchenopa binotata* Say. Studies such as this have provided strong evidence for sympatric speciation resulting from host specialization by phytophagous insects (Bush, 1994). Menken and Ulenberg (1987) list many more examples of the usefulness of allozymes for separating sibling species. Among Hymenoptera, aphid parasitoids can be difficult to identify morphologically. However, Tomiuk and Wöhrmann (1980) were able to separate two aphidiid parasitoids of *Macrosiphum rosae* (L.), by their malate dehydrogenase patterns, while Castañera *et al.* (1983) identified diagnostic esterase patterns for five species of parasitoid attacking *Sitobion avenae* (F.). In some important instances, however, allele patterns may not be entirely reliable for identification where some overlap exists between closely related species, as for example between *Anopheles gambia* (Giles) and *A. arabiensis* (Patton) (Hemingway *et al.*, 1996).

The abilities of different enzyme systems to distinguish between species and genera within a taxonomic group vary considerably, and need to be investigated in advance. For example, Tomiuk *et al.* (1979) studied the electrophoretic patterns for 23 aphid species using 20 enzyme systems, 4 of which proved useful for separating species and genera. Similarly, Loxdale *et al.* (1983) used 14 enzymes to study differences between 6 species of cereal aphid, and found 10 to be effective for taxonomic classification. Difficulties can arise where extremely small arthropods need to be studied; conventional PAGE and starch gel electrophoresis frequently fails to reveal some or all of the most active enzymes in mites and parasitoids such as *Trichogramma* spp. (Menken and Raijmann, 1996). Techniques to overcome this problem were developed by Kazmar (1991), using isoelectric focusing and ultra-thin agarose gels or cellulose acetate membranes. Several *Trichogramma* spp. have proved to be valuable biocontrol agents, and species identification is therefore clearly of vital importance. Equally important is separation of pest from non-pest, at all developmental stages; Menken and Raijmann (1996) developed electrophoretic procedures that can identify *Liriomyza* leafminers, some of which are subject to quarantine regulations within the European Community.

There is an unresolved problem as to what constitutes a species, sibling species, subspecies, host races, metapopulation or local population of insects, when in practice all conceivable intermediate states exist in nature.

Given the dynamics of evolutionary processes, population geneticists using electrophoretic techniques have developed a number of measures to quantify the amount of genetic differentiation observed. This is achieved by measuring the similarity (S), identity (I) or genetic distance (D) between two populations over a set of enzyme loci (Nei, 1972; Thorpe, 1982), or by simply calculating the frequency with which a particular allele occurs within different populations. The degree of variation within a species provides some indication of its evolutionary potential. Of equal importance, however, is the rate of gene flow through a population, which can determine the speed with which, for example, insecticide resistance genes can spread (Devonshire and Moores, 1982) (see Section 12.5). As the majority of insecticides are esters, and cleavage of ester bonds by insect esterases is one of the primary detoxification methods used by insects developing resistance, electrophoresis to detect amplified and/or modified esterase activity has proven to be a powerful method of detecting resistance (Devonshire, 1989). Gene flow can be estimated from allozyme studies by calculating $N_e m$, the average number of migrants exchanged between two populations per generation. This has been calculated from $F_{ST} = 1/(1 + 4\ N_e m)$, where F_{ST} is the among-population variance in allele frequency (Wright, 1943), or from the logarithm of $p(1)$, the average frequency of 'private alleles', which are alleles found in one particular population (Slatkin, 1981, 1985). A comparison between these methods was made by Slatkin and Barton (1989). A good example of a long-term electrophoretic study of gene flow, sympatric speciation, and biochemical systematics in general is that of the *Yponomeuta* (Lepidoptera: Yponomeutidae) (Menken *et al.*, 1992; Menken and Raijmann, 1996), a complex of species and host races.

DETECTION OF PARASITISM

Integrated pest management systems sometimes rely upon monitoring rates of parasitism in the field in order to forecast the ability of parasitoids to control the growth of pest populations. Traditional methods involving rearing parasitoids from their hosts are both labour intensive and slow, too slow in many cases to allow effective modifications to cultural operations to be undertaken. Dissection of hosts to measure rates of parasitism is generally inaccurate, because simply finding early instars and eggs, even with the aid of appropriate stains, can be a problem (M.P. Walton, Victoria, Australia, personal communication). Morphological identification to species of the early stages of parasitoids is also often difficult or impossible (Walton *et al.*, 1990a). Several workers have therefore developed electrophoretic methods to improve speed and accuracy. These techniques depend upon the existence of diagnostic banding patterns that are consistent within a species, preferably at all developmental stages, which can distinguish the parasitoid from the host enzymes and which can identify different species of parasitoid attacking the target host. As in all such studies, preliminary work is required to find the best enzyme systems. Castañera *et al.* (1983) found 3 out of 14 enzyme systems useful for examining parasitism by *Aphidius uzbek-*

istanicus Luzhetzki of the cereal aphid *S. avenae*, one of which, malate dehydrogenase, was used to investigate rates of parasitism for this host–parasitoid relationship using isoelectric focusing (Höller and Braune, 1988). In the majority of such studies, esterase appears to be one of the most useful enzyme systems for detecting parasitism. Examples include the detection of the braconid wasp *Aphidius matricariae* Haliday in the aphid *Myzus persicae* (Sulzer) (Wool *et al.*, 1978); parasitism of the aphid *Macrosiphum rosae* (L.) in the field (Tomiuk and Wöhrmann, 1980); parasitism by the endoparasitic ichneumonid wasp *Glypta fumiferanae* (Viereck) of the western spruce budworm *Choristoneura occidentalis* Freeman (Castrovillo and Stock, 1981); parasitism of small ermine moths (Yponomeutidae) by a number of endoparasitoids (Menken, 1982); and parasitism of the whitefly *B. tabaci* by two endoparasitoids, *Encarsia lutea* and *Eretmocerus mundus* (Wool *et al.*, 1984). Walton *et al.* (1990a) developed electrophoretic keys for diagnostic banding patterns representing esterase activity for six species of parasitoid attacking *S. avenae*. These were later tested in the analysis of parasitism rates in the field, and gave results comparable with those obtained by live rearing (contrary to the findings of Höller and Braune, 1988), both in terms of percentage parasitism and parasitoid species identification (Walton *et al.*, 1990b). Electrophoretic analysis of parasitized aphids can involve processing in the order of 100–150 insects per day by skilled personnel (Powell *et al.*, 1996), allowing realistically large field samples to be taken.

DETECTION OF PREDATION

Prior to the development of monoclonal antibodies for species-specific detection of predation in the field, electrophoretic techniques were found to be an effective method for investigating the dietary range of polyphagous predators and may still have a valuable role to play. Whereas monoclonal antibodies are probably the most efficient method for identifying the predators of a particular target prey, zymograms can be used to detect a wide range of prey species consumed by particular polyphagous predators, when suitable electrophoretic keys are available and where clear diagnostic bands can be found. Electrophoretic techniques can also be undertaken immediately, without the need to undertake the lengthy procedures required to produce and characterize specific monoclonal antibodies against each target species. When very small prey, such as mites, are the target, the difficulty of gathering sufficient prey material for immunizations prior to antibody production has been seen as an argument in favour of the electrophoretic approach (Solomon *et al.*, 1996).

Murray and Solomon (1978) investigated the ability of staining for esterase activity to detect predation on a range of prey, including two mites, *Panonychus ulmi* (Koch) and *Tetranychus urticae* (Koch), plus several insects such as the aphid *Rhopalosiphum insertum* (Walker) and the psyllid *Psylla simulans* Förster. The system proved to be capable of detecting the remains of a single red spider mite, *P. ulmi*, within the predatory mite, *Typhlodromus pyri* (Scheuten) (mean weight 7 μg), at least 31 h after consumption. They were able to demonstrate that the alder sawfly *Fenusa dohrnii* (Tisch.) was not a significant prey item for the anthocorid predator *Anthocoris nemoralis* (F.) in the field. Fitzgerald *et al.* (1986) went on to develop methods of quantifying predation on *P. ulmi* and *T. urticae* on gradient gels, including dilution series of prey material from which to establish calibration curves. Giller (1982, 1984, 1986) used an esterase system to investigate the prey range of waterbugs, *Notonecta* spp. Characteristic banding patterns were found for 33 aquatic prey species, and the dietary range of different instars of two species of *Notonecta* were investigated by assays of field samples. In general, prey could only be identified when the foregut of the predators contained clearly observable prey remains. Van der Geest and Overmeer (1985) used the methods developed by Murray and Solomon (1978) to detect predation on spider mites by predatory phytoseiid mites. They found that they could identify four species of spider mites (*Tetranychus viennensis* Zacher, *T. urticae*, *P. ulmi* and *Eotetranychus pruni* (Oudemans)) from their banding patterns post-consumption by the predators. However, detection of eriophyid mites proved more problematic, and could only be effective where more than 30 individuals had been consumed, while the banding patters of tydeids sometimes overlapped with those of the predators. Dicke and De Jong (1988) also investigated predation by *T. pyri* on two species of phytophagous mites, *P. ulmi* and *Aculus schlechtendali* (Napela) in an orchard, finding a preference for the former. Lovei (1986) demonstrated that enzyme electrophoresis had potential for the investigation of the diets of carabid beetles, and this technique has recently been used to investigate predation on slugs (W. Paill, Graz, Austria, personal communication). Lister *et al.* (1987) used an esterase system, based upon the methods of Murray and Solomon (1978), to conduct a study of predation by an Antarctic mite, *Gamasellus racovitzai* (Trouessart), upon a range of microarthropods in the field. The results were analysed qualitatively, to determine the prey range at different sites, and quantitatively to determine predation rates from proportional analyses. Quantities of target prey remains in predatory mites were determined by transmission densitometry, which allowed quantified esterase bands to be compared with exponential decay curves derived from prepared standards. Attack rates could then be calculated from a model based upon meal size (Sections 11.4 and 12.2.5).

12.4. DNA Techniques

The range of new molecular techniques is expanding every year, but to date interest has primarily been directed towards two areas of entomological research. The first of these broadly covers studies of insect systematics, taxonomy and populations genetics, while the second has been that of detecting and monitoring insecticide resistance.

12.4.1. *Molecular techniques used in studies of insect taxonomy and population genetics*

Numerous molecular techniques are now available for looking at different taxonomic problems in insects. These techniques are rapidly replacing biochemical assays, primarily because the quantity and accuracy of information which can be obtained are orders of magnitude greater than for isoenzyme assays. An excellent review comparing the merits of DNA and enzyme electrophoretic systems for entomological research can be found in Loxdale *et al.* (1996). The optimum choice of molecular method depends on the taxonomic question to be tackled and the amount of data available on the target DNA from the insects of interest. The former will determine the necessary level of sequence divergence needed between DNA samples, and hence will guide not only method choice but also the target DNA. If sequence data are already available an approach using species-specific probes can be made directly. Where no sequence is available, random primers can be used to generate species-specific probes, or universal or degenerate primers can be targeted at specific DNA sequences using a variety of PCR-based techniques.

MITOCHONDRIAL DNA

The complete sequence of the mitochondrial DNA of four insect species (*Drosophila yakuba* Burla, *Apis mellifera, Anopheles gambiae* and *An. quadrimaculatus* Say) is now known (Beard *et al.*, 1993). The sequences of various loci within the MtDNA are known for a larger range of insects including several locusts (McCraken *et al.*, 1987; Haucke and Gellissen, 1988).

Because of its ease of purification, relatively rapid rate of change, simple organization and maternal inheritance, MtDNA has found many uses in phylogenetic studies (Liu and Bechenbach, 1992). It is particularly suited to studies of sibling or closely related species, but of less use for studying distantly related species. Comparisons can be made either directly at the sequence level or at the cruder level of gene order. In insects there is significant variation in the position and orientation of a number of the tRNA genes (Pruess *et al.*, 1992; Crozier and Crozier, 1993). Comparisons of these genes and their genomic organization have been very useful in resolving phylogenetic relationships of a number of Diptera.

Intact mitochondria can be recovered from insect homogenates and purified simply from other organelles by CsCl density gradient centrifugation, due to its unusual buoyant density. An excellent review of techniques involved in extraction and purification of MtDNA is given in Lansman *et al.* (1981). Once clean DNA has been recovered it can be used for restriction digestion, hybridization studies and probing as for all other DNA samples.

RIBOSOMAL DNA (rDNA)

rDNA has found extensive and increasing use in taxonomic studies primarily using the ribosomal RNA gene cluster. Sequence data, which are now available for this region for a number of insects, have been used to infer phylogenetic relationships (Paskewitz and Collins, 1989; Fang *et al.*, 1993). It is possible to use the rDNA sequence itself, or to look at insertion sites of retrotransposons within the rDNA sequence to infer relationships (Besansky *et al.*, 1992). The major advantages of using the rDNA for phylogenetic studies are its high to moderate repeat nature and high quantity, both of which simplify hybridization work. The rDNA genes are present in hundreds of tandem copies per cell nucleus in most multicellular organisms, and present at more than 500 copies per diploid genome in most Diptera (Scott *et al.*, 1993). The highly conserved nature of the ribosomal genes results in two further benefits. Firstly, their rate of change is slow, making them ideal for taxonomic studies covering large time intervals. Secondly, their DNA sequence in new species can be obtained relatively simply, by taking advantage of short, highly conserved or invariable sequences within the ribosomal genes. Primers to these sequences will bind to the ribosomal genes of any species, allowing a simple PCR, subcloning and sequencing experiment to be undertaken to obtain the sequence of the more degenerate sequence.

Clones of rDNA can be obtained from standard genomic libraries using probes to specific coding regions. Alternatively, a PCR approach using degenerate primers to conserved regions can be used to obtain regions of interest. Digestion, hybridization and probing are then as for any DNA sample.

12.4.2. *General method for handling DNA samples*

The methods given here are basic ones which will work with most DNA samples. There are now numerous manuals providing details of molecular techniques which expand on the methods given here. Berger and Kimmel (1987) and Sambrook *et al.* (1989) are particularly recommended.

Electrophoresis of DNA samples is described in Appendix D. This allows separation of DNA fragments of different lengths. It is used after restriction enzyme digestion, or following purification of DNA samples, to check purity and quality.

Southern blotting allows you to detect sequences in your target DNA which are homologous to sequences in a probe of interest. The probe and the Southern conditions determine the specificity, such that it may be species- or enzyme-specific or may detect the members of a species family. Southern blotting is a powerful tool in understanding basic problems such as genome organization and gene structure. It plays a role in the diagnosis of heritable disease and the detection of viral and microbial pathogens, as well as insecticide resistance due to gene amplification. The basic steps are the isolation of the DNA, which is then cut with restriction endonucleases, subjected to electrophoresis and blotted on to a membrane. A cloned, labelled piece of DNA from the organism, or a related organism, can then be used as a probe for hybridization to the membrane. Initially, labelling of the DNA probe was done with ^{32}P-labelled nucleotides. This method is still used when high levels of sensitivity are needed, but it is increasingly being replaced by non-radioactive methods, which are now approaching the sensitivity of radiolabelling for most applications. Non-radioactive systems include horseradish peroxidase, digoxigenin, and biotin–streptavidin. In the horseradish peroxidase system a chemical reaction is used to label the DNA fragment. Horseradish peroxidase is covalently linked to polyethyleneimine, and this is mixed with denatured DNA and 1% glutaraldehyde. The glutaraldehyde is a bifunctional cross-linking reagent which covalently binds the marker enzyme to the DNA probe.

The basic steps in Southern blotting are described in Appendix E. With the streptavidin–biotin, radiolabelled and digoxigenin systems the DNA probe is labelled either by DNA nick translation or by random primer labelling.

Nick translation uses the enzymes DNAase I and DNA polymerase I. The DNAase I randomly nicks the double stranded DNA fragment to be labelled. The DNA polymerase I then adds nucleotides to the 3'- OH created by the nicking activity of DNAase I, while the 5' to 3' exonuclease activity of DNA polymerase I simultaneously removes nucleotides from the 5' side of the nick. If labelled nucleotides are present in the reaction mixture, then the pre-existing nucleotides are replaced by radiolabelled ones. Kits for nick translation labelling reactions are commercially available. As little as 20 ng of probe DNA can be used in the reaction, but generally 1 μg of DNA is used. The stability, storage conditions and length of viability of the probe depend on the label used.

In random priming short oligonucleotides of random sequence are hybridized to your single stranded DNA fragment. The hybridized oligonucleotides serve as primers for the initiation of DNA synthesis by a DNA polymerase. The DNA polymerase used is the Klenow fragment of DNA polymerase I (this is a cleaved form of the enzyme used in the nick translation system, which has no 5' to 3' exonuclease activity). The use of this fragment ensures that the DNA synthesis occurs only by primer extension and that the bound primers are not degraded. The size of the probe produced is inversely proportional to the concentration of the primers used. At low primer concentrations, the probe may be up to 50% of the length of the original DNA template, while at high concentrations the probe will be ~400 bp. Kits with all the components for random prime labelling are available from a number of manufacturers.

Southern blotting is straightforward, but necessitates having a probe. Molecular techniques can still be used for some purposes when no sequence data or probe are available. The rapid amplification of polymorphic DNA (RAPDs) technique was first reported 5 years ago (Welsh and McClelland, 1990). It is a simple technique for amplifying up non-specific fragments of genomic DNA for which no sequence data are available. The technique utilizes short (8–12 base pair) random primers. Banks of RAPD primers are now commercially available. Its advantages over work on MtDNA or rDNA are that no prior knowledge of insect sequence is necessary, and it is cheap and quick to undertake the reactions and analyses. The disadvantages are that the results can be difficult to analyse, they do not reflect genotype and the technique can be subject to artefacts unless carefully controlled. The major application of the technique is in looking at closely related species, or even different populations of the same species.

The RAPD technique can be compared in many ways to the earlier electrophoretic enzyme isozyme systems for looking at populations, and they have many similar applications. RAPDs have been used, for example, to investigate taxonomic differences between different aphid species (Cenis *et al.*, 1993; Robinson *et al.*, 1993), and they have also been widely used to look at clonal variation within single species of aphid (Black *et al.*, 1992; Puterka *et al.*, 1993). They proved to be an effective tool for investigating the population genetics of arionid slugs (Noble and Jones, 1996). RAPDs have the advantage over isozyme variants in producing greater levels of variability. Hence insect populations which are invariant for enzyme markers may show a measurable level of RAPD polymorphism.

RAPDs are essentially a form of PCR whose specificity is determined by the random primer sequence used and the exact reaction conditions set. The process of PCR, and its associated enzyme DNA polymerase, were named by *Science* as the 1989 'Molecule of the Year' due to their likely effect on future science. The Nobel Prize in Chemistry was also awarded to the inventor of the PCR process in 1993. The basic steps in any PCR experiment are the same. These are:

1. Denaturation of the DNA template by heating.
2. The annealing of primers to the DNA template.
3. Copying of the DNA template by extension from the primers by the DNA polymerase.

These three steps are then repeated many times (usually through 20–40 cycles) to amplify the DNA fragments for subsequent RAPD analysis on agarose gels. At each cycle there is a doubling of the target DNA. Conditions can be modified by altering the temperature in step 2, altering the time in step 3, changing the $MgCl_2$ concentration of the buffers used or altering the primer concentration. The skill in using RAPD analysis is largely in determining the initial PCR conditions. These need to be set so that variant bands are visualized, but need to be sufficiently controlled to avoid artefacts.

Two DNA techniques, isolation of genomic DNA and the use of DNA dot-blots for esterase gene quantification, are relevant to the detection of insecticide resistance, or have been specifically modified for that purpose. These are described below.

12.5. Biochemical and Molecular Methods for Detection of Insecticide Resistance

Insecticide resistance is an increasing problem in the efficient control of agricultural pests. Resistance monitoring programmes should no longer rely on testing the response to one insecticide, with the intention of switching to another chemical when resistance levels rise above the threshold which affects disease control. Effective resistance management depends on early detection of the problem and rapid assimilation of information on the resistant insect population, so that rational pesticide choices can be made.

The correct use of biochemical, immunological and/or molecular methods for resistance detection at a mechanistic level can provide a powerful tool for analysing field and laboratory populations, with the aim of improving resistance detection and management. These assays provide more information about the insect population being controlled.

Where a mechanistic approach to resistance detection is being undertaken, the investigator needs to have a basic understanding of the possible resistance mechanisms likely to be encountered. This chapter deals only with the four major groups of insecticides: the organochlorines, organophosphates, carbamates and pyrethroids. New compounds are clearly coming on to the market, such as insect growth regulators, but a mechanistic detection of resistance to these compounds has by necessity to be reactive rather than proactive (i.e. methods can only be developed for resistance detection when we know the range of mechanisms selected in different insect populations). Hence monitoring for resistance to compounds outside the four well characterized pesticide groups still relies heavily on standard bioassays. There are four possible types of resistance mechanism to the main insecticide groups in all insects analysed to date. These are:

1. Increased metabolism to non-toxic products.
2. Decreased target site sensitivity.
3. Decreased rates of insecticide penetration.
4. Increased rates of insecticide excretion.

Of these four categories the first two are by far the most important. Penetration rate changes in isolation generally produce insignificant ($<$ five-fold) levels of resistance, and only become important when found in combination with other resistance mechanisms. Increased rates of insecticide excretion are very uncommon, producing low levels of resistance, and are only included in this list for completeness.

The first two categories of resistance mechanisms, which are the most common and produce the highest levels of resistance, can be subdivided further. The enzyme groups involved in insecticide metabolism are esterase, multifunction oxidases and glutathione-*S*-transferases. The target sites involved are the sodium (Na^+) channels for the pyrethroids and DDT, acetylcholinesterase for the organophosphates and carbamates, and the GABA receptors for cyclodienes.

There are two major ways that the metabolic enzymes can produce resistance: (i) overproduction of the enzyme, leading to increased metabolism or sequestration; (ii) an alteration in the catalytic centre activity of the enzyme, increasing the rate at which an enzyme unit metabolizes the insecticide. These two routes are not mutually exclusive, and an enzyme may be both physically changed and overproduced. When an enzyme is overproduced, but the pesticide is only slowly metabolized by that enzyme, the cause of resistance may be considered to be sequestration rather than metabolism, with the increased enzyme levels acting as a means of holding the pesticide and preventing it from reaching the target site within the insect. The level of resistance conferred is then roughly proportional to the increase in the quantity of enzyme produced.

It should be stressed that at present simple field biochemical assays do not exist for all resistance mechanisms, the one major deficiency being the altered sodium channels. Hence, when initially surveying an insect population for resistance, it is important that the biochemical assays and a limited number of bioassays are run together to avoid missing any resistance. When a population is well characterized, some of the biochemical assays can be used in isolation to measure changes in resistance gene frequencies in field populations under different selection pressures.

12.5.1. *The biochemical assays*

Because the resistance mechanisms detected by these methods are common to all insects they are applicable across the range of pests. However, it is important to note that the baselines may differ between insects, and where possible a known susceptible strain of the same species as the field population being tested should be analysed at the same time.

Two main variants of two of the assays are in use. To save repeating the arguments for and against each twice, these are outlined here, although both methods are later given in detail. One variant of the assays uses filter paper or another solid support medium, the second variant is run in microtitre plates. The 'filter' paper or nitrocellulose membrane assays generally use one insect per assay and are quantified visually or using a densitometer, but provide a permanent record which can be rechecked in the future. The microtitre plate tests allow the same insect to be used for all assays and are quantified visually or with a spectrophotometer. A permanent record can be made on paper by simply using a transfer plate, but this is not an automatic result of the test. Space allows only details of the microtitre plate assays to be given, but details of the filter paper assays may be found in Pasteur and Georghiou (1981), Georghiou and Pasteur (1989) and Dary *et al.* (1991).

THE MICROTITRE PLATE TESTS

It is practical to run assays for altered acetylcholinesterase, elevated esterase, glutathione-*S*-transferase, monooxygenase and protein from the same insect. As 96 insects can be run per microtitre plate, a system for grinding this number of insects rapidly is beneficial (for details of one system see ffrench-Constant and Devonshire, 1987). Alternatively, insects can be ground directly in the microcentrifuge tubes ready for centrifugation.

All reagents listed in the following experiments (Appendices F–I) are basic grade chemicals and are widely available from a range of suppliers. Unless otherwise stated, phosphate buffer refers to a sodium phosphate buffer system (NaH_2PO_4 + Na_2HPO_4 solutions mixed in the correct molarities and ratios).

A standard method is given here for the preparation of adult planthoppers. For other insects the initial homogenization volume may need to be altered depending on the size of the insect. Any stage of the insect can be used, but ideally this should be the life stage of the insect that you wish to control with the pesticide, as resistance mechanisms do not necessarily operate throughout all stages. With some insects, for example house flies, the head should be removed and used for the acetylcholinesterase assay, as the large amounts of pigment released from the eyes on homogenization interfere with the other assays.

Homogenize individual insects in 250 µl of distilled water. Take 2×25 µl of homogenate for the acetylcholinesterase assay and 2×20 µl for the monooxygenase assay and then spin the remainder of the homogenate at 14K for 30 s in a microfuge. The microfuge step is essential only for the glutathione-S-transferase assay, as particulate matter in the crude homogenate has a significant impact on the absorption of light at 340 nm. The samples for some assays are removed before spinning, as being membrane bound much of this enzyme may be pelleted and lost during centrifugation.

Live or frozen insects can be used for these assays. Whole insects stored at $-20°C$ are viable for several months, while those stored at $-70°C$ or in liquid nitrogen remain viable for years.

In the acetylcholinesterase assay (Appendix F) each insect acts as its own control. A bright yellow colour should appear in all the control wells, i.e. those with insect homogenate + ASCHI but not propoxur. A carbamate is used in this assay, but the oxon analogues of any of the OPs could be substituted for this. The concentration of propoxur is set so that >70% of the total acetylcholinesterase activity is inhibited in the susceptible insect. This gives a clear distinction between resistant and susceptible insects both visually and spectrophotometrically. If an end point assay is used, then susceptible (SS) individuals can be easily differentiated from heterozygote (RS) and homozygote (RR) resistants, but differentiating some RS and RR is difficult. If a rate assay is used, then all three genotypes can be clearly distinguished. For other insects, the concentration of propoxur may need to be adjusted slightly, by experimentation with the susceptible strain of the test insect to achieve >70% inhibition of acetylcholinesterase activity.

Resistant insects should have a percentage value greater than 30%. The accuracy of your results should be evident if the data for your population is expressed graphically (Hemingway *et al.*, 1986; ffrench-Constant and

Bonning, 1989). Values for homozygous resistant insects may be substantially higher than 100%. This is partially due to the absorbance of propoxur in the microtitre plate well and is normal in resistant strains.

The naphthyl acetate end point assay (Appendix G) measures esterase activity directly, and linking this to resistance requires confirmation via synergist and/or electrophoretic work. The assay as detailed above uses two substrates, in most insects elevated esterases involved in resistance are generally active with both substrates, but in other insects, such as *Blattella germanica*, there can be a marked preference for one of the two substrates used.

Individuals with non-elevated levels of esterase activity should have a pale blue or pink colour with α-NA or β-NA respectively. Individuals with elevated esterase activity show an intense blue/black or pink/red colour.

PNPA rate reaction esterase assay (Appendix H) is simpler and quicker than the naphthyl acetate assays and equally accurate. However, it cannot be used as an end point assay. If your plate reader will not do kinetic measurements, it is preferable to use the naphthyl acetate assay. This assay also measures only elevation of esterase activity and the association of this with resistance still needs confirming, as with the naphthyl acetate assays, by electrophoresis and/or synergist work.

The glutathione-*S*-transferase assay (Appendix I) must be read spectrophotometrically as the change in absorbance is in the UV range (i.e. you will not see a colour change in the wells). Ideally, the assay should be read as a rate, as this gives a much better differentiation between resistant and susceptible insects with this mechanism.

Dichloronitrobenzene can be used rather than chlorodinitrobenzene. However, the amount of insect homogenate will need to be increased, as the baseline activity with DCNB is much lower than that with CDNB for most insects.

Where a simple protein assay is required, this can be performed using widely available commercial kits. Alternatively the methods of Lowry *et al.* (1951) or Bradford (1976) can be followed.

With many insects it is not possible to measure cytochrome P^{450} activity rates in individuals. The amount of haem containing enzyme (the bulk of which is P^{450}) can however be measured by a simple haem titration. This differentiates between resistant insects with elevated P^{450} and susceptible insects. See Brogdon *et al.* (1996) for details of the monooxygenase assay.

Currently immunological detection methods are available only for specific elevated esterases in collaboration with laboratories which have access to the antiserum. There are no monoclonal antibodies, as yet, available for this purpose.

An antiserum has been prepared against the E4 carboxylesterase in the aphid *M. persicae*. An affinity-purified IgG fraction from this antiserum has been used in a simple immunoplate assay to discriminate between the three common resistant variants of *M. persicae* found in UK field populations (Devonshire *et al.*, 1986). The sensitivity of this assay is such that it gives a clearer differentiation of resistant phenotypes than the esterase microplate assay (Devonshire *et al.*, 1992).

Various antisera have been raised to the A_2, B_1 and B_2 elevated esterases in *Culex*. The literature suggests that sera raised to any of the 'B' esterases will detect other 'B', but not 'A' esterases and vice versa (Hemingway *et al.*, 1986; Mouches *et al.*, 1987; Beyssat-Arnaouty *et al.*, 1989). This has recently been shown to be incorrect, as sera raised to purified A_2 will in fact bind to B_2, although the sensitivity is 60-fold lower against the latter enzyme (Karunaratne *et al.*, 1995). The aphid IgG fraction used in the immunoplate assay is specific for the E_4 enzyme and the closely related FE_4, and shows no cross-reactivity with any of the elevated *Culex* esterases. The antiserum raised to the B_1 esterase also detects other immunologically related proteins in *Aedes aegypti* (L.) and *Musca domestica* L. (Mouches *et al.*, 1987).

Detection of oxidase (cytochrome P^{450})-based insecticide resistance can be undertaken using synergist pre-exposure. The levels of oxidase activity in individual mosquitoes are relatively low, and no reliable microtitre plate or dot-blot assay has yet been developed to detect the presence of this mechanism. The P^{450}s are also a complex family of enzymes, and it appears that different cytochrome P^{450}s produce resistance to different insecticides. The simplest way currently to obtain an indication of the presence of an oxidase-based mechanism is to use pre-exposure to the synergist piperonyl butoxide (PB), followed by exposure to an insecticide within the class against which you wish to detect resistance (e.g. a pyrethroid, carbamate or organophosphorus insecticide). The dosage of synergist used should be the maximum that results in no mortality after exposure to the synergist alone. This dose varies between species and insect groups, and should be determined for your species of interest by pre-exposing a sample of insects to a range of different concentrations.

The interpretation of results for organophosphorus insecticides can be difficult, as the oxidases are involved in both activation and detoxication of these compounds. Hence, piperonyl butoxide can lower the toxicity of OPs to both susceptible and resistant strains without an oxidase-based OP resistance mechanism.

ELECTROPHORESIS (PAGE) FOR IDENTIFICATION OF SPECIFIC ESTERASES

Detection of specific esterases can be performed using the polyacrylamide gel electrophoresis protocol given earlier (Appendix C). Details of a starch gel method can be found in Georghiou and Pasteur (1978). With the method described in Appendix C, very intensely stained red/purple or purple/blue bands should appear on the gel within minutes if esterases are elevated.

Individual bands can be identified by measuring their mobility from the origin of the gel relative to a xylene cyanol marker. Ideally, insect strains with known esterase bands, e.g. A_2/B_2 or B_1 in *Culex* or E_4 in aphids, should be run on the same gel if these bands are to be accurately identified.

12.5.2. *Molecular methods for resistance detection*

ISOLATION OF GENOMIC DNA

All the DNA resistance detection systems require access to genomic DNA. This can be obtained from groups of insects or from individuals. The techniques for isolating DNA differ. With larger amounts of starting material it is more important to remove the majority of RNA and protein from the preparation, as these will inhibit future cutting of the DNA with restriction enzymes and/or complicate the interpretation of results. It should also be noted that although the steps and basic principles for DNA isolation are the same for all insects, modification of the methods may be necessary for each species depending on the quantity of pigments, DNAases or inhibitors found in the insect. The methods given in Appendix J work well for mosquitoes which have naturally high levels of polyphenols, which bind to the DNA unless removed rapidly (Hemingway *et al.*, 1995; Vaughan *et al.*, 1995). Protocols are given for large-scale isolation of DNA from 0.1 to 1 g of insects, and for extracting DNA from single insects or fractions of an insect.

DNA DOT-BLOTS FOR ESTERASE GENE QUANTIFICATION

DNA-based detection systems are not yet generally available. However, the recent emphasis on molecular studies of all the major insecticide resistance genes suggests that these are likely to become more accessible over the next few years. A dot-blot detection system, combined with an immunoassay to identify resistant aphids which have reverted to susceptibility, has been described by Field *et al.* (1989). Probes are now available for all the elevated esterases commonly found in the mosquitoes where a similar system can be employed. The one resistance mechanism which can, however, be detected only by a molecular method at present is the altered GABA receptor. The changes in the GABA receptor associated with cyclodiene resistance appear to involve the same amino acid residue in all insects tested to date. Hence various PCR techniques can be used to accurately score genotypes. Details are given in ffrench-Constant *et al.* (1993a, b, 1994), Aronstein *et al.* (1994) and Steichen and ffrench-Constant (1994).

It is likely in the next few years that a PCR-based molecular technique will become available for 'field' detection of 'kdr' in individual insects, as the relevant proteins have now been cloned from a number of insects.

12.5.3. *Case studies*

Molecular and biochemical assays can often be combined on the same sample of insects to investigate a range of biological problems. For example a study was undertaken in The Gambia (Hemingway *et al.*, 1995) to assess the efficacy of pyrethroid-impregnated bednets against local vectors. This study posed several problems. Firstly, the vectors were part of a group of sibling species, which could not be differentiated visually. Second, information was needed on whether the insects were infected with parasites and, thirdly, their insecticide resistance status needed to be determined. The study util-ized a PCR approach with rDNA to identify each insect to species level. A PCR approach with primers specific for the parasite DNA was used to determine the infection rates and biochemical assays were used to determine the resistance status of the same insects (Hemingway *et al.*, 1995). Similarly a study in Mexico on insecticide resistance management is using both biochemical and molecular techniques. The biochemical techniques are again being used to identify resistance gene frequencies, while RAPDs are being used on the same insects to estimate the baseline level of gene flow between populations (Penilla *et al.*, 1996).

12.6. Other Biochemical Techniques

Many other biochemical tests are used in entomology labs to provide information relevant to ecological investigations, but which can be mentioned only briefly here. Some, such as the use of rare elements or radioactive labels, are not strictly biochemical or molecular, and have been well reviewed by Southwood (1978) and Sunderland (1988).

Analysis of cuticular hydrocarbons can provide data similar in many ways to that of isozymes. Many of these hydrocarbons are known to have a semiochemical role (Stanley-Samuelson and Nelson, 1993), and thus their production and composition are often precisely controlled for specific functions. These and other components of the cuticular lipid layer are thought to be important, for example, in short-range mate recognition, and thus the development of differences between populations or sibling species may provide a mechanism by which sympatric speciation can occur and, subsequently, hybridization be prevented (Phillips *et al.*, 1990). The technique involves extraction of the hydrocarbons with a solvent, such as hexane, and subsequent analysis by gas chromatography (Phillips *et al.*, 1988). The analysis generates hydrocarbon profiles which can be compared statistically using similar methods as for allozymes (Section 12.3.2.), particularly Nei's genetic distance coefficients, and identify values between populations (Nei, 1972), or discriminant analyses (e.g. Phillips *et al.*, 1990). Cuticular hydrocarbons have been used recently to separate closely related species (Puttler and Sacks, 1993; Anyanwu *et al.*, 1994; Richmond and Page, 1995) and populations (Kruger and Pappas, 1993; Byrne *et al*, 1995; Estradapena *et al.*, 1995), to study variation across hybrid zones (Neems and Butlin, 1994) and to examine caste and colony recognition signals in social insects (Bonavita-Cougourdan *et al.*, 1993; Heinze *et al.*, 1994; Butts *et al.*, 1995).

A variety of techniques have been developed to answer very specific questions. For example, bloodmeals in haematophagous insects can be quantified in tests that measure haemoglobin (Randolph and Rogers, 1978; Langley, 1996). In this test the dried abdomens of the insects are crushed and the haemoglobin extracted with sodium hydroxide and ethanol. The resulting alkaline haematin is then mixed with pyridine and sodium dithionite to produce pyridine haemochromagen, a coloured product that can be quantified in a spectrophotometer. The age of certain flies can be determined by measuring accumulations of pteridine waste products in the head capsule (Lehane and Mail, 1985). The pteridine flourescence technique involves the extraction of pteridine, which is then quantified in a flourimeter, test samples being compared with standards of known age. This technique has been used recently to study the age spectra of, for example, tsetse flies, *Glossina* spp. (Langley *et al.*, 1988), house flies, *Musca domestica* L. (McIntyre and Gooding, 1995), melon flies, *Bactrocera curcurbitae* (Mochizuki *et al.*, 1993) and blowflies, *Lucilia sericata* Meigen (Wall, 1993).

12.7. Acknowledgements

We would like to thank Dr J.E. Liddell and Dr M.P. Walton for their useful comments on the antibody and electrophoresis sections respectively.

12.8. References

Allen, W.R. and Hagley, E.A.C. (1982) Evaluation of immunoelectroosmophoresis on cellulose polyacetate for assessing predation on Lepidoptera (Tortricidae) by Coleoptera (Carabidae) species. *Canadian Entomologist* 114, 1047–1054.

Anyanwu, G.I., Phillips, A. and Molyneux, D.H. (1994) Variation in the cuticular hydrocarbons of larvae of *Anopheles gambiae* and variation in the cuticular hydrocarbons of larvae of *Anopheles gambiae* and *A. arabiensis. Insect Science and its Applications* 15, 117–122.

Aronstein, K., Ode, P. and ffrench-Constant, R.H. (1994) Direct comparison of PCR-based monitoring for cyclodiene resistance in *Drosophila* populations with insecticide bioassay. *Pesticide Biochemistry and Physiology* 48, 229–233.

Ashby, J.W. (1974) A study of arthropod predation of *Pieris rapae* L. using serological and exclusion techniques. *Journal of Applied Ecology* 11, 419–425.

Baker, J.P., Maynard-Smith, J. and Strobeck, C. (1975) Genetic polymorphism in the bladder campion, *Silene maritima. Biochemical Genetics* 13, 393–410.

Beard, C.B., Hamm, D.M. and Collins, F.H. (1993) The mitochondrial genome of the mosquito *Anopheles gambiae*: DNA sequence, genome organization, and comparisons with mitochondrial sequences of other insects. *Insect Molecular Biology* 2, 103–124.

Benjamin, D.C., Berzofsky, J.A., East, I.J., Gurd, F.R.N., Hannum, C., Leach, S.J., Margoliash, E., Michael, J.G., Miller, A., Prager, E.M., Reichlin, M., Sercaz, E.E., Smith-Gill, S.J., Todd, P.E. and Wilson, A.C. (1984) The antigenic structure of proteins. *Annual Review of Entomology* 2, 67–101.

Berger, S.L. and Kimmel, A.R. (1987) *Methods in Enzymology. Volume 152. Guide to Molecular Cloning Techniques.* Academic Press, London.

Berkelhamer, R.C. (1983) Intraspecific genetic variation and haplodiploidy, eusociality and polygyny in the Hymenoptera. *Evolution* 37, 540–545.

Berlocher, S.H. (1984) Insect molecular systematics. *Annual Review of Entomology* 29, 403–433.

Besansky. N.J., Paskewitz, S.M., Hamm, D.M. and Collins, F.H. (1992) Distinct families of site-specific retrotransposons occupy identical positions in the rRNA genes of *Anopheles gambiae*. *Molecular and Cellular Biology* 12, 5102–5110.

Beyssat-Arnaouty, V., Mouches, C., Georghiou, G.P. and Pasteur, N. (1989) Detection of organophosphate detoxifying esterases by dot-blot immunoassay in *Culex* mosquitoes. *Journal Of The American Mosquito Control Association* 5, 196–200.

Billiald, P. and Goyffon, M. (1992) Hemocyanin and immunoelectrophoretic criteria in the systematics of the scorpions. *Belgian Journal of Botany* 125, 282–289.

Black, W.C. (IV), DuTeau, N.M., Puterka, G.J., Nechols, J.R. and Pettorini, J.M. (1992) Use of the random amplified polymorphic DNA polymerase chain reaction (RAPD-PCR) to detect DNA polymorphisms in aphids (Homoptera: Aphididae). *Bulletin of Entomological Research* 82, 151–159.

Blackwell, A., Mordue (Luntz), A.J. and Mordue, W. (1994) Identification of bloodmeals of the Scottish biting midge, *Culicoides impunctatus*, by indirect enzyme-linked immunosorbent assay (ELISA). *Medical and Veterinary Entomology* 8, 20–24.

Blackwell, A., Brown, M. and Mordue, W. (1995) The use of an enhanced ELISA method for the identification of *Culicoides* bloodmeals in host-preference studies. *Medical and Veterinary Entomology* 9, 214–218.

Bonavita-Cougourdan, A., Clements, J.L. and Lange, C. (1993) Functional subcaste discrimination (foragers and brood-tenders) in the ant *Componotus vagus* Scop – polymorphism of cuticular hydrocarbon patterns. *Journal of Chemical Ecology* 19, 1461–1477.

Boorman, J., Mellor, P.S., Boreham, P.F.L. and Hewett, R.S. (1977) A latex agglutination test for the identification of blood-meals of *Culicoides* (Diptera: Ceratopogonidae). *Bulletin of Entomological Research* 67, 305–311.

Boreham, P.F.L. and Ohiagu, C.E. (1978) The use of serology in evaluating invertebrate predator–prey relationships: a review. *Bulletin of Entomological Research* 68, 171–194.

Bradford, M. M. (1976) A rapid and sensitive method for the quantitation of microgram quantities of protein utilizing the principle of protein dye binding. *Analytical Biochemistry* 72, 248–254.

Brogdon, W.G., McAllister, J.C. and Vulule J. (1996) Association of heme peroxidase activity measured in single mosquitoes with insecticide resistance oxidase levels. *Journal of the American Mosquito Control Association* (in press).

Bull, C.G. and King, W.V. (1923) The identification of the blood meal of mosquitoes by means of the precipitin test. *American Journal of Hygiene* 3, 491–496.

Burkhart, B.D., Montgomery, E., Langley, C.H. and Voelker, R.A. (1984) Characterization of allozyme null and low activity alleles from two natural populations of *Drosophila melanogaster*. *Genetics* 107, 295–306.

Bush, G.L. (1994) Sympatric speciation in animals: new wine in old bottles. *Trends in Ecology and Evolution* 9, 285–288.

Butts, D.P., Camann, M.A. and Espelie, K.E. (1995) Workers and queens of the European hornet *Vespa crabro* L. have colony-specific cuticular hydrocarbon profiles (Hymenoptera: Vespidae). *Insects Sociaux* 42, 45–55.

Byrne, A.L., Camann, M.N., Cyr, T.L., Catts, E.P. and Espelie, K.E. (1995) Forensic implications of biochemical differences among geographic populations of the black blow fly, *Phormia regina* (Meigen). *Journal of Forensic Sciences* 40, 372–377.

Cameron, E.A. and Reeves, R.M. (1990) Carabidae (Coleoptera) associated with Gypsy moth, *Lymantria dispar* (L.) (Lepidoptera: Lymantriidae), populations, subjected to *Bacillus thuringiensis* Berliner treatments in Pennsylvania. *Canadian Entomologist* 122, 123–129.

Castañera, P., Loxdale, H.D. and Nowak, K. (1983) Electrophoretic study of enzymes from cereal aphid populations. II. Use of electrophoresis for identifying aphidiid parasitoids (Hymenoptera) of *Sitobion avenae* (F.) (Hemiptera: Aphididae). *Bulletin of Entomological Research* 73, 659–665.

Castrovillo, P.J. and Stock, M.W. (1981) Electrophoretic techniques for detection of *Glypta fumiferanae*, an endoparasite of western spruce budworm. *Entomologia Experimentalis et Applicata* 30, 176–180.

Catty, D. and Raykundalis, C. (1988) Production and quality control of polyclonal antibodies. In: Catty, D. (ed.) *Antibodies: A Practical Approach*, Vol. 1. IRL Press, Oxford, pp. 19–79.

Cenis, J.L., Perez, P. and Fereres, A. (1993) Identification of aphid (Homoptera: Aphididae) species and clones by random amplified polymorphic DNA. *Annals of the Entomological Society of America* 86, 545–550.

Chippendale, G.M. and Beck, S.D. (1966) Haemolymph proteins of *Ostrinia nubilarlis* during diapause and prepupal differentiation. *Journal of Insect Physiology* 12, 1629–1638.

Chiverton, P.A. (1987) Predation of *Rhopalosiphum padi* (Homoptera: Aphididae) by polyphagous predatory arthropods during the aphids' pre-peak period in spring barley. *Annals of Applied Biology* 111, 257–269.

Chrambach, A. and Rodbard, D. (1971) Polyacrylamide gel electrophoresis. *Science* 172, 440–451.

Crook, N.E. and Payne, C.C. (1980) Comparison of three methods of ELISA for baculoviruses. *Journal of General Virology* 46, 29–37.

Crook, N.E. and Sunderland, K.D. (1984) Detection of aphid remains in predatory insects and spiders by ELISA. *Annals of Applied Biology* 105, 413–422.

Crowle, A.J. (1980) Precipitin and microprecipitin reactions in fluid medium and in gels. In: Rose, N.R. and Friedman, H. (eds) *Manual of Clinical Immunology*. American Society for Microbiology, Washington, pp. 3–14.

Crozier, R.H. and Crozier, Y.C. (1993) The mitochondrial genome of the honeybee *Apis mellifera*: complete sequence and genome organization. *Genetics* 133, 97–117.

Dahlman, D.L. and Greene, J.R. (1981) Larval hemolymph protein patterns in tobacco hornworm parasitized by *Apanteles congregatus*. *Annals of the Entomological Society of America* 74, 130–133.

Dary, O., Georghiou, G.P., Parsons, E. and Pasteur, N. (1991) Dot-blot test for identification of insecticide-resistant acetylcholinesterase in single insects. *Journal of Economic Entomology* 84, 28–33.

Davis, B.J. (1964) Disk electrophoresis II. Method and application to human serum proteins. *Annals of the National Academy of Sciences* 121, 404–427.

Dempster, J.P. (1960) A quantitative study of the predators of the eggs and larvae of the broom beetle, *Phytodecta olicavea* (Forster), using the precipitin test. *Journal of Animal Ecology* 29, 149–167.

Dennison, D.F. and Hodkinson, I.D. (1983) Structure of the predatory beetle community in a woodland soil ecosystem. I. Prey selection. *Pedobiologia* 25, 109–115.

Devonshire, A.L. (1989) The role of electrophoresis in the biochemical detection of insecticide resistance. In: Loxdale, H.D. and den Hollander, J. (eds)

Electrophoretic Studies on Agricultural Pests. Systematics Association Special Volume 39, Clarendon Press, Oxford, pp. 363–374.

Devonshire, A.L. and Moores, G.D. (1982) A carboxylesterase with broad substrate specificity causes organophosphorus, carbamate and pyrethroid resistance in peach-potato aphids (*Myzus persicae*). *Pesticide Biochemistry and Physiology* 18, 235–246.

Devonshire, A.L., Moores, G.D. and ffrench-Constant, R.H. (1986) Detection of insecticide resistance by immunological estimation of carboxylesterae activity in *Myzus persicae* (Sulzer) and cross reaction of the antiserum with *Phorodon humuli* (Shrank) (Hemiptera:Aphididae). *Bulletin of Entomological Research* 76, 97–107.

Devonshire, A.L., Devine, G.J. and Moores, G.D. (1992) Comparison of microplate esterase assays and immunoassay for identifying insecticide resistant variants of *Myzus persicae* (Homoptera: Aphidae). *Bulletin of Entomological Research* 82, 459–464.

Dicke, M. and De Jong, M. (1988) Prey preference of the phytoseiid mite *Typhlodromus pyri*. 2. Electrophoretic diet analysis. *Experimental and Applied Acarology* 4, 15–25.

Doane, J.F., Scotti, P.D., Sutherland, O.R.W. and Pottinger, R.P. (1985) Serological identification of wireworm and staphylinid predators of the Australian soldier fly (*Inopus rubriceps*) and wireworm feeding on plant and animal food. *Entomologia Experimentalis et Applicata* 38, 65–72.

DuDevoir, D.S. and Reeves, M.R. (1990) Feeding activity of carabid beetles and spiders on gypsy moth larvae (Lepidoptera: Lymantriidae) at high density prey populations. *Journal of Entomological Science* 25, 341–356.

Ernsting, G. and van der Werf, D.C. (1988) Hunger, partial consumption of prey and prey size preference in a carabid beetle. *Ecological Entomology* 13, 155–164.

Estradapena, A., Dusbabek, F. and Castella, J. (1995) Cuticular hydrocarbon variation and progeny phenotypic similarity between laboratory breeds of allopatric populations of *Argas* (*Persicargas*) *persicus* (Oken). *Acta Tropica* 59, 309–322.

Fang, Q., Black, W.C., Blocker, H.D. and Whitcomb, R.F. (1993) A phylogeny of New World *Deltocephalus*-like leafhopper genera based on mitochondrial 16S ribosomal DNA sequences. *Molecular Phylogenetics and Evolution* 2, 119–131.

Ferguson, A. (1980) *Biochemical Systematics and Evolution.* Blackie, London.

Ferkovich, S.M., Greany, P.D. and Dillard, C. (1983) Changes in haemolymph proteins of the fall armyworm, *Spodoptera frugiperda* (J.E. Smith), associated with parasitism by the braconid parasitoid *Cotesia marginiventris* (Cresson). *Journal of Insect Physiology* 29, 933–942.

ffrench-Constant, R.H. and Bonning, B.C. (1989) Rapid microtitre plate test distinguishes insecticide resistant acetylcholinesterase genotypes in the mosquitoes *Anopheles albimanus*, *An. nigerrimus* and *Culex pipiens*. *Medical and Veterinary Entomology* 3, 9–16.

ffrench-Constant, R.H. and Devonshire, A.L. (1987) A multiple homogenizer for rapid sample preparation in immunoassays and electrophoresis. *Biochemical Genetics* 25, 493–499.

ffrench-Constant, R.H., Rocheleau, T.A., Steichen, J.C. and Chalmers, A.E. (1993a) A point mutation in a *Drosophila* GABA receptor confers insecticide resistance. *Nature* 363, 449–451.

ffrench-Constant, R.H., Steichen, J.C., Rocheleau, T.A., Aronstein, K. and Roush, R.T. (1993b) A single-amino acid substitution in a gamma-aminobutyric acid subtype A receptor locus is associated with cyclodiene insecticide resistance in *Drosophila* populations. *Proceedings of The National Acadamy of Sciences of The United States of America* 90, 1957–1961.

ffrench-Constant, R.H., Steichen, J.C. and Shotkoski, F. (1994) Polymerase chain reaction diagnostic for cyclodiene insecticide resistance in the mosquito *Aedes aegypti*. *Medical and Veterinary Entomology* 8, 99–100.

Fichter, B.L. and Stephen, W.P. (1981) Time related decay in prey antigens ingested by the predator *Podisus maculiventris* (Hemiptera: Pentatomidae) as detected by ELISA. *Oecologia* 51, 404–407.

Fichter, B.L. and Stephen, W.P. (1984) Time related decay of prey antigens by arborial spiders as detected by ELISA. *Environmental Entomology* 13, 1583–1587.

Field, L.M., Devonshire, A.L., ffrench-Constant, R.H. and Forde, B.G. (1989) The combined use of immunoassay and a DNA diagnostic technique to identify insecticide-resistant genotypes in the peach-potato aphid, *Myzus persicae* (Sulz.). *Pesticide Biochemistry and Physiology* 34, 174–178.

Fitzgerald, J.D., Solomon, M.G. and Murray, R.A. (1986) The quantitative assessment of arthropod predation rates by electrophoresis. *Annals of Applied Biology* 109, 491–498.

Fuerst, P.A. and Ferrell, R.E. (1980) The stepwise mutation model: an experimental evaluation utilising hemoglobin variants. *Genetics* 94, 185–201.

Gardener, W.A., Shepard, M. and Noblet, T.R. (1981) Precipitin test for examining predator–prey interactions in soybean fields. *Canadian Entomologist* 113, 365–369.

Georghiou, G.P. and Pasteur, N. (1978) Electrophoretic esterase patterns in insecticide resistant and susceptible mosquitoes. *Journal of Economic Entomology* 71, 201–205.

Georghiou, G.P. and Pasteur, N. (1989) Novel tests for organophosphate insecticide resistance in single mosquitoes: an overview of recent progress and outline of filter paper test. *57th Annual Conference of the California Mosquito and Vector Control Association*, pp. 174–178.

Giller, P.S. (1982) The natural diet of waterbugs (Hemiptera–Heteroptera): electrophoresis as a potential method of analysis. *Ecological Entomology* 7, 233–237.

Giller, P.S. (1984) Predator gut state and prey detectability using electrophoretic analysis of gut contents. *Ecological Entomology* 9, 157–162.

Giller, P.S. (1986) The natural diet of the Notonectidae: field trials using electrophoresis. *Ecological Entomology* 11, 163–172.

Goding, J.W. (1986) *Monoclonal Antibodies: Principles and Practice.* Academic Press, San Diego.

Gordon, A.H. (1975) *Electrophoresis of Proteins in Polyacrylamide and Starch Gels.* North Holland, Amsterdam.

Greenstone, M.H. (1977) A passive haemagglutination inhibition assay for the identification of the stomach contents of invertebrate predators. *Journal of Applied Ecology* 14, 457–464.

Greenstone, M.H. (1983) An enzyme-linked immunosorbent assay for the *Amblyospora* of *Culex salinarius* (Microspora: Amblyosporidae). *Journal of Invertebrate Pathology* 14, 250–255.

Greenstone, M.H. (1995) Bollworm or budworm? Squashblot immunoassay distinguishes eggs of *Helicoverpa zea* and *Heliothis virescens* (Lepidoptera: Noctuidae). *Journal of Economic Entomology* 88, 213–218.

Greenstone, M.H. (1996) Serological analysis of arthropod predation: past, present and future. In: Symondson, W.O.C. and Liddell, J.E. (eds) *The Ecology of*

Agricultural Pests: Biochemical Approaches. Systematics Association Special Volume 53. Chapman & Hall, London, pp. 265–300.

Greenstone, M.H. and Morgan, C.E. (1989) Predation on *Heliothis zea*: an instar-specific ELISA for stomach analysis. *Annals of the Entomological Society of America* 84, 457–464.

Greenstone, M.H. and Trowell, S.C. (1994) Arthropod predation: a simplified immunodot format for predator gut analysis. *Annals of the Entomological Society of America* 87, 214–217.

Greenstone, M.H., Stuart, M.K. and Haunerland, N.H. (1991) Using monoclonal antibodies for phylogenetic analysis: an example from the Heliothinae (Lepidoptera: Noctuidae). *Annals of the Entomological Scoiety of America* 84, 457–464.

Guttman, S.I., Wood, T.K. and Karlin, A.A. (1981) Genetic differentiation along host plant lines in the sympatric *Enchenopa binotata* Say complex (Homoptera: Membracidae). *Evolution* 35, 205–217.

Hagler, J.R. and Cohen, A.C. (1990) Effects of time and temperature on digestion of purified antigen by *Geocoris punctipes* (Hemiptera: Lygaeidae) reared on artificial diet. *Annals of the Entomological Society of America* 83, 1177–1180.

Hagler, J.R. and Naranjo, S.E. (1994a) Qualitative survey of two coleopteran predators of *Bemisia tabaci* (Homoptera: Aleyrodidae) and *Pectinophora gossypiella* (Lepidoptera: Gelechiidae) using a multiple prey gut content ELISA. *Environmental Entomology* 23, 193–197.

Hagler, J.R. and Naranjo, S.E. (1994b) Determining the frequency of heteropteran predation on sweetpotato whitefly and pink bollworm using multiple ELISAs. *Entomologia Experimentalis et Applicata* 72, 63–70.

Hagler, J.R., Cohen, A.C., Enriquez, F.J. and Bradley-Dunlop, D. (1991) An egg-specific monoclonal antibody to *Lygus hesperus*. *Biological Control* 1, 75–80.

Hagler, J.R., Cohen, A.C., Bradley-Dunlop, D. and Enriquez, F.J. (1992) Field evaluation of predation on *Lygus hesperus* (Hemiptera: Miridae) using a species- and stage-specific monoclonal antibody. *Environmental Entomology* 21, 896–900.

Hagler, J.R., Brower, A.G., Zhijian Tu, Byrne, D.N., Bradley-Dunlop, D. and Enriques, F.J. (1993) Development of a monoclonal antibody to detect predation of the sweetpotato whitefly, *Bemisia tabaci* (Gennadius). *Entomologia Experimentalis et Applicata* 68, 231–236.

Hagler, J.R., Naranjo, S.E., Bradley-Dunlop, D., Enriquez, F.J. and Henneberry, T.J. (1994) A monoclonal antibody to pink bollworm (Lepitoptera: Gelechiidae) egg antigen: a tool for predator gut analysis. *Annals of the Entomological Society of America* 87, 85–90.

Hames, B.D. and Rickwood, D. (1981) *Gel Electrophoresis of Proteins: a Practical Approach.* IRL Press, Oxford.

Hance, T. and Renier, L. (1987) An ELISA technique for the study of the food of carabids. *Acta Phytopathologica et Entomologicae Hungaricae* 22, 363–368.

Hance, T. and Rossignol, P. (1983) Essai de quantification de la prédation des Carabidae par le test ELISA. *Mededelingen van de Fakulteit Landbouwwetenschappen Rijksuniversiteit Gent* 48, 475–485.

Haucke, H.R. and Gellissen, G. (1988) Different mitochondrial gene orders among insects: exchanged tRNA gene positions in the COII/COIII region between an orthopteran and a dipteran species. *Current Genetics* 14, 471–476.

Heinze, J., Ortius, D., Holldobler, B. and Kaib, M. (1994) Interspecific aggression in colonies of the slave-making ant *Harpagoxenus sublaevis*. *Behavioural Ecology and Sociobiology* 35, 75–83.

Hemingway, J., Smith, C., Jayawardena, K.G.I. and Herath, P.R.J. (1986) Field and laboratory detection of the altered acetylcholinesterase resistance genes which confer organophosphate and carbamate resistance in mosquitoes (Diptera: Culicidae). *Bulletin of Entomological Research* 76, 559–565.

Hemingway, J., Lindsay, S.W., Small, G.J., Jawara, M. and Collins, F.H. (1995) Insecticide susceptiblity status in individual species of the *Anopheles gambiae* complex where pyrethroid impregnated bednets are used extensively for malaria control. *Bulletin of Entomological Research* 85, 229–234.

Hemingway, J., Small, G.J., Lindsay, S.W. and Collins, F.H. (1996) Combined use of biochemical, immunological and molecular assays for infection, species identification and resistance detection in field populations of *Anopheles* (Diptera: Culicidae). In: Symondson, W.O.C. and Liddell, J.E. (eds) *The Ecology of Agricultural Pests: Biochemical Approaches.* Systematics Association Special Volume 53. Chapman & Hall, London, pp. 31–49.

Hillis, D.M. and Moritz, C. (eds) (1990) *Molecular Systematics.* Sinauer, Sunderland.

Höller, C. and Braune, H.J. (1988) The use of isoelectric focusing to assess percentage hymenopterous parasitism in aphid populations. *Entomologia Experimentalis et Applicata* 47, 105–114.

Hsiao, T.H. (1989) Estimation of genetic variability amongst coleoptera. In: Loxdale, H.D. and den Hollander, J. (eds) *Electrophoretic Studies on Agricultural Pests.* Systematics Association Special Volume 39. Clarendon Press, pp. 143–180.

Hudson, L. and Hay, F.C. (1989) *Practical Immunology.* Blackwell, London.

Kapuge, S.H., Danthanarayana, W. and Hoogenraad, N. (1987) Immunological investigation of prey–predator relationships for *Pieris brassicae* (L.) (Lepidoptera: Pieridae). *Bulletin of Entomological Research* 77, 247–254.

Karunaratne, S.H.P.P., Jayawardena, K.G.I. and Hemingway, J. (1995) The cross-reactivity spectrum of a polyclonal antiserum raised against the A_2 esterase involved in insecticide resistance. *Pesticide Biochemistry and Physiology* 53, 75–83.

Kazmar, D.J. (1991) Isoelectric focusing procedures for the analysis of allozyme variation in minute arthropods. *Annals of the Entomological Society of America* 84, 332–339.

Köhler, G. and Milstein, C. (1975) Continuous cultures of fused cells secreting antibody of predefined specificity. *Nature* 256, 495–497.

Kruger, E.L. and Pappas, C.D. (1993) Geographic variation of cuticular hydrocarbons amongst 14 populations of *Aedes albopictus* (Diptera: Culicidae). *Journal of Medical Entomology* 30, 544–548.

Kuperstein, M.L. (1979) Estimating carabid effectiveness in reducing the Sunn pest, *Eurygaster integriceps* Puton (Heteroptera: Scutelleridae) in the USSR. *Entomological Society of America Miscellaneous Publications* 11, 80–84.

Langley, P.A. (1996) Practical applications for techniques to determine the nutritional state and age of field caught tsetse flies. In: Symondson, W.O.C. and Liddell, J.E. (eds) *The Ecology of Agricultural Pests: Biochemical Approaches.* Systematics Association Special Volume 53. Chapman & Hall, London, pp. 479–497.

Langley, P.A., Hall, M.J.R. and Felton, T. (1988) Determining the age of tsetse flies, *Glossinia* spp. (Diptera: Glossinidae): an appraisal of the pteridine fluorescence technique. *Bulletin of Entomological Research* 78, 387–395.

Lansman, R.A., Shade, R.O., Shapira, J.F. and Avise, J.C. (1981) The use of restriction endonucleases to measure mitochondrial DNA sequence relatedness in natural populations III. Techniques and potential applications. *Journal of Molecular Evolution* 17, 214–226.

Leach, E.H. (1947) Bismarck brown as a stain for mucoproteins. *Stain Technology* 22, 73–76.

Lehane, M.J. and Mail, T.S. (1985) Determining the age of adult male and female *Glossina morsitans morsitans* using a new technique. *Ecological Entomology* 10, 219–224.

Lesiewicz, D.S., Lesiewicz, J.L., Bradley, J.R. and van Duyn, I.W. (1982) Serological determination of carabid (Coleoptera: Adephaga) predation on corn earworm (Lepidoptera, Noctuidae) in field corn. *Environmental Entomology* 11, 1183–1186.

Leslie, G.W. and Boreham, P.F.L. (1981) Identification of arthropod predators of *Eldana saccharina* Walker (Lepidoptera: Pyralidae) by cross-over electrophoresis. *Journal of the Entomological Society of South Africa* 44, 381–388.

Liddell, J.E. and Cryer, A. (1991) *A Practical Guide to Monoclonal Antibodies*. Wiley, Chichester.

Liddell, J.E. and Symondson, W.O.C. (1996) The potential of combinatorial gene libraries in pest–predator relationship studies. In: Symondson, W.O.C. and Liddell, J.E. (eds), *The Ecology of Agricultural Pests: Biochemical Approaches*. Systematics Association Special Volume 53, Chapman & Hall, London, pp. 347–366.

Liddell, J. and Weeks, I. (1995) *Antibody Technology*. BIOS, Oxford.

Lister, A., Usher, M.B. and Block, W. (1987) Description and quantification of field attack rates by predatory mites: an example using electrophoresis method with a species of Antarctic mite. *Oecologia (Berlin)* 72, 185–191.

Liu, H. and Bechenbach, A.T. (1992) Evolution of the mitochondrial cytochrome oxidase II gene among ten orders of insects. *Molecular Phylogenetics and Evolution* 1, 41–52.

Lovei, G.L. (1986) The use of biochemical methods in the study of carabid feeding: the potential of isozyme analysis and ELISA. In: den Boer, P.J., Grum, L. and Szyszko, J. (eds) *Feeding Behaviour and Accessibility of Food for Carabid Beetles*. Warsaw Agricultural University Press, Warsaw, pp. 21–27.

Lovei, G.L., Monostori, E. and Andó, I. (1985) Digestion rate in relation to starvation in the larva of a carabid predator, *Poecilus cupreus*. *Entomologia Experimentalis et Applicata* 37, 123–127.

Lovei, G.L., Sopp, P.I. and Sunderland, K.D. (1987) The effect of mixed feeding on the digestion of the carabid *Bembidion lampros*. *Acta Phytopathologica et Entomologica Hungarica* 22, 403–407.

Lovei, G.L., Sopp, P.I. and Sunderland, K.D. (1990) Digestion rate in relation to alternative feeding in three species of polyphagous predators. *Ecological Entomology* 15, 293–300.

Lowry, O.H., Rosebrough, N.J., Farr, A.L. and Randall, R.J. (1951). Protein measurement with the folinphenol reagent. *Journal of Biological Chemistry* 193, 265–275.

Loxdale, H.D. (1994) Isozyme and protein profiles of insects of agricultural and horticultural importance. In: Hawksworth, D.L. (ed.) *The Identification and Characterisation of Pest Organisms*. CAB International, Wallingford, pp. 337–375.

Loxdale, H.D. and den Hollander, J. (eds) (1989) *Electrophoretic Studies on Agricultural Pests*. Systematics Association Special Volume 39. Clarendon Press.

Loxdale, H.D., Castañera, P. and Brookes, C.P. (1983) Electrophoretic study of enzymes from cereal aphid populations. I. Electrophoretic techniques and staining systems for characterising isoenzymes from six species of cereal aphids (Hemiptera: Aphididae). *Bulletin of Entomological Research* 73, 645–657.

Loxdale, H.D., Brookes, C.P. and de Barro, P.J. (1996) Application of novel molecular markers (DNA) in agricultural entomology. In: Symondson, W.O.C. and Liddell, J.E. (eds) *The Ecology of Agricultural Pests: Biochemical Approaches*. Systematics Association Special Volume 53. Chapman & Hall, London, pp. 149–198.

Lucas, J.R. and Graften, A. (1985) Partial prey consumption by ambush predators. *Journal of Theoretical Biology* 113, 455–473.

Lund, R.D. and Turpin, F.T. (1977) Serological investigation of black cutworm larval consumption by ground beetles. *Annals of the Entomological Society of America* 70, 322–324.

McCraken, A., Uhlenbrusch, I. and Gellissen, G. (1987) Structure of the cloned *Locusta migratoria* mitochondrial genome: restriction mapping and sequence of its ND1 (URF1) gene. *Current Genetics* 11, 1625–1630.

McIntyre, G.S. and Gooding, R.H. (1995) Pteridine accumulation in *Musca domestica*. *Journal of Insect Physiology* 41, 357–368.

McIver, J.D. (1981) An examination of the utility of the precipitin test for evaluation of arthropod predator–prey relationships. *Canadian Entomologist* 113, 213–222.

McIver, J.D. and Tempelis, C.H. (1993) The arthropod predators of anti-mimetic and aposematic prey: a serological analysis. *Ecological Entomology* 18, 218–222.

Menken, S.B.J. (1982) Enzyme characterisation of nine endoparasite species of small ermine moths (Yponomeutidae). *Experimentia* 38, 1461–1462.

Menken, S.B. and Raijmann, L.E.L. (1996) Biochemical systematics: principles and perspectives for pest management. In: Symondson, W.O.C. and Liddell, J.E. (eds) *The Ecology of Agricultural Pests: Biochemical Approaches*. Systematics Association Special Volume 53. Chapman & Hall, London, pp. 7–29.

Menken, S.B.J. and Ulenberg, S.A. (1987) Biochemical characters in agricultural entomology. *Agricultural Zoology Reviews* 2, 305–360.

Menken, S.B.J., Herrebout, W.M. and Wiebes, J.T. (1992) Small ermine moths (*Yponomeuta*): their host relations and evolution. *Annual Review of Entomology* 37, 41–66.

Menschik, Z. (1953) Nile blue histochemical method for phospholipids. *Stain Technology* 28, 13–18.

Miller, M.C. (1981) Evaluation of enzyme-linked immunosorbent assay of narrow- and broad-spectrum anti-adult southern pine beetle serum. *Annals of the Entomological Society of America* 74, 279–282.

Mochizuki, A., Shiga, M. and Imura, O. (1993) Pteridine accumulation for age determination in the melon fly, *Bactrocera* (*Zeugodacus*) *cucurbitae* (*coquillett*) (Diptera: Tephritidae). *Applied Entomology and Zoology* 28, 584–586.

Mouches, C., Magnin, M., Berge, J.B., Desilvestri, M., Beyssat, V., Pasteur, N. and Georghiou, G. P. (1987) Overproduction of detoxifying esterases in organophosphate-resistant *Culex* mosquitoes and their presence in other insects. *Proceedings of the National Acadamy of Sciences of the United States of America* 84, 2113–2116.

Murray, R.A. and Solomon, M.G. (1978) A rapid technique for analysing diets of invertebrate predators by electrophoresis. *Annals of Applied Biology* 90, 7–10.

Nakamura, K. (1977) A model for the functional response of a predator to varying prey densities, based on the feeding ecology of wolf spiders. *Bulletin of the National Institute of Agricultural Sciences (Japan), series C* 31, 28–89.

Nakamura, M. and Nakamura, K. (1977) Population dynamics of the chestnut gall wasp (Cynipidae: Hym.). 5. Estimation of the effect of predation by spiders on the mortality of imaginal wasps based on the precipitin test. *Oecologia* 27, 97–116.

Neems, R.M. and Butlin, R.K. (1994) Variation in cuticular hydrocarbons across a hybrid zone in the grasshopper *Chorthippus parallelus*. *Proceedings of the Royal Society of London Series B – Biological Sciences* 257, 135–140.

Nei, M. (1972) Genetic distance between populations. *American Naturalist* 106, 283–292.

Nevo, E., Beiles, A. and Ben-Chlomo, R. (1984) The evolutionary significance of genetic diversity: ecological, demographic and life history correlates. *Lecture Notes in Biomathematics* 53, 13–213.

Ngumbi, M.P., Lawyer, P.G., Johnson, R.D., Kiilu, G. and Asiago, C. (1992) Identification of phlebotomine sandfly bloodmeals from Baringo district, Kenya, by direct enzyme-linked immunosorbent assay (ELISA). *Medical and Veterinary Entomology* 6, 385–388.

Nichols, W.S. and Nakamura, R.M. (1980) Agglutination and agglutination inhibition assays. In: Rose, N.R. and Friedman, H. (eds) *Manual of Clinical Immunology*. American Society for Microbiology, Washington, pp. 15–22.

Noble, L.R. and Jones, C.S. (1996) A molecular and ecological investigation of the large arionid slugs of North-Western Europe: the potential for new pests. In: Symondson, W.O.C. and Liddell, J.E. (eds) *The Ecology of Agricultural Pests: Biochemical Approaches*. Systematics Association Special Volume 53. Chapman & Hall, London, pp. 93–131.

Nuttall, G.H.F. (1904) *Blood Immunity and Blood Relationships*. Cambridge University Press, Cambridge.

Oien, C.T. and Ragsdale, D.W. (1992) A species-specific enzyme-linked immunosorbent assay for *Nosema furnicalis* (Microspora: Nosematidae). *Journal of Invertebrate Pathology* 60, 84–88.

Olsen, P.E., Grant, G.A., Nelson, D.L. and Rice, W.A. (1990) Detection of American foulbrood disease of the honeybee, using a monoclonal antibody specific to *Bacillus larvae* in an enzyme-linked immunosorbent assay. *Canadian Journal of Microbiology* 36, 732–735.

Paskewitz, S.M. and Collins, F.H. (1989) Site-specific ribosomal DNA insertion elements in *Anopheles gambiae* and *Anopheles arabiensis* – nucleotide-sequence of gene-element boundaries. *Nucleic Acids Research* 17, 8125–8133.

Pasteur, N. and Georghiou, G.P. (1981) Filter paper test for rapid determination of phenotypes with high esterase activity in organophosphate resistant mosquitoes. *Mosquito News* 41, 181–183.

Pasteur, N., Pasteur, G., Bonhomme, F., Catalan, J. and Britton-Davidian, J. (1988) *Practical Isozyme Genetics*. Ellis Horwood, Chichester.

Penilla, R.P., Rodriguez, A.D., Hemingway, J., Estrada, J.L.T., Jimenez, J.I.A. and Rodriguez, M.H. (1996) Rotational and mosaic strategies for delaying the development of insecticide resistance in mosquitoes – baseline data for a large scale field trial in Southern Mexico. In: Wildey, K.B. and Robinson, W.H. (eds) *Proceedings of the 2nd International Conference on Insect Pests in the Urban Environment*. BPCC Wheatons Ltd. Exeter.

Pettersson, J. (1972) Technical data of a serological method for quantitative predator efficiency studies on *Rhopalosiphum padi* (L.). *Swedish Journal of Agricultural Research* 2, 65–69.

Phillips, A., Milligan, P.J.M., Broomfield, G. and Molyneux, D.H. (1988) Identification of medically important Diptera by analysis of cuticular hydrocarbons. In: Service, M.W. (ed.) *Biosystematics of Haematophagous Insects*. Systematics Association Special Volume 37, Ch. 4. Clarendon Press, Oxford.

Phillips, A., Sabatini, A., Milligan, P.J.M., Boccolini, D., Broomfield, G. and

Molyneux, D.H. (1990) The *Anopheles maculipinnis* complex (Diptera: Culicidae): comparison of the cuticular hydrocarbon profiles determined in adults of five Palaearctic species. *Bulletin of Entomological Research* 80, 459–464.

Powell, W. and Walton, M.P. (1995) Populations and communities. In: Jervis, M. and Kidd, N. (eds) *Insect Natural Enemies*. Chapman & Hall, London, pp. 223–292.

Powell, W., Walton, M.P. and Jervis, M.A. (1996) Populations and communities. In: Jervis, M.A. and Kidd, N.A.C. (eds) *Insect Natural Enemies: Practical Approaches to their Study and Evaluation*. Chapman & Hall, London, pp. 223–292.

Pruess, K.P., Zhu, X. and Powers, T.O. (1992) Mitochondrial transfer RNA genes in blackfly, *Simulium vittatum* (Diptera: Simulidae), indicate long divergence from mosquitoes (Diptera: Culicidae) and fruitfly (Diptera: Drosophilidae). *Journal of Medical Entomology* 29, 644–651.

Puterka, G.J., Black, W.C. (IV), Steiner, W.M. and Burton, R.L. (1993) Genetic variation and phylogenetic relationships amongst worldwide collections of Russian wheat aphid, *Diuraphis noxia* (Mordvilko), inferred from allozyme and RAPD-PCR markers. *Heredity* 70, 604–618.

Puttler, B. and Sacks, J.M. (1993) Morphological and chemical differences between the ectoparasitoids *Euplectrus comstockii* and *E. plathypenae* (Hymenoptera: Eulophidae). *Annals of the Entomological Society of America* 86, 551–559.

Quicke, D.L.J. (1993) *Principles and Techniques of Contemporary Taxonomy*. Blackie, London.

Ragsdale, D.W. and Kjer, K.M. (1989) Diagnosis of tracheal mite (*Acarapis woodi* Rennie) parasitism of honey bees using a monoclonal based enzyme-linked immunosorbent assay. *American Bee Journal* 129, 550–552.

Ragsdale, D.W. and Oien, C.T. (1996) An environmental risk assessment for release of an exotic microsporidium for European corn borer control in North America. In: Symondson, W.O.C. and Liddell, J.E. (eds) *The Ecology of Agricultural Pests: Biochemical Approaches*. Systematics Association Special Volume 53, Chapman & Hall, London, pp. 401–417.

Ragsdale, D.W., Larson, A.D. and Newsom, L.D. (1981) Quantitative assessment of the predators of *Nezara viridula* eggs and nymphs within a soybean agroecosystem using an ELISA. *Environmental Entomology* 10, 402–405.

Randolf, S.E. and Rogers, D.J. (1978) Feeding cycles and flight activity in field populations of tsetse (Diptera: Glossinidae). *Bulletin of Entomological Research* 68, 655–671.

Richardson, B.J., Baverstock, P.R. and Adams, M. (1986) *Allozyme Electrophoresis*. Academic Press, London.

Richmond, J. and Page, M. (1995) Genetic and biochemical similarities among four species of pine coneworms (Lepidoptera: Pyralidae). *Annals of the Entomological Society of America* 88, 271–280.

Robinson, J., Fischer, M. and Hoisington, D. (1993) Molecular characterisation of *Diuraphis* spp. using random amplified polymorphic DNA. *Southwestern Entomologist* 18, 121–127.

Roitt, I.M., Brostoff, J. and Male, D.K. (1989) *Immunology*. Gower Medical Publishing, London.

Sambrook, J., Fritsch, E.F. and Maniatis, T. (1989) In: Nolan C. (ed.) *Molecular Cloning: a Laboratory Manual. Volumes I–III*. Cold Spring Harbor Laboratory Press, US.

Sargent, J.R. and George, S.G. (1975) *Methods in Zone Electrophoresis*. BDH Chemical Ltd, Poole, Dorset.

Sasaki, H., Kang'ethe, E.K. and Kaburia, F.A. (1995) Blood meal sources of *Glossina pallidipes* and *G. longipennis* (Diptera: Glossinidae) in Nguruman, Southwest Kenya. *Journal of Medical Entomology* 32, 390–393.

Schoof, D.D., Palchick, S. and Tempelis, C.H. (1986) Evaluation of predator–prey relationships using an enzyme immunoassay. *Annals of the Entomological Society of America* 79, 91–95.

Schwartz, M. and Sofer, W. (1976) Diet induced alterations in distribution of multiple forms of alcohol dehydrogenase in *Drosophila. Nature* 263, 129–131.

Scott, J.A., Brogdon, W.G. and Collins, F.H. (1993) Identification of single specimens of the *Anopheles gambiae* complex by the polymerase chain reaction. *American Journal of Tropical Medicine and Hygiene* 49, 520–529.

Sergeeva, T.K. (1975) Use of cross immunoelectrophoresis for increasing sensitivity of serological reactions in studying trophic relations in insects. *Zoologicheski Zhurnal* 54, 1014–1019.

Service, M.W. (1973) Study of the natural predators of *Aedes cantans* (Meigen) using the precipitin test. *Journal of Medical Entomology* 10, 503–510.

Service, M.W., Voller, A. and Bidwell, D.E. (1986) The enzyme-linked immunosorbent assay (ELISA) for the identification of blood-meals of haematophagous insects. *Bulletin of Entomological Research* 76, 321–330.

Sigsgaard, L. (1996) Serological analysis of predators of *Helicoverpa armigera* Hübner (Lepidoptera: Noctuidae) eggs in sorghum–pigeonpea intercropping at ICRASAT, India: a preliminary field study. In: Symondson, W.O.C. and Liddell, J.E. (eds) *The Ecology of Agricultural Pests: Biochemical Approaches*. Systematics Association Special Volume 53. Chapman & Hall, London, pp. 367–381.

Slatkin, M. (1981) Estimating levels of gene flow in natural populations. In: Real, L.A. (ed.), *Ecological Genetics*. Princeton University Press, Princeton, pp. 3–17.

Slatkin, M. (1985) Rare alleles as indicators of gene flow. *Evolution* 39, 53–65.

Slatkin, M. and Barton, N. (1989) A comparison of three indirect methods for estimating average levels of gene flow. *Evolution* 43, 1349–1368.

Smilowitz, Z. and Smith, C.L. (1977) Hemolymph proteins of developing *Pieris rapae* larvae parasitized by *Apanteles glomeratus. Annals of the Entomological Society of America* 70, 447–454.

Solomon, M.G., Fitzgerald, J.D. and Murray, R.A. (1996) Electrophoretic approaches to predator–prey interactions. In: Symondson, W.O.C. and Liddell, J.E. (eds) *The Ecology of Agricultural Pests: Biochemical Approaches*. Systematics Association Special Volume 53, Chapman & Hall, London, pp. 457–468.

Sopp, P.I. and Sunderland, K.D. (1989) Some factors affecting the detection period of aphid remains in predators using ELISA. *Entomologia Experimentalis et Applicata* 51, 11–20.

Sopp, P.I., Sunderland, K.D., Fenlon, J.S. and Wratten, S.D. (1992) An improved quantitative method for estimating invertebrate predation in the field using an enzyme-linked immunosorbent assay (ELISA). *Journal of Applied Ecology* 79, 295–302.

Southwood, T.R.E. (1978) *Ecological Methods*, 2nd edn. Chapman & Hall, London, 524pp.

Stanley-Samuelson, D.W. and Nelson, D.R. (eds) (1993) *Insect Lipids: Chemistry, Biochemistry and Biology*. University of Nebraska Press, Nebraska.

Steichen, J.C. and ffrench-Constant, R.H. (1994) Amplification of specific cyclodiene insecticide resistance alleles by the polymerase chain reaction. *Pesticide Biochemistry and Physiology* 48, 1–7.

Stock, M.W., Amman, G.D. and Higby, P.K. (1984) Genetic variation among

Mountain Pine Beetle (*Dendroctonus ponderosae*) populations from 7 Western states. *Annals of the Entomological Society of America* 77, 760–764.

Stuart, M.K. and Greenstone, M.H. (1990) Beyond ELISA: a rapid, sensitive, specific immunodot assay for identification of predator stomach contents. *Annals of the Entomological Society of America* 83, 1101–1107.

Stuart, M.K. and Greenstone, M.H. (1996) Serological diagnosis of parasitism: a monoclonal antibody-based immunodot assay for *Microplitis croceipes* (Hymenoptera: Braconidae). In: Symondson, W.O.C. and Liddell, J.E. (eds) *The Ecology of Agricultural Pests: Biochemical Approaches*. Systematics Association Special Volume 53. Chapman & Hall, London, pp. 300–321.

Stuart, M.K., Barak, A.V. and Burkholder, W.E. (1994) Immunological identification of *Trogoderma granarium* Everts (Coleoptera: Dermestidae). *Journal of Stored Products Research* 30, 9–16.

Sunderland, K.D. (1988) Quantifying methods for detecting invertebrate predation occurring in the field. *Journal of Applied Biology* 112, 201–224.

Sunderland, K.D. (1996) Progress in quantifying predation using antibody techniques. In: Symondson, W.O.C. and Liddell, J.E. (eds) *The Ecology of Agricultural Pests: Biochemical Approaches*. Systematics Association Special Volume 53, Chapman & Hall, London, pp. 419–455.

Sunderland, K.D., Crook, N.E., Stacey, D.L. and Fuller, B.J. (1987) A study of feeding by polyphagous predators on cereal aphids using ELISA and gut dissection. *Journal of Applied Ecology* 24, 907–933.

Sunderland, K.D., De Snoo, G.R., Dinter, A., Hance, T., Helenius, J. Jepson, P., Kromp, B., Samu, F., Sotherton, N.W., Toft, S. and Ulber, B. (1995) Density estimation for invertebrate predators in agroecosystems. In: Toft, S. and Riedel, W. (eds) *Arthropod Natural Enemies in Arable Land. Density, Spatial Heterogeneity and Dispersal. Acta Jutlandica 70*. Aarhus University Press, Denmark, pp. 133–164.

Sutton, S.L. (1970) Predation on woodlice: an investigation using the precipitin test. *Entomologia Experimentalis et Applicata* 13, 279–285.

Symondson, W.O.C. and Liddell, J.E. (1993a) The development and characterisation of an anti-haemolymph antiserum for the detection of mollusc remains within carabid beetles. *Biocontrol Science and Technology* 3, 261–275.

Symondson, W.O.C. and Liddell, J.E. (1993b) A monoclonal antibody for the detection of arionid slug remains in carabid predators. *Biological Control* 3, 207–214.

Symondson, W.O.C. and Liddell, J.E. (1993c) Differential antigen decay rates during digestion of molluscan prey by carabid predators. *Entomologia Experimentalis et Applicata* 69, 277–287.

Symondson, W.O.C. and Liddell, J.E. (1993d) The detection of predation by *Abax parallelepipedus* and *Pterostichus madidus* (Coleoptera: Carabidae) on Mollusca using a quantitative ELISA. *Bulletin of Entomological Research* 83, 641–647.

Symondson, W.O.C. and Liddell, J.E. (1995) Decay rates for slug antigens within the carabid predator *Pterostichus melanarius* monitored with a monoclonal antibody. *Entomologia Experimentalis et Applicata* 75, 245–250.

Symondson, W.O.C. and Liddell, J.E. (1996a) A species-specific monoclonal antibody system for detecting the remains of field slugs, *Deroceras reticulatum* (Müller) (Mollusca: Pulmonata), in carabid beetles (Coleoptera: Carabidae). *Biocontrol Science and Technology* 6, 91–99.

Symondson, W.O.C. and Liddell, J.E. (1996b) Immunological approaches to the detection of predation upon New Zealand flatworms (Tricladida: Terricola): problems caused by shared epitopes with slugs (Mollusca: Pulmonata).

International Journal of Pest Management 42, 95–99.

Symondson, W.O.C. and Liddell, J.E. (eds) (1996c) *The Ecology of Agricultural Pests: Biochemical Approaches*. Chapman & Hall, London.

Symondson, W.O.C. and Walton, M.P. (1993) Electrophoretic separation of pulmonate haemocyanins: a simple taxonomic tool. *Journal of Molluscan Studies* 60, 351–354.

Symondson, W.O.C., Mendis, V.W. and Liddell, J.E. (1995) Monoclonal antibodies for the identification of slugs and their eggs. *EPPO Bulletin* 25, 377–382.

Symondson, W.O.C., Glen, D.M., Wiltshire, C.W., Langdon, C.J. and Liddell, J.E. (1996) Effects of cultivation techniques and methods of straw disposal on predation by *Pterostichus melanarius* (Coleoptera: Carabidae) upon slugs (Gastropoda: Pulmonata) in an arable field. *Journal of Applied Ecology* 33, 741–753.

Tempelis, C.H. (1975) Host-feeding patterns of mosquitoes, with a review of advances in analysis of blood meals by serology. *Journal of Medical Entomology* 11, 635–653.

Thorpe, J.P. (1982) The molecular clock hypothesis: biochemical evolution, genetic differentiation and systematics. *Annual Review of Ecology and Systematics* 13, 139–168.

Tijssen, P. (1985) *Practice and Theory of Enzyme Immunoassays*. Elsevier, Oxford.

Tod, M.E. (1973) Notes on beetle predators of molluscs. The *Entomologist* 106, 196–201.

Tomiuk, J. and Wöhrmann, K. (1980) Population growth and population structure of natural populations of *Macrosiphum rosae* (L.) (Hemiptera: Aphididae). *Zeitschrift für Angewandte Entomologie* 90, 464–473.

Tomiuk, J., Wöhrmann, K. and Eggers-Schumacher, H.A. (1979) Enzyme patterns as a characteristic for the identification of aphids. *Zeitschrift für Angewandte Entomologie* 88, 440–446.

Vachon, M. and Goyffon, M. (1978) Chiniotaxonomie: valeur et limites des caractères tirés du protéinogramme de l'hémolymphe en gel de polyacrylamide de différentes espèces de scorpions (Arachnides). *Symposia. Zoological Society of London* 42, 317–325.

Van der Geest, L.P.S. and Overmeer, W.P.J. (1985) Experiences with polyacrylamide gel gradient gel electrophoresis for the detection of gut contents of phytoseiid mites. *Mededelingen van de Faculteit voor Landbouwwetenschappen Rijksuniversiteit Gent* 50, 469–471.

Vaughan, A., Rodriguez, M. and Hemingway, J. (1995) The independent gene amplification of indistinguishable esterase B electromorphs from the insecticide resistant mosquito *Culex quinquefasciatus*. *Biochemical Journal* 305, 651–658.

Voller, A. and Bidwell, D.E. (1986) Enzyme-linked immunosorbent assay. In: Rose, N.R., Friedman, H. and Fahley, J.L. (eds) *Manual of Clinical Laboratory Immunology*. American Society for Microbiology, Washington, pp. 99–109.

Voller, A., Bidwell, D.E. and Bartlett, A. (1979) *The Enzyme-Linked Immunosorbent Assay (ELISA): a Guide with Abstracts of Microplate Applications*. Dynatech Europe, Guernsey.

Wall, R. (1993) The reproductive output of the blowfly *Lucilia sericata*. *Journal of Insect Physiology* 39, 743–750.

Walton, M.P., Loxdale, H.D. and Allen-Williams, L. (1990a) Electrophoretic 'keys' for the identification of parasitoids (Hymenoptera: Braconidae: Aphelinidae) attacking *Sitobion avenae* (F.) (Hemiptera: Aphididae). *Biological Journal of the Linnean Society* 40, 333–346.

Walton, M.P., Powell, W., Loxdale, H.D. and Allen-Williams, L. (1990b) Electrophoresis as a tool for estimating levels of hymenopterous parasitism in field populations of the cereal aphid, *Sitobion avenae*. *Entomologia Experimentalis et Applicata* 54, 271–279.

Wang, C. and Patton, R.L. (1968) The separation and characterisation of the haemolymph proteins of several insects. *Journal of Insect Physiology* 14, 1068–1075.

Washino, R.K. and Tempelis, C.H. (1983) Mosquito host bloodmeal identification: methodology and data analysis. *Annual Review of Entomology* 28, 179–201.

Weitz, B. (1956). Identification of blood meals of blood-sucking arthropods. *Bulletin WHO* 15, 473–490.

Weitz, B. (1960) Feeding habits of bloodsucking arthropods. *Experimental Parasitology* 9, 63–82.

Welsh, J. and McClelland, M. (1990) Fingerprinting genomes using PCR with arbitrary primers (RAPDs). *Nucleic Acids Research* 18, 7213–7218.

Whallon, M.E. and Parker, B.L. (1978) Immunological identification of tarnished plant bug predators. *Annals of the Entomological Society of America* 71, 453–456.

Williams, D.E. and Reisfeld, R.A. (1964) Disk electrophoresis in polyacrylamide gels: extension to new conditions of pH and buffer. *Annals of the New York Academy of Sciences* 121, 373–381.

Williamson, R.A., Burioni, R., Sanna, P.P., Partridge, L.J., Barbas, C.F. iii and Burton, D.R. (1993) Human monoclonal antibodies against a plethora of viral pathogens from single combinatorial libraries. *PNAS* 90, 4141–4145.

Winter, G., Griffiths, A.D., Hawkins, R.E. and Hoogenboom, H.R. (1994) Making antibodies by phage display technology. *Annual Review of Immunology* 12, 433–455.

Wool, D., van Emden, H.F. and Bunting, S.D. (1978) Electrophoretic detection of the internal parasite *Aphidius matricariae* in *Myzus persicae*. *Annals of Applied Biology* 90, 21–26.

Wool, D., Gerling, D. and Cohen, I. (1984) Electrophoretic detection of two endoparasite species, *Encarsia lutea* and *Eretmocerus mundus* in the whitefly *Bemisia tabaci* (Genn.) (Hom.: Aleurodidae). *Zeitschrift für Angewandte Entomologie* 98, 276–279.

Wright, S. (1943) Isolation by distance. *Genetics* 28, 114–138.

Young, S.Y and Kring, T.J. (1991) Selection of healthy and nuclear polyhedrosis virus infected *Anticarsia gemmatalis* (Hubner) (Lep.: Noctuidae) as prey by nymphal *Nabis roseipennis* Reuter (Hemiptera: Nabidae) in laboratory and on soybean. *Entomophaga* 36, 265–273.

12.9. Appendices: Experimental Techniques

Appendix A: Immunization and antiserum production

Antigen (approx. 10–50 µg per dose)
Freund's complete adjuvant
Freund's incomplete adjuvant
1 ml syringes plus 21-g needles

1. Prepare an emulsion of 0.5 ml antigen solution with 0.5 ml Freund's complete adjuvant, by gradual addition of the antigen to the adjuvant, with vortex mixing at each stage.
2. Inject the mixture intradermally, or subcutaneously, at several sites into a large rabbit.
3. After 3 to 4 weeks inject a secondary dose using Freund's incomplete adjuvant.
4. One week later take approximately 20 ml of blood from the marginal ear vein. Allow the blood to clot, then centrifuge to separate sera.
5. Aliquot and store antiserum at $-20°C$.

Appendix B: Enzyme-linked immunosorbent assays

96-well microtitre plate
Phosphate buffered saline (PBS)
    PBS (1 litre $\times$ 10 stock):
    NaCl   80 g
    KCl   2 g
    Na_2HPO_4 11.5 g
    KH_2PO_4 2 g
    Made up to 1 litre in distilled water
Carbonate–bicarbonate buffer, pH 9.6 (double antibody sandwich only)
    Na_2CO_3 1.59 g
    $NaHCO_3$ 2.93 g
    Made up to 1 litre in distilled water
Tween 20
Antigen samples
Polyclonal or monoclonal antibodies
Commercial anti-species horseradish peroxidase conjugate
Enzyme substrate – orthophenylenediamine (ODP) in a citrate–phosphate buffer:
    24.3 ml 0.1 M citric acid
    25.7 ml 0.2 M Na_2HPO_4
    50 ml distilled water
    Immediately before use add:

40 mg ODP
40 µl H_2O_2
2.5 M H_2SO_4

METHOD ONE – INDIRECT ELISA

The letters in parenthesis refer to those in Fig. 12.1.

1. Dilute homogenized antigen samples appropriately in PBS. If a block is not used then a dilution rate of 1:20,000 in PBS may be found to be generally appropriate for invertebrate homogenates. Add 100–200 µl per well on the microtitre plate, and include positive and negative controls. Incubate overnight at room temperature to allow antigens to bind to the plate (A).
2. Wash wells three times with PBS/Tween (250 µl Tween in 500 ml PBS). Add antibodies diluted in PBS/Tween. The rate of dilution will probably be 1:1000–1:10,000 for a polyclonal antiserum, or 1:5000–1:50,000 for monoclonal antibodies as ascitic fluid. Incubate for 2 h at room temperature, allowing antibody–antigen binding to occur (B).
3. Wash three times with PBS/Tween. Add anti-species enzyme conjugate, e.g. an anti-rabbit IgG/horseradish peroxidase conjugate if a rabbit polyclonal antiserum was used in step 2. Dilute according to concentration and manufacturer's recommendations. Incubate for 1 h to allow binding with rabbit antibodies (C).
4. Wash three times with PBS/Tween. Add the enzyme substrate and place in the dark to allow colour development. Reagents in the previous steps should be adjusted to allow adequate development within 20–30 min. Stop the reaction by adding 50 µl of 2.4 M sulphuric acid.
5. Obtain absorbance reading at 492 nm. Store readings as a paper printout or on disk.

METHOD TWO – DOUBLE ANTIBODY SANDWICH ELISA

1. Dilute purified antibodies from a rabbit polyclonal antiserum, as appropriate, in carbonate buffer, and add 100–200 µl/well on a microtitre plate (A). Incubate overnight at room temperature.
2. Wash three times in PBS/Tween. Add antigen samples diluted approx. 1:20,000 and incubate for 2 h (B).
3. Wash three times in PBS/Tween. Add mouse monoclonal antibodies in the form of diluted ascitic fluid (diluted 1:5000–1:50,000), and incubate for 2 h (C).
4. Wash three times in PBS/Tween. Add an anti-mouse enzyme conjugate, diluted as recommended by the manufacturers, and incubate for 1 h (D).
5. Wash three times in PBS/Tween. Add the enzyme substrate and place in the dark to allow colour development. Reagents in the previous steps should be adjusted to allow adequate development within 20–30 min. Stop the reaction by adding 50 µl of 2.4 M sulphuric acid.
6. Obtain absorbance reading at 492 nm. Store readings as a paper printout or on disk.

Blocks are sometimes used to prevent non-specific binding of antibodies to the polystyrene. If this proves to be a problem, antibodies can be diluted, for example, in 1% bovine serum albumin, or a block can be added as an additional step, for 15–30 min, after binding antigen or antibodies to the plate in step one. This usually reduces the sensitivity of the assay, and may require higher concentrations of antigen and antibodies to overcome this.

The methods described above are the simplest, but a number of variants exist. For example, the double antibody sandwich may be performed in a direct, rather than indirect manner, by conjugating enzyme with the same specific purified antibody as used in step one. If a monoclonal antibody is used, this will only work if the same target epitope is duplicated on antigen molecules.

A regression equation, derived from absorbance readings for a dilution series of known antigen concentrations on the same ELISA plate as test samples, will allow antigen concentrations to be calculated from absorbance readings for such samples (Sopp and Sunderland, 1989; Symondson *et al.*, 1996).

Appendix C: Polyacrylamide gel electrophoresis to detect esterase activity (Adapted from Davis, 1964)

REAGENTS

Gel buffer – Tris borate/EDTA, pH 8.6
12.11 g	Tris (Tris (hydroxymethyl) aminomethane)
0.93 g	EDTA (di-sodium ethylenediaminetetra acetic acid) (0.0025 M)
2.47 g	Boric acid (0.04 M)
	Distilled water (make up to 250 ml total)

7.5 % running gels
Solution (a):
2.25 g	Methyl bis acrylamide	(1.5 g for a 5% gel)
75 g	Acrylamide	(50 g for a 5% gel)
500 ml	Distilled water	

Solution (b):
25 g	Sucrose
250 ml	Gel buffer

Solution (c):
0.16%	Ammonium persulphate in distilled water

Electrode buffer – 0.1 M TEB, pH 8.0
42.39 g	Tris
2.79 g	EDTA
14.85 g	Boric acid
3.5 l	Distilled water

Staining buffer (×10)
   1 M potassium phosphate (KH$_2$PO$_4$)
   1 M sodium phosphate (Na$_2$HPO$_4$)

Additional reagents
   0.75 μl/ml of gel solution TEMED (N,N,N1,N1-tetramethylethylene-diamine)
   Saturated butanol
   Triton X-100
   Sucrose
   Acetic acid
   α-naphthyl acetate
   β-naphthyl acetate
   Fast blue RR stain

PROTOCOL

1.   Make up the gel buffer to a total volume of 250 ml.
2.   Filter solution (a) through a Whatman No. 1 filter. Mix the running gel solutions in the ratio of 2(a):1(b):1(c) and de-gas for 5 min.
3.   Add TEMED (0.75 μl/ml) and immediately pour gel.
4.   If thicker gels required (e.g. 1 cm) use a 5% homogeneous gel. Thin gels, generally giving higher resolution, can be made using a 7.5% running gel and a 4% stacking gel.
5.   Exclude oxygen by layering the top of the gel with saturated butanol.
6.   Mix electrode buffer and adjust pH to 8.0 with saturated boric acid.
7.   Pre-run the gel for at least 20 min at 150 volts.
8.   Homogenize samples in electrode buffer plus 1% Triton X-100 (v/v). Centrifuge to remove particulate material.
9.   Add 15% sucrose (w/v). A few crystals of bromophenol blue may be added (optional).
10.   Load samples with micropipette with ultrapipette tip, or microsyringe.
11.   Include channel with an appropriate standard in order to calculate relative mobility ratios. If bromophenol blue has not been included as a tracking dye, run, for example, 20 μl of a separate blue marker (100 μl 1% xylene cyanol in 5 ml 15% glycerol in electrode buffer) in a separate channel.
12.   Run the samples into the gel at 200 volts for 5 min, then wash wells with electrode buffer.
13.   Reduce voltage to 150 volts (varies with gel size and equipment) and run until marker approaches the end of the gel.
14.   Add 4 ml α-naphthyl acetate, 4 ml β-naphthyl acetate (30 mM in acetone) and 25 mg Fast Blue RR stain to 100 ml staining buffer (diluted 1:9). Stain gels in this mixture.
15.   When bands have developed sufficiently, de-stain and fix with 7% acetic acid for 5 min.
16.   Photograph gels. Vacuum seal in polythene and refrigerate.

Many other buffering systems can be used with equal success, including Tris/barbitone and Tris/citrate buffers widely used in entomological studies (Williams and Reisfeld, 1964; Baker *et al.*, 1975; Loxdale *et al.*, 1983).

Magnesium chloride and NADP⁺ (β-nicotimamide adenine dinucleotide phosphate) are sometimes added to enhance isoenzyme resolution (Loxdale *et al.*, 1983).

Great care should always be taken with toxic materials such as acrylamide and bis-acrylamide, and indeed with many of the dyes used in electrophoresis. Always follow suppliers' safety information.

Appendix D: Electrophoresis of DNA samples

REAGENTS

DNA grade agarose
10 × TBE (0.89 M Trizma base, 0.89 M boric acid, 0.02 M EDTA)
Gel loading buffer (50% glycerol, 0.1% bromophenol blue, 0.7% SDS)

METHOD

1. Make up an 0.8% agarose solution by mixing agarose and 1 × TBE.
2. Microwave this on full power until the solution goes clear.
3. Allow the solution to cool to ~45°C and pour into the gel cassette (agarose gels are used in flat bed systems). Ethidium bromide (1 μl of a 20 mg/ml solution) can be added to the gel at this stage, or alternatively the gel can be immersed in a solution containing EthBr at the end of the electrophoresis step.
4. Once the gel has solidified remove the gel comb and immerse the gel in 1 × TBE in the gel tank.
5. Load the DNA samples to be analysed.
6. Connect up the gel tank and run at an appropriate voltage (5 V cm⁻¹ measured between the electrodes) until the marker reaches the midpoint of the gel. NB DNA migrates as a polyanion.
7. Visualize the DNA on a UV transilluminator.

Appendix E: Southern blotting

1. After isolating your genomic, MtDNA or rDNA of interest, first check that it will cut well with your chosen endonuclease(s). Common reasons for DNA not cutting include high residual salt content, residual traces of phenol or protein contamination.
2. Cut 5 μg aliquots of your DNA with 50 units of a restriction endonuclease such as *Eco*RI, *Hind*III or *Hinc*III.
3. Run the samples out on an agarose gel, stain and photograph the gel.
4. Denature the DNA in the gel by immersing in denaturing solution (0.5 M NaOH, 1 M NaCl) for about 1 h. (Some people also like to put a dilute

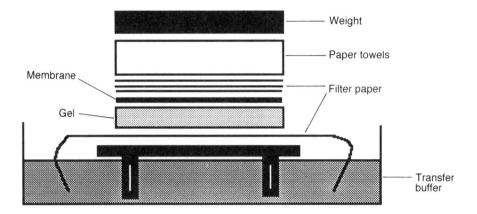

Fig. 12.4. System for conducting Southern blotting. The DNA migrates overnight from the gel to the membrane by capillary action.

HCl immersion before this step. This nicks and exposes the DNA 'backbone' of the template.)

5. Wash the gel with distilled water.

6. Immerse the gel in neutralizing solution (0.5 M TrisHCl, pH 7, 1.5 M NaCl) for 1 h.

7. Cut the nitrocellulose membrane to the gel size and soak in distilled water then transfer to SSC (standard sodium citrate) for 10–15 min before use (NB The concentration of SSC will depend on the brand of membrane used, and manufacturer's recommendations should be followed for this.)

8. The gel and membrane can then be placed in a blotting apparatus or a home-made blotting system can be made as illustrated (Fig. 12.4). The DNA is then allowed to blot overnight. (During this period the DNA migrates out of the gel and into the membrane by capillary action.)

9. The membrane can then be baked at high temperature for several hours or UV cross-linked for 1–2 min to bind the DNA irreversibly to the membrane.

10. The exact membrane hybridization procedure will then depend on the type of probe to be used. For radiolabelled probes, the membrane is immersed in prehybridization solution overnight. The labelled probe is then added to the solution and the hybridization continued for a further 1–4 h (or overnight).

11. The hybridized membrane is then washed to remove unbound probe. The stringency of the wash will determine how similar the probe needs to be to the target DNA to remain bound. The lower the concentrations of SSC and SDS in the wash the more homologous the probe will need to be to remain bound.

12. Detection of the radiolabelled target DNA is then by autoradiography on X-ray films. Blots are usually left in contact with the film overnight at −70ºC before bands are visualized on the X-ray film by standard development techniques.

APPENDIX F: Acetylcholinesterase assay

REAGENTS

0.01 M Dithiobis 2-nitrobenzoic acid in 0.1 M phosphate buffer pH 7.0
0.01 M Acetylthiocholine iodide in distilled water
0.1 M Propoxur in acetone
Split the ASCHI solution into two 10 ml aliquots, to one aliquot add 20 μl of 0.1 M propoxur
1% Triton X-100 in 0.1 M phosphate buffer pH 7.8

METHOD

1. Take the 2×25 μl of crude insect homogenate and place in separate wells of a microtitre plate.
2. Add 145 μl of Triton phosphate buffer and 10 μl DTNB solution to each replicate.
3. Add 25 μl of ASCHI to *one* replicate and 25 μl of ASCHI + propoxur to the *other* replicate.
4. Blanks should contain 25 μl distilled water, 10 μl DTNB solution, 25 μl ASCHI solution and 145 μl Triton buffer but *no* insect homogenate.
5. Read continuously at 405 nm for 5 min or leave for 1 h and read as a fixed point assay at 405 nm. If a fixed point is used the same time needs to be used for the inhibited and uninhibited wells for the same insect. If a continuous reading is used this must be completed no more than 20 min after the addition of the ASCHI solutions.

NOTES

All solutions for this assay, with the exception of the 0.1 M propoxur, should be made up freshly and used within 4 h if left at room temperature or 4°C. The propoxur stock solution can be kept at 4°C in a tightly stoppered bottle for several months.

Appendix G: Naphthyl acetate end point assays (esterase assay 1)

REAGENTS

30 mM α-Naphthyl acetate (NA) in acetone.
30 mM β-Naphthyl acetate made as above.
Both solutions can be made up and stored separately in a tightly stoppered bottle at 4°C for several months.
Working naphthyl acetate solutions 1 ml of 30 mM stock in 99 ml of phosphate buffer 0.02 M pH 7.2. Solution should be made up freshly and used within 1–2 h of preparation.
Stain (150 mg Fast Blue B salt dissolved in 15 ml distilled water, then add

35 ml of 5% sodium lauryl sulphate (SDS)). Solution should be made up freshly and used within 1–2 h of preparation.

METHOD

1.   Take 2×20 μl replicates of spun homogenate and place in separate wells in a microtitre plate.
2.   Add 200 μl of α-NA working solution to one replicate and 200 μl of β-NA to the second replicate. Leave at room temperature for 15 min.
3.   Add 50 μl of Fast Blue stain solution.
4.   One or more plate blanks should be included per plate. This should contain 20 μl distilled water, 200 μl of NA solution and 50 μl of stain.
5.   Read at 570 nm as an end point and convert values to nmoles of product produced by analysing against α-naphthol or β-naphthol standard curves.

Notes. The 15 min time point is an arbitrary one. However, extending the time to more than 15 min may produce such an intense colour in insects which have a highly amplified esterase that spectrophotometric readings become inaccurate. Occasionally, where esterases are greatly overproduced, even a time of 10 min has given a production of naphthol too dark to read on a plate reader. This problem can be overcome by shortening the incubation time or alternatively by diluting the homogenate or taking a smaller volume (e.g. 5 μl) of homogenate.

Appendix H: PNPA rate reaction esterase assay (esterase assay 2)

REAGENTS

100 mM *p*-Nitrophenyl acetate stock solution in acetonitrile (can be stored in a tightly stoppered bottle at 4°C for 1–2 months).
Working solution dilute stock solution PNPA 1:100 with 50 mM sodium phosphate buffer pH 7.4.

METHOD

1.   Take 2×10 μl replicates of homogenate and put into separate wells of a microtitre plate.
2.   Add 200 μl of PNPA working solution.
3.   One or more plate blanks should be included per microtitre plate. These should include 10 μl of distilled water and 200 μl working PNPA solution.
4.   Read at 405 nm continuously for 2 min.

Appendix I: Glutathione-S-transferase assay

REAGENTS

10 mM GSH (reduced glutathione) in 0.1 M phosphate buffer pH 6.5.
63 mM Chlorodinitrobenzene (CDNB) in methanol.
Working solution: add 125 μl of CDNB solution to 2.5 ml GSH solution.

All solutions should be prepared freshly and used within 1–2 h.

METHOD

1.   Take 2×10 μl replicates of homogenate and place in separate wells of a microtitre plate.
2.   Add 200 μl of the GSH/CDNB working solution.
3.   One or more plate blanks should be used per microtitre plate. These should contain 10 μl distilled water + 200 μl of the GSH/CDNB working solution.
4.   Read at 340 nm continuously for 5 min or leave at room temperature for 20 min and then read at 340 nm as an end point.

Appendix J: DNA isolation

A. GENERAL METHOD FOR LARGE-SCALE ISOLATION OF DNA FROM 1 TO 0.1 g OF INSECTS

1.   Up to 1 g wet weight of insects should be ground under liquid nitrogen. (NB If more than 1 g of material is used recovery of DNA from this method tends to be poor.)
2.   The homogenate is added to extraction buffer (10 mM Tris HCl, pH 8, 0.1 M EDTA, 0.5% (w\v) sodium lauryl sulphate (SDS), 20 μg ml^{-1} pancreatic RNAase).
3.   The mixture should be incubated at 37°C for 1 h, to break open the cells and degrade the RNA.
4.   Proteinase K is added to a final concentration of 100 μg ml^{-1} and the mixture is incubated at 50°C for 3 h to break down the protein.
5.   The mixture is cooled on ice for 10 min and 0.35 volumes of saturated NaCl are added to precipitate the protein. The solution should be mixed well and stood on ice for a further 5 min before centrifuging at 16,000 g for 20 min to remove the protein.
6.   The supernatant is taken and the DNA precipitated from it by adding an equal volume of propan-2-ol. This step removes the degraded RNA which remains in solution.
7.   The DNA pellet is resuspended in 7.5 ml of Tris EDTA buffer (TE), pH 8.0, containing 20 μg ml^{-1} RNAase, and incubated at 37°C for 1 h to remove the last traces of RNA.
8.   The solution is then extracted with phenol, 1:1 phenol:chloroform, and finally chloroform to remove the last traces of protein.
9.   The DNA is precipitated from solution by the addition of 2 volumes of ethanol.
10.   The DNA pellet is washed in 80% ethanol to remove salts, briefly air dried and resuspended in a small volume of TE. DNA extracted in this manner can be stored at 4°C without signs of degradation.

B. ISOLATION OF DNA FROM SINGLE INSECTS OR FRACTIONS OF AN INSECT

This method will work with small whole insects, insect legs or abdomens or other body parts.

1. The insect is placed in an Eppendorf tube and homogenized in 100 µl of 10% SDS.
2. The homogenate is left on ice for 30 min and then spun at 8000g for 10 min.
3. The supernatant is transferred to a fresh Eppendorf tube and 1 ml of 1:1 phenol:chloroform added to precipitate protein.
4. The upper aqueous layer is transferred to a fresh tube and 10 µl of 5 M sodium acetate plus 2 volumes of ethanol are added to precipitate the DNA.
5. Where very small samples of insect are used it may be necessary to add 1 µl of 1 mg ml^{-1} tRNA at this stage to facilitate the DNA pelleting. DNA is pelleted by spinning the ethanol solution at 14,000g for 10 min. The pellets are washed with 80% ethanol and briefly air dried before being taken up again in a small volume of TE.

Samples prepared in this way are good for use in PCR reactions. Where larger insects are used, or cleaner DNA is needed, it may be necessary to include an RNAase and/or Proteinase K step to the method.

13 Modelling

J. Holt and R.A. Cheke
Natural Resources Institute, Central Avenue, Chatham Maritime, Chatham, Kent ME4 4TB, UK

13.1. Introduction

A model is a simplified representation of a system or process. In ecological and agricultural entomology, models are used: (i) to seek general conclusions about how organisms interact with each other and their environment; and (ii) to predict how populations or communities will behave in relation to some expected changes in their existing relationships or in the prevailing environmental conditions. The latter will often include man's interventions with biocides or introductions of exotic organisms.

At one extreme a model may consist of a very complicated mathematical representation. At the other end of the spectrum, there may be no mathematics at all other than the 'true' or 'false' of Boolean logic or simple inequalities based on qualitative judgements. Central to model building are the decisions to be made about both the nature of the representation to be adopted and the degree of simplification. The reasons for developing the model should determine its structure. The activity of model-building helps to organize ideas and improve understanding. The implications of the improved understanding may be testable when models lead to the development of new hypotheses which can be investigated experimentally. Useful models are effective tools for thought, not inscrutable 'black boxes' for processing data and delivering 'answers'.

13.2. System Definition and Model Design

The design of a model depends upon careful definition of the system or systems involved, the questions to be asked and whether these are specific and detailed or general and qualitative. Many of the problems encountered in entomology involve both incomplete knowledge and incomplete data. Regardless of the type of model built, it will be required to perform two main tasks: to assist in decision-making about (i) the most effective use of the existing data and (ii) what new data need to be collected. Before these tasks can be tackled, decisions need to be taken about both the nature of the model to be built and its construction. It is the objective of this chapter to suggest some of the ways these decisions can be made.

An initial decision concerns the scope of the model and the number of variables to be included, which will depend upon whether the system involves, for instance, several herbivorous insects, several predators, the crop, the soil or several weed species. Clearly, the number of interactions rapidly becomes very large with an increasing number of variables. The behaviour of unwieldy models is difficult to understand and the intellectual control of such models becomes tenuous, even though they may possess more realism than simple representations: there is a trade-off in model-building between detail and comprehensibility. A plethora of parameters and variables complicates validation and sensitivity analyses.

Time spent on careful definition of the system to be modelled and the variables to be included, with due regard to the questions to be posed and the purpose of the investigation, will be time well spent. A range of techniques for structuring this descriptive phase (Norton and Mumford, 1993) can improve the chances that the design of the model will match the task. It is usually easier to see when a model is too simple rather than too complex. If the initial attempt is judged too simple, the effect of elaborations can be examined.

The following sections discuss a number of issues which need to be considered when embarking on model building.

13.2.1. Stochastic or deterministic?

Deterministic models produce identical results each time (for the same starting state and set of parameter values). Variability is fundamental to biological processes, however, and parameter values may be better represented as having a mean and variance, not simply a fixed value. For example, daily fecundity for the individuals in a population may be somewhere in the range 10–30 eggs with a mean of 20. In a deterministic model, fecundity might be set at 20. In a stochastic model, explicit account is taken of the variability. The actual fecundity used is determined by selecting a value from the frequency distribution of possible values. Thus, each time a stochastic model is run, a different, unpredictable result is obtained. The range of outcomes from many runs can then enable an estimate to be made of the likely range of outcomes in nature. Results, of course, are critically dependent on the assumptions incorporated in the model regarding parameter variance. It is therefore not only necessary to decide which parameters to treat stochastically but also to have estimates of their variances.

Stochastic processes can be important to the outcome of a model when dealing with small populations. Here the probability of individual events can alter the course of a simulation, perhaps determining survival or extinction. Simple models in population genetics provide classic examples involving gene frequency shifts (Roughgarden, 1979). When questions involve detailed prediction of numbers over time it may be essential to incorporate stochasticity. For example, timing of emergence of a damaging life-stage of a pest may be critically dependent on the distribution of development times. The magnitude of the variance in development time has a direct bearing on patterns of abundance.

Unpredictability is not entirely the preserve of stochastic models. Under some circumstances, deterministic models are very sensitive to their initial conditions with results which are chaotic (May, 1974a; Logan and Hain, 1991).

13.2.2. Space

Most models are concerned with the dynamics of processes over time but spatial dynamics are also important. By choosing a non-spatial model it is assumed that the use of mean values for the whole space will give a fair representation of the dynamics of the system. However, if it is known that different parts of the system are not spatially homogeneous, then some partitioning of the model to represent this heterogeneity may be useful. Space may be represented in various ways such as cells in a grid, as in cellular automaton models (Comins *et al.*, 1992), or as the realizations of a statistical model such as the negative binomial distribution (Broadhead and Cheke, 1975). Empirical models concerned with aggregation and spatial heterogeneity are discussed in Chapter 4.

13.2.3. Discrete time or continuous time?

If the model variables can be represented as a continuous process, a model in continuous time can be appropriate. Such models lend themselves to representation as systems of differential equations. Typically, for each variable in the model, an equation is specified for its rate of change. It is sometimes possible to find solutions to such models algebraically and they would then properly be called analytical models. Frequently, however, models comprising systems of rate equations cannot be solved analytically, though it is generally still possible to draw some conclusions from the mathematics *per se*, rather than from numerical simulation. Models of this type have the advantage that they are usually sufficiently simple for the connection between the form of the equations and the behaviour of the model to be clear. When it is the global behaviour of the system which is of interest, formulation of a model as a set of differential equations is particularly useful as this facilitates a more general qualitative approach to the analysis.

If the model concerns generation changes or annual population estimates, difference equations with discrete time steps may be more appropriate. The choice of the size of these steps is important. They should be short enough to keep track of changes in the values of the variables: for example, given an insect with a 6 day egg stage and a 9 day nymph stage, a time step of 3 days would be convenient with the eggs lasting two steps and the nymphs 3. If it is only the change in the insect population from one generation to the next which is of interest, then a time step equal to the generation time, say 3 weeks, would be suitable. The time step to be chosen depends on the resolution with which the variables are to be described.

13.2.4. Units

The question of variable resolution also relates to the units of population size to be used. All the individuals in a field could be treated as a single population. Alternatively, the individuals on each plant, or even each individual, could be modelled separately. If the population of interest is modelled as sub-units, this almost inevitably introduces a spatial dimension. Spatial models involving individuals or sub-populations may sound very complex but they often only involve multiple versions of one simple model.

13.2.5. Deductive or inductive?

Analytical models are usually deductive in the sense that the model is specified on arguments from the general to the particular. The, often complementary, inductive approach is to argue from the particular to the general by starting with the available data and attempting to induce patterns and relationships. Empirical models based on statistical relationships are usually inductive, whereas simulation models are usually built using a mixture of deductive and inductive methods. The overall model structure may be reached deductively, although the details of individual functional relationships may be determined empirically from experimental results.

13.3. Modelling Techniques

With analytical models, whether complete solutions are possible or not, useful deductions can be made from the equations without actually solving them, as already mentioned. It is often also useful to try to solve the equations for the steady state, i.e. when all the equations equal zero in a set of n equations with n unknowns. The steady state solution is only biologically meaningful in certain systems – for a pest population in a particular field of an annual crop a steady state is unlikely except, of course, at the end of the season.

Numerical methods also allow simulations of models expressed in differential equations. Phase diagrams, in which one variable is plotted against another (e.g. prey numbers vs. predator numbers), derived from such simulations are a convenient way of investigating the equilibrium behaviour of such models. Trajectories over time can then be compared for different starting conditions.

Analytical models have most frequently been used for systems where discrete events in time do not occur and where the same equations hold, unaltered, throughout. With the increasing availability of computer algorithms for numerical integration, such constraints have been relaxed and it is possible to manipulate, not just the starting states but even events during the simulation. A useful comparison between analytic and simulation approaches has been published by Jeger (1986).

With more complicated models, and especially complex simulations, it is useful to summarize the model in the form of a flow diagram. This acts as a check that the intended structure of the model is biologically sensible. It also acts as the specification for conversion of the algorithm to computer language code and finally as a reference to check that the algorithm is correct.

Selecting appropriate values for the parameters in a model is often difficult. Good data may exist in order to estimate some, but probably not all, of the parameters and so it is worth tabulating the likely ranges of each parameter. Later, the effects of varying these values can be examined. If a model proves sensitive to a parameter for which there are few data available to estimate a value, this may reveal a critical information gap.

When a model is intended to reflect the dynamics in a specific situation, comparison of model output with relevant data is essential. The more sets of data, exhibiting differing dynamics, that are available, the better. Some form of model calibration may be warranted when discrepancies are found between simulated results and observed events, but it is important to be as systematic as possible during calibration. A possible approach involves reviewing both the structure of the model and each of the parameter estimates, identifying those aspects which may be suspect and listing a set of possible modifications. A combination of several changes may need to be considered but it is useful first to re-examine the fit of the model for each of these modifications in turn.

Once a model, parametrized to give a reasonable fit to the sets of test data, has been created then a start can be made on using it. If, for example, a model has been built to investigate insecticide spraying strategies against a pest, simulations can be made of sprays at different crop stages or at different levels of persistence, or with different pest thresholds. Simulation models allow abstract experimentation with different options and so may reduce the number of trials to a manageable set for field testing.

13.3.1. Developing a simple population model in discrete time

Imagine an annual crop with a population of an insect pest, N, which produces a offspring per day per individual, and in which a proportion, b, die each day. This can be represented with a model which shows the change in population from one day, t, to the next, $t+1$:

$$N_{t+1} = N_t + aN_t - bN_t \qquad (13.1)$$

The population will grow exponentially if $a > b$ or decline exponentially if $a < b$. A population growing indefinitely is not realistic, so let us assume that the insect prefers younger foliage, and that its fecundity declines as the crop ages. Further, assume that the fecundity is simply inversely proportional to plant age, reaching zero in the later stages of the crop, at a plant age of d days. The model incorporating this effect becomes:

$$N_{t+1} = N_t + aN_t \left(1 - \frac{t}{d}\right) - bN_t \qquad \text{if } t <= d,$$

$$N_{t+1} = N_t - bN_t \qquad \text{otherwise} \qquad (13.2)$$

As specified in Eqn (13.2), the occurrence of negative birth rates when $t > d$ is avoided. Suppose that the control options to be investigated affect the different growth stages of the pest in different ways. Then it is important to distinguish the different growth stages. If the insect is an homopteran for instance, it is possible to distinguish: eggs, X, nymphs, Y, and adults, Z. Each day, a maximum of a eggs are produced per adult but fecundity is also dependent on plant age. A proportion of eggs die per day (due, for example, to predation), b_1, and a proportion hatch to become nymphs, s_1. The change in eggs from one day to the next is therefore given by:

$$X_{t+1} = X_t + aZ_t \left(1 - \frac{t}{d}\right) - s_1 X_t - b_1 X_t \qquad (13.3)$$

Nymph numbers are added to as eggs hatch, b_1, but are reduced by a mortality rate, b_2, and the proportion moulting to the adult stage, s_2:

$$Y_{t+1} = Y_t + s_1 X_t - b_2 Y_t - s_2 Y_t \qquad (13.4)$$

Similarly, adult numbers increase as nymphs moult and decline according to a daily mortality, b_3:

$$Z_{t+1} = Z_t + s_2 Y_t - b_3 Z_t \qquad (13.5)$$

The flow chart (Fig. 13.1) summarizes the model, which was implemented using microcomputer spreadsheet software. Mumford and Holt (1993) and Cartwright (1994) give worked examples of modelling with spreadsheets. Figure 13.2(a) shows sample output of simulated egg, nymph and adult numbers. It was assumed that the population was initiated by the entry of a low number of adults to the crop on the day of crop emergence. The numbers of eggs, nymphs and adults increase, reach peaks, then decline. The

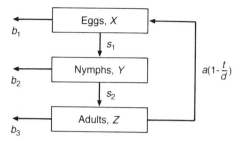

Fig. 13.1. Summary of a simple insect population model. Arrows indicate the transitions. See text for explanation of symbols.

decline is caused by the assumption of reduced fecundity with increasing crop age.

Figures 13.2(b) and (c) show how the model can be used to examine the potential impact of insecticide applications on the pest population. As an example, an insecticide applied at 25 days after emergence was assumed to kill all nymphs and adults present that day, but to have no residual effects on subsequent days. The result, Fig. 13.2(b), was a dramatic decline in numbers with the population barely recovering to pre-spray levels later in the season. Perhaps suspecting that the spray might also have a deleterious effect particularly on egg predators, Fig. 13.2(c) shows the result of also decreasing egg mortality for a period of one week following the spray. Using the assumption that it might take 1 week for the egg predators to recover, therefore, was sufficient to almost nullify the positive effects of the pesticide.

To examine the robustness of the model, the extent to which the conclusion about spray-induced resurgence is sensitive to model parameter changes can be tested. As an example, the sensitivity to the decline in fecundity with plant age was investigated. Different values of d changed the timing of the population peak but this did not affect the relative degree of population resurgence following the spray. Therefore, it is reasonable to conclude that the relative lack of knowledge in this respect is not too important in the context of the question posed.

Examples of models of the type described in this section include one on the problem of insecticide-induced resurgence of *Nilaparvata lugens* (Holt *et al.*, 1992), and another on the timing of insecticide sprays to control *Helicoverpa armigera* (Holt *et al.*, 1990).

13.3.2. A model in continuous time

The above model could also be expressed in continuous time as differential equations. The parameters, rather than being a net value by which the system changes from one step to the next, are now rates. For example, in the discrete model above, a is the number of new eggs laid per adult between the times t and $t + 1$ (the unit is number). In the continuous model below, a

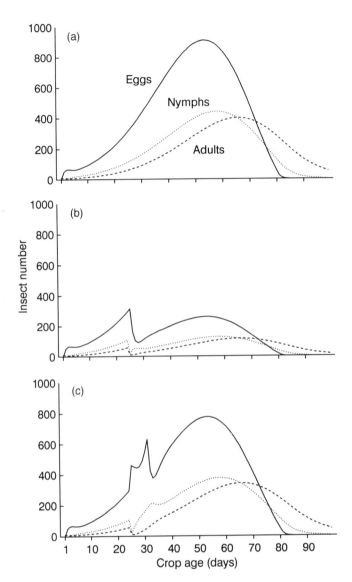

Fig. 13.2. Simulated insect numbers over time: (a) with no intervention; (b) and (c) with a simulated insecticide application having two different effects (see text for details).

is the egg-laying rate per unit time (the unit is number/time). With this proviso, the parametrization is the same for both models.

Continuous models are most useful for systems where the variables have equilibrium values and the stabilities of the equilibria are of interest to managers. In the above model of an annual crop, the only stable equilibrium is $x = y = z = 0$, reached at the end of the season. The hypothetical model is of more interest if it is assumed that the insect is a pest of a perennial crop and a fourth variable, the number of host plants, U, is introduced. Thus a set of equations for the rate of change of each of the variables in the model can be written as:

$$\frac{dX}{dt} = aZU - s_1X - b_1X \tag{13.6}$$

$$\frac{dY}{dt} = s_1X - b_2Y - s_2Y \tag{13.7}$$

$$\frac{dZ}{dt} = s_2Y - b_3Z \tag{13.8}$$

$$\frac{dU}{dt} = rU - mU(Y + Z) \tag{13.9}$$

Equations (13.7) and (13.8) are analogous to (13.4) and (13.5), respectively. In Eqn (13.6) it is assumed that the rate of egg laying is proportional to host plant availability, instead of egg laying being dependent on plant age. In Eqn (13.9), the planting rate of the host plant, r, is taken as proportional to the amount of host material present. Such an assumption may be appropriate for a crop like cassava, which is propagated from material cut from old plants. It is also now assumed that the crop is eaten by the pest at a rate, m, proportional to the amount available and to the numbers of the pest (nymphs and adults being counted as equal in this respect). The model is summarized in Fig. 13.3.

What can be gleaned from these equations without actually solving

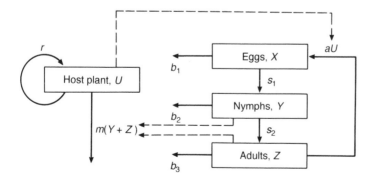

Fig. 13.3. Summary of a simple herbivorous insect–plant model. Solid arrows indicate transitions and broken arrows indicate effects. See text for explanation of symbols.

them? From Eqn (13.9), dU/dt will be 0 when

$$r = m(Y + Z)$$ (13.10)

and from Eqns (13.6), (13.7) and (13.8), $d(X + Y + Z)/dt$ will be 0 when

$$aUZ = b_1X + b_2Y + b_3Z$$ (13.11)

For crop abundance to be stable, therefore, re-planting must balance consumption by the herbivore, and for the insect herbivore population to be stable, fecundity must balance mortality. This highlights a possible flaw in the model: if the herbivore is not present, then crop abundance will increase indefinitely. A modification of the model to address this problem is discussed later. Assuming a situation where the herbivore is present, it is useful to determine the equilibrium values, i.e. can the crop and pest coexist in acceptable numbers? The stability of the equilibrium is also important, i.e. will crop abundance fluctuate wildly or be constant?

In this example, it is possible to solve Eqns (13.6) to (13.9) for the equilibrium state, i.e. derive algebraic expressions for the values of X, Y, Z and U when Eqns (13.6) to (13.9) all equal zero at the same time. Two sets of equilibria exist, one of them is the case where all variables equal zero (crop is destroyed and pest becomes extinct). In the other, the equilibrium level of the crop is given by:

$$U = \frac{b_3(b_1 + s_1)(b_2 + s_2)}{s_1 s_2 a}$$ (13.12)

From this equation it can be seen that all the pest mortality terms appear in the numerator only so, not surprisingly, high pest mortality leads to a higher equilibrium crop level. Conversely, pest fecundity appears only in the denominator, so low fecundity also leads to a higher crop equilibrium. Of more interest is that m, the *per capita* consumption rate of the insect, does not affect the equilibrium abundance of the crop. This implies that it is the rate of increase, not the damage capability of a pest species, which determines sustainable crop levels.

The stability properties of models of this type and many others can be investigated by rigorous mathematical methods (Rosen, 1970), especially when a neighbourhood stability analysis also describes the global stability of the model and a Liapunov function exists (May, 1974b). Roughgarden (1979) provides an appendix on stability theory. Alternatively, a graphical way of examining whether the equilibrium is stable is to look at a phase diagram for the variables in the system. To do this for the current example, simulations were run using a numerical integration package. It was found that the starting values of the variables were critical. Typically, pest and crop abundance oscillated through a sequence as follows: pest numbers increase, crop abundance declines, pest numbers decline, crop abundance increases, pest numbers increase, etc. In some simulations, the amplitude of the oscillations reduced with time until the equilibrium was reached. In others the amplitude increased until the crop was effectively destroyed (Fig. 13.4). Divergent oscillations occurred when there were large differences between the starting values and the equilibrium values. One practical implication of this might be that a policy to maintain a high cropping density would be inherently less sustainable than one involving less intensive cropping.

This model is of similar form to the classic Lotka–Volterra predator–prey (or herbivore-plant) model:

$$\frac{dN}{dt} = \alpha N - \beta PN \qquad (13.13)$$

$$\frac{dP}{dt} = \gamma NP - \lambda P \qquad (13.14)$$

where N is the number of prey, α their intrinsic rate of increase, β the capture rate of prey per predator (P), λ the death rate per individual of the predators in the absence of prey, and γ is the predator's rate of increase per individual prey present.

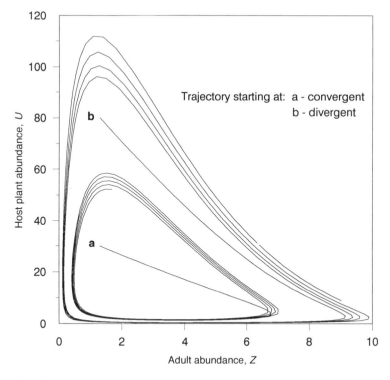

Fig. 13.4. Examples of two phase trajectories show the importance of starting state: (a) leads to a stable equilibrium, and (b) shows oscillations of increasing amplitude occurring.

Rather than a two-equation model with a single variable to represent the herbivore, we have three life-stages. It is interesting to note that while the Lotka–Volterra model has neutral stability, i.e. produces stable limit cycles of constant amplitude dependent on starting state (May 1972), this age-structured model has quite different stability properties. Some of the consequences of introducing age-structure, in the form of different insect stages, into models of this type are discussed by Bernstein (1985) who modelled mite predator–prey interactions including the prey's effects on the host plant. Crawley (1983) also describes some modifications of the basic Lotka–Volterra model which are appropriate for plant–herbivore interactions.

A modification to deal with the problem mentioned earlier, of unlimited growth in the absence of the herbivore, involves the introduction of a crop carrying capacity, K. Equation (13.9) can be rewritten as:

$$\frac{dU}{dt} = rU\left(1 - \frac{U}{K}\right) - mU(Y + Z) \tag{13.15}$$

As crop abundance approaches K, the replanting rate now drops towards zero. The equation defining the equilibrium value of U (13.12) is unchanged from the original model but there is a qualitative difference in the stability properties. In common with the Lotka–Volterra equations, (13.13) and (13.14), introduction of density dependence stabilizes the model so that an equilibrium is reached in all cases (i.e. divergent oscillations do not occur). Such a stepwise approach to model development allows consideration of the importance of the various functional relationships on the global behaviour of the system.

Other examples of the use of differential equations for modelling ecological interactions involving insects include research on gypsy moth *Lymantria dispar* (Foster *et al.*, 1992; Wilder *et al.*, 1994), biological control (Hearne *et al.*, 1994), vector-borne disease (Dye and Hasibeder, 1986), vector-borne plant viruses (Chan and Jeger, 1994; Holt *et al.*, 1995), and insect pathology (Begon *et al.*, 1992).

The kinds of conclusions we have described here illustrate how continuous models are well suited to thinking about general questions of system behaviour; the equivalent discrete model showed the usefulness of that approach for 'experiments' with management interventions.

13.3.3. *Inclusion of behaviour into predator–prey and other models*

Holling (1963) devised an approach to modelling known as experimental components analysis, which allows the introduction of behavioural effects. Holling's work involved series of experiments to build up relationships between identified components of a system, in his case a predator–prey one. Realizing that the behaviour of predators would influence the rate at which they consumed prey, he effectively parametrized the Lotka–Volterra model using experimental results to augment the information subsumed under the constants in the Lotka–Volterra equations, (13.13) and (13.14). A classic experiment involved a blindfolded subject seeking randomly distributed sandpaper discs. The resultant formulation is known as the disc equation (Holling, 1959):

$$\frac{Ne}{P} = \frac{aT_tN}{1 + aT_hN} \qquad (13.16)$$

where Ne/P = the number of prey encountered per predator (P). P is the number of predators, a is a constant encounter coefficient, T_t is the time the prey are available, T_h is the predator's handling time to deal with a victim, and N is the number of prey. The disc equation lacked realism as there was no explicit 'exploitation', i.e. the effect of the discs being consumed was not accounted for, but it included the effects of the predators' prey-handling times. Later elaborations involving experiments with praying mantises incorporated hunger and learning (Holling, 1965, 1966), provided further formulations and a classification of functional responses (the relationship between the number of prey consumed per predator and prey density) into Type I (linear), Type II (curvilinear) and Type III (sigmoid). A clear account of functional responses in such models and related ones for host–parasitoid systems is given by Hassell (1978).

One aspect of behaviour which is seldom modelled satisfactorily is migration. Although many insects may only move short distances during their lifetimes, others migrate very long distances. Often such phenomena are subsumed into a simple emigration–immigration function whereby a percentage of insects are assumed to move in or out of the modelled arena. These percentages may be expressed as functions of a density-dependent process or as responses to a shift in morph frequencies from brachypterous to macropterous forms or vice versa (e.g. Holt *et al.*, 1987). Metapopulation models or those using cellular automata take specific account of insect movements. One such model to investigate screwworm invasion in Australia also highlights the importance of stochastic processes in insect dispersal (Mayer *et al.*, 1993). Long-range migrations such as those of locusts (Steedman, 1990), the *Simulium damnosum* species complex (Garms *et al.*, 1979) or many of the numerous examples given by Johnson (1969) require a different approach. Johnson and Johnson (1994) describe a model of long distance movement by *S. damnosum* s.l. from source areas to successively more distant sites, in which it is shown that the models with predictions closest to the observed events were those with the smallest proportion of laggards: cohesion of the day-to-day patterns at the sources was maintained over long distances (> 300 km) and for up to three weeks.

One method for modelling long-range migration is to use trajectory models, in which the direction and distances of movements are simulated in response to specific climatic variables such as wind speed at the heights at which the insects in question are thought to fly when travelling long distances (Chapter 5). The subject is, however, complicated by the concentration of flying insects by wind convergence zones (Rainey, 1951, 1989), but Riley and Reynolds (1990) have devised a model to represent the proportion of a migrant population likely to be concentrated. If there is sufficient information on the migratory pathways of an insect, probabilistic models of likely trajectories are possible as discussed for the African armyworm, *Spodoptera exempta* (Cheke and Tucker, 1995).

13.3.4. Matrix models

Population models in discrete time can be represented using a matrix to define the transitions from one time interval to the next. If the transition values are constants, then matrix algebra can be used to derive various quantities which may be of interest, including the equilibrium age structure of the population and its rate of increase. In a model considered earlier (Eqns (13.3) to (13.5)), all parameters except fecundity were constants. In a matrix representation of the same model it is assumed that fecundity is also constant (Fig. 13.5). The cells of the matrix are daily transition values. For example, the proportion, $1 - (s_1 + b_1)$ of eggs at time t, become eggs at $t + 1$, while the proportion, s_1 become nymphs. This is a stage-structured matrix. If stage-specific survival and fecundity have been determined from lifetable studies (as described in Chapter 4), or stage-specific survival rates obtained from analysis of insect stage-frequency data (Manly, 1974, 1990), they could provide the values for such matrices.

   In Fig. 13.5 (and its earlier form as Eqns (13.3) to (13.5)), no account was taken of the age structure of the individuals within each life-stage; in fact the implicit assumption was that, within a life-stage, the age structure is constant. For modelling the age structure of the population in more detail, more categories can be defined. One way to do this is by using an age-structured matrix. For example, daily age classes could be used, so successive classes of the egg stage would be eggs aged 1 day, eggs aged 2 days, etc. In an example of an age-structured matrix, Fig. 13.6, weekly age classes were adopted for convenience. The egg stage is assumed to have a period of 1 week and the nymphs and adults, 2 weeks each. The model therefore has

Time t (days)

		Eggs X	Nymphs Y	Adults Z
Time $t + 1$ (days)	Eggs, X	$1-(s_1 + b_1)$		a
	Nymphs, Y	s_1	$1-(s_2 + b_2)$	
	Adults, Z		s_2	$1 - b_3$

Fig. 13.5. A stage-structured transition matrix for an insect population.

five age classes: eggs aged <1 week, nymphs aged <1 week, nymphs aged 1–2 weeks, adults aged <1 week, adults aged 1–2 weeks. As indicated in the matrix, it is assumed that eggs (aged <1 week) in week t, become nymphs aged <1 week in week $t+1$. Given information about the variance of development times, a model employing daily age classes might define transitions to the nymph stage for eggs of a range of ages. Variance can be incorporated into the matrix in Fig. 13.6 by assuming that the transition, $X_{1,t}$ to $X_{1,t+1}$ is greater than zero, i.e. that a proportion of nymphs do not hatch after 1 week but remain as eggs.

To determine population change, the transition matrix is multiplied by the population vector at time t, to obtain the new population vector at $t+1$. According to the rules of matrix algebra the multiplication for a 3×3 matrix (e.g. Fig. 13.5) is given by:

$$\begin{bmatrix} u_{1,1} & u_{1,2} & u_{1,3} \\ u_{2,1} & u_{2,2} & u_{2,3} \\ u_{3,1} & u_{3,2} & u_{3,3} \end{bmatrix} \begin{bmatrix} X_t \\ Y_t \\ Z_t \end{bmatrix} = \begin{bmatrix} u_{1,1}X_t + u_{1,2}Y_t + u_{1,3}Z_t \\ u_{2,1}X_t + u_{2,2}Y_t + u_{2,3}Z_t \\ u_{3,1}X_t + u_{3,2}Y_t + u_{3,3}Z_t \end{bmatrix} = \begin{bmatrix} X_{t+1} \\ Y_{t+1} \\ Z_{t+1} \end{bmatrix} \quad (13.17)$$

where, $u_{i,j}$ is the value in the ith row and jth column of the matrix. Repeated application of (13.17) allows the population dynamics to be simulated. Depending on the transition values, an equilibrium may be reached, when the ratio of the numbers in each stage is constant (i.e. $X_{t+1}:Y_{t+1}:X_{t+1} = X_t:Y_t:Z_t$). This equilibrium age structure is equivalent to the eigenvector of the matrix. The rate at which the population is increasing when this equilibrium is reached (i.e. $X_{t+1}/X_t = Y_{t+1}/Y_t = Z_{t+1}/Z_t$) is equivalent to the dominant eigenvalue of the matrix.

A comprehensive treatment of matrix models is given by Caswell (1989). An entomological example was provided by Lefkovitch (1966) who used matrices to model movements of the cigarette beetle *Lasioderma serricorne*

Time t (weeks)

	Eggs Age <1 week	Nymphs Age <1 week	Nymphs Age 1–2 weeks	Adults Age <1 week	Adults Age 1–2 weeks
Eggs age <1 wk				a	a
Nymphs age <1 wk	$1 - b_1$				
Nymphs age 1–2 wks		$1 - b_2$			
Adults age <1 wk			$1 - b_2$		
Adults age 1–2 wks				$1 - b_3$	

(row labels span Time $t + 1$ (weeks))

Fig. 13.6. An age-structured transition matrix for an insect population.

(see also Takada and Nakajima, 1992). Cheke (1978) used matrix models to evaluate rates of population increase in different phases of *Schistocerca gregaria* and to model Markov chain descriptions of sequences of a parasitic wasp's grooming behaviour (Cheke, 1977).

13.3.5. A simulation model with spatial and stochastic elements

In this example, the population dynamics of a pest in relation to both the pattern of cultivation of different crop varieties and the variance of planting times is examined. The model is designed to represent a square grid of fields and simulates the movement of pests between crops. The model is general but could reflect the conditions in a tropical, irrigated, contiguous rice system. Each field in the grid can be planted with a variety which is either susceptible or resistant to the pest in question. It was assumed that a crop is present for a standard time interval and that two crops per year are grown in one field. When a crop is harvested, the varietal resistance of the next crop is chosen randomly. The proportion of the fields with resistant varieties is a model parameter, p. To choose the resistance of the field, p is compared with a uniformly distributed random variate in the range 0 to 1. If the variate is less than or equal to p then the field is allocated as being resistant; if greater than p, then it is allocated susceptible.

Changes in the pest population in a particular field from one month to the next were simulated using a discrete form of the logistic equation. In an earlier example, Eqn (13.1), we started with a simple model of exponential growth. This model can also be expressed as

$$N_{t+1} = N_t + RN_t \qquad (13.18)$$

where R is the net growth $(a - b)$. In Eqn (13.2), population growth was limited by declining fecundity with increasing plant age. Here, in a similar way to Eqn (13.15), it is assumed that competition constrains the population by making R decline in proportion to how close numbers are to a maximum supportable population size, the carrying capacity, K.

$$N_{t+1} = N_t + RN_t\left(1 - \frac{N_t}{K}\right) \qquad (13.19)$$

To represent the pest population increasing more slowly on the resistant variety, R was made variety specific. The value of R also allows for the loss of a proportion of the population from each field as emigrants, which may move to other fields in the grid.

To describe the movement of insects between crops, the simplest possible assumption is made: that insects leaving the source crop settle on new crops at a constant rate with respect to distance. This assumption results in an exponential decline in the number settling with increasing distance. It is also assumed that there is an equal probability of travelling in any direction. The immigration to a field, therefore, is the sum of the movement from all other fields. The total input of immigrants, Z, to the field, j, is given by the following equation:

$$Z_j = \sum_{i=1}^{m} \left[N_i u_{i,j} Exp^{-sd_{i,j}} \right]$$
(13.20)

where:

m is the number of fields in the grid.

$u_{i,j}$ specifies movement potential from i to j, dependent on the age of the source field, i, and the age and variety of the destination field, j:

    $u_{i,j} = v$, if age of $i<=h$ months and age of $j<=h$ months and variety of j is susceptible,

    $u_{i,j} = wv$, if age of $i<=h$ months and age of $j<=h$ months and variety of j is resistant,

    $u_{i,j} = 0$, otherwise,

        where v and w are parameters determining movement potential due to host age and host resistance respectively, and h is the age of the crop at harvest.

s determines the distribution of settling with respect to distance (the larger the value, the nearer to the source is the distribution of settling).

d is the distance between the source and destination fields.

    At each step of the model the calculation (13.20) is repeated for all fields (i.e. from $j = 1$ to m) to determine the immigration to each field in the grid. In this example, it was assumed that the crop is harvested at the end of the fifth month ($h = 5$), so the potential for movement, to or from the crop, becomes zero. It was also assumed that movement to resistant varieties was less than to susceptible varieties by the factor, w. The movement process is illustrated in Fig. 13.7. According to Eqn (13.20), the closest cells contribute most to $Z_{j,t}$, with the largest contribution being a component of local movement within the cell in question. It is a feature of this representation of the spatial dynamics that the cells nearer the edge of the grid have fewer neighbours and therefore receive fewer immigrants. If our grid of fields is isolated from external influences then this may be reasonable.

    The growth of the population in each field is calculated taking into account the new arrivals. The population size is also scaled to a maximum value of unity, so K (Eqn 13.19), is omitted from the equation below.

$$N_{j,t+1} = (N_{j,t} + Z_{j,t})(1 + R\,(1 - (N_{j,t} + Z_{j,t})))\ \text{ if age of } j <= h,$$

$$N_{j,t+1} = 0 \qquad\qquad\qquad\qquad\qquad \text{otherwise} \qquad (13.21)$$

As indicated in Eqn (13.21), the population in each field is set to 0 when the crop age exceeds h months. After one month's fallow, it was assumed that a new crop was planted and the crop age was reset to 1.

    By varying the proportion of fields occupied by resistant varieties, p, the dynamics of the pest within the system can be altered. There appears to be a

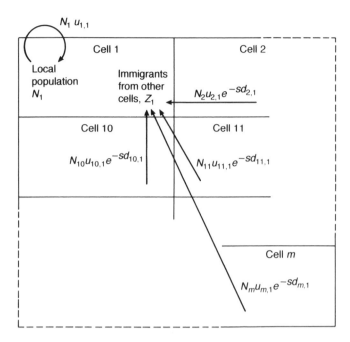

Fig. 13.7. Depiction of the metapopulation model comprising a square grid of cells. Movement into the first cell of the grid is shown. A similar movement occurs into each cell as specified in Eqn 13.20.

threshold below which the pest can persist, and above which it gradually dies out. This implies that small changes in the mix of varieties might have a large impact on long-term pest dynamics. Similarly, alteration of the proportion of fields in each age category at any one time has similar effects. It can also be inferred that small changes in the degree of planting time asynchrony can have profound effects.

The stochastic elements in the model reside in the allocation of crop age and variety to each cell. The proportions of each category in the grid as a whole are specified as parameters, but positions within the grid are allocated at random. In some simulations, therefore, the proximity of susceptible varieties to each other may allow pests to transfer in large numbers and perpetuate the population. In others, the spatial distribution may be less 'helpful'. The impact of such stochasticity in the spatial structure of the model leads to dramatic differences between each run of the model. The effect is most marked for high values of s: when short-range movement predominates, the identity of neighbouring fields becomes critical.

The above is an example of a metapopulation model using difference equations. Metapopulation models, which may be formulated in different ways including differential equations, involve discrete but interacting

populations and have led to new insights and controversies in insect population dynamics (Gilpin and Hanski, 1991; Nee and May, 1992).

13.3.6. Models using physiological time

The development rate of insects is usually dependent upon temperature.
Thus it is possible to relate insect development not to calendar time, as we
have done in the examples so far, but to temperature. It is a common
assumption in such models that above a certain temperature threshold,
development rate is proportional to temperature. Below this threshold,
development is assumed to be zero. For example, if the development threshold is 3°C and the mean temperature on a particular day was 15°C, then the
physiological time elapsed during that day would be 12 degree-days.
Average temperature is frequently calculated from the approximation (daily
max + daily min) / 2. The unit, degree-days, is probably the most common
in such models but there is no reason why other units such as degree-hours
should not be used instead.

It is often possible to conduct experiments in controlled temperature
conditions in order to determine the physiological time period of each of the
life-stages of the species in question, although quiescence may complicate
matters in some instances (Matthee, 1951).

Relating the passing of physiological time to the development of the
population can be executed by considering single age cohorts of individuals. Each cohort progresses through its development at a different rate (in
terms of calendar time) dependent upon the temperatures it experiences.

An illustration is given in Fig. 13.8. Here development is considered
from one week to the next according to the number of day-degrees accumulated in that week. Starting with a newly laid cohort of eggs, C_1, at the start

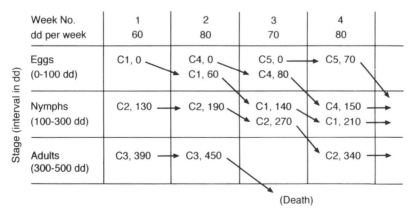

Week No.	1	2	3	4
dd per week	60	80	70	80
Eggs (0-100 dd)	C1, 0	C4, 0 / C1, 60	C5, 0 → C5, 70 / C4, 80	C5, 70
Nymphs (100-300 dd)	C2, 130 →	C2, 190	C1, 140 / C2, 270	C4, 150 → / C1, 210 →
Adults (300-500 dd)	C3, 390 →	C3, 450		C2, 340 →

(Death)

Fig. 13.8. Representation of a model using physiological time. Development
of cohorts of insects, $C_{i,n}$, where i = cohort number, n = number of day-
degrees accumulated by the cohort at the start of each week.

of Week 1, this cohort accumulates 60 day-degrees in that week. In Week 2, a further 80 day-degrees are accumulated. The egg stage is assumed to last 100 day-degrees, so at the start of Week 3, C_1 become nymphs, with 140 day-degrees accumulated. In this example, C_3 is assumed to start Week 1 with 390 days accumulated. During that week this adult cohort lays eggs. The eggs produced by C_3 in Week 1 are assumed to constitute a new cohort of eggs, C_4, which starts in Week 2 with 0 day-degrees accumulated. Clearly, a simplification is used here in combining egg production by C_3 for the entire week into a single cohort. This of course may be too crude for some applications – daily or hourly time steps could be employed instead at the expense of a larger model.

Working in physiological time obviously means that a more complicated algorithm is necessary to execute the calculations. This is often necessary when the effects of temperature change are central to the problem being studied. Examples include models of *Spodoptera frugiperda* (Labatte, 1994), the cabbage aphid *Brevicorne brassicae* (Hughes and Gilbert, 1968) and the cereal leaf beetle *Oulema melanopus* (Gutierrez *et al.*, 1974).

13.3.7. *Population genetics models*

Roughgarden (1979) gives an introduction to this extensive field from an ecological standpoint. In entomology, simple models of population genetics have been used quite extensively to investigate the dynamics of insecticide resistance (Comins, 1984). Here we describe the essential features of a single gene model. Most such models are based on the Hardy–Weinberg equilibrium: an expression which gives the proportion of each genotype in a population, for a given gene frequency. For example, in a situation where resistance to an insecticide in an insect population is controlled by a single gene, the proportions at time t, of resistant homogygotes, R_t, heterogygotes, H_t, and susceptible homogygotes, S_t, would be:

$$R_t = p_t^2, \quad H_t = 2p_t(1 - p_t), \quad S_t = (1 - p_t)^2 \qquad (13.22)$$

where, p_t = frequency of the resistance gene at time t. It should be noted that these equations only hold when the population is mating at random.

If the resistance gene confers no advantage or disadvantage, then Eqn (13.22) represents an equilibrium. What is more likely in our example is that when the population is exposed to the insecticide in question, the R genotype would be at an advantage relative to the S genotype. This advantage can be expressed as a fitness and indicates the relative contribution each genotype makes to the next generation. Depending on the dominance of the gene, the heterozygote may have a fitness equal to the R type, the S type, or an intermediate value. The proportions of each genotype following selection pressure are therefore:

$$R_t = \frac{R_t f_R}{T_t}, \quad H_t = \frac{H_t f_H}{T_t}, \quad S_t = \frac{S_t f_S}{T_t} \qquad (13.23)$$

where f = fitness for each genotype, and $T_t = R_t f_R + H_t f_H + S_t f_S$.

It is necessary to divide by the total, T_t, so that following selection, values are again expressed as proportions. The frequency of the resistance gene, prior to selection in the next generation, depends on the proportion of homozygous resistant and heterozygous insects entering the next generation as follows:

$$p_{t+1} = R_{t+1} + \frac{H_{t+1}}{2} \tag{13.24}$$

The sequence of calculation (13.22) to (13.24) can now be repeated at the next iteration of the model. If the resistance gene confers no disadvantage through reduced fitness (in the absence of insecticide exposure), then at any level of insecticide use, the model predicts that the resistance gene frequency will increase following an S-shaped curve, a rate dependent on the level of exposure, until all individuals are resistant. Resistance dynamics can be complicated in various ways. This is well illustrated by *Helicoverpa armigera* in India, where models were used to investigate two possible means by which some of the population may escape insecticide exposure: through the colonization of unsprayed hosts (Madden *et al.*, 1995) and through diapause (De Souza *et al.*, 1995).

13.3.8. Rule-based models

It is possible to represent a system as a set of linked logical relationships. In examples of such models (e.g. Geurrin, 1991; Knight and Mumford, 1992; Holt and Day, 1993), the relationships have taken the form of rules of the structure:

'IF some condition(s) THEN some conclusion(s) ELSE other conclusions(s).'

Such Boolean conditions were incorporated into conventional models in Eqns (13.2), (13.20) and (13.21). A rule-based model is composed entirely of Boolean logic. Rule-based models are the qualitative equivalent of simulation models and provide a means by which quite crude or subjective knowledge about a system can be used. In rule-based models, system components are represented by a small set of discrete states and changes are described by rules.

The population dynamics of the whitefly, *Bemisia tabaci*, on cassava provides a useful case study to illustrate the approach, as a variety of factors and interactions are thought to be involved (Legg, 1995). In the model, most of the components of the system were defined as being in one of three possible states, low (L), medium (M) or high (H). Although qualitative categories were used, in formulating the rules, it was vital that, for example, 'low rain' had a consistent meaning throughout. A time step of one month was chosen since this is approximately the period for one whitefly generation. The first set of rules defined water availability for cassava plant growth in terms of rainfall and the water retention properties of the soil:

IF rainfall in a given month is high, THEN water availability is high.
IF rainfall is medium, THEN water availability is equal to soil water retention ability.
IF rainfall is low and soil water retention ability is low or medium, THEN water availability is low.
IF rainfall is low and soil water retention ability is high, THEN water availability is medium.

The rules can easily be coded using any spreadsheet, database or other computer language software that contains the necessary logical operators (if, and, or, greater than, etc.). A simple example using a spreadsheet is given in Knight and Mumford (1992). The above set of rules are summarized (Fig. 13.9(a)). Figures 13.9(b)–(d) define how water availability and temperature determine leaf production, how leaf production and plant age determine host suitability for whitefly, and how suitability for whitefly and temperature determine whitefly fecundity. The rate of whitefly population change depends on the balance between fecundity and mortality (Fig. 13.9(e)). In the absence of information about mortality in different months, initially, it was assumed that mortality was 'M' throughout. Figures 13.9(f) and (g) show the definition of the plant age states, and the rainfall and temperature states for an example location, respectively. The water retention property of

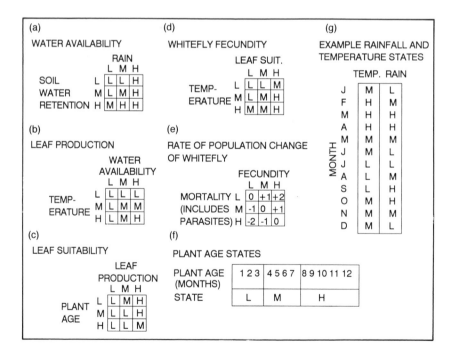

Fig. 13.9. Summary of a rule-based model of whitefly population dynamics on cassava.

the soil was assumed to be constant for a particular location and set at 'H' in this example.

The whitefly population was allowed to be in one of five states from very low to very high (1–5). At each monthly step, the rules determined changes to the population density. Because the model predicts the rate of change of the population, the successive states of the whitefly population are better regarded as steps on a logarithmic scale rather than a linear scale. An example of output from the model (Fig. 13.10) shows the predicted whitefly populations on crops planted in different months. Plotted as cumulative number for clarity, the figure shows that material planted between October and February was more likely to suffer a high whitefly burden than at other times. The implications of a qualitative perception of the workings of this system can thus be reviewed in the light of comparisons with field observations of whitefly numbers.

13.3.9. Computer software and artificial intelligence

An expert system is a model which mimics the decision-making of a human expert. Although developments such as fuzzy set theory have been considered in ecological applications for a number of years (Bosserman and Ragade, 1982), most applications in entomology which use the term expert

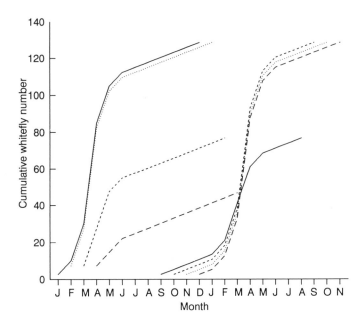

Fig. 13.10. Cumulative whitefly abundance predicted for cassava planted in different months illustrated by different lines. Values remained low on plantings from May to August inclusive; these are not shown.

system refer to rule-based models. A rule-based model was used above to simulate the dynamics of processes over time. In an expert system, rules are used to represent the decision-making associated with a particular problem. A wide variety of software is now available for construction of expert systems on PCs. Edwards-Jones (1992) reviewed applications of expert systems in pest management. They have been directed at problems of diagnosis, treatment prescription, or information management.

In diagnosis, for example, a diagnostic key can be converted into a set of rules for coding as a computerized expert system. There is some advantage in the greater flexibility and ease of use provided by a computerized key. Such systems are particularly powerful when they incorporate techniques from numerical taxonomy (Sands, 1992). Treatment prescription often poses a more difficult problem because the decisions may not be as well defined: the act of attempting to build the expert system can in fact expose failings in current understanding and practice. A major output from expert system construction can be that the decision problem has been specified rigorously (Mumford and Norton, 1989). A worked example of expert system construction for the problem of desert locust control can be found in Mumford and Norton (1993), based on Steedman (1990, Table 19). Expert systems form a central component of many decision support systems, which can also include simulation models and databases. For example the aphid forecasting package, FLYPAST (Knight *et al.*, 1992), combines expert system and database.

Software packages are becoming available, which offer a simple means of experimenting with simulated insect populations using generic population models (Sutherst, 1993). Spreadsheets, already mentioned, are also rapid development tools for smaller models but where large matrices or complex iteration are involved, they can become rather clumsy. In such cases, it may be more efficient to construct the model of a programming language such as Basic or Pascal. A growing number of commercially available, general simulation packages allow models to be specified, on screen, in the form of flow charts (similar to those in Figs 13.1 and 13.2), and are particularly useful for processes that can be described using differential equations. A range of more mathematically-orientated packages are also available with facilities such as symbolic algebra, useful for qualitative analysis of analytical models. In general, models for research can be constructed quickly with minimum attention to the user interface, but when a model is intended as a software product, sophisticated and elaborate programming is essential. This can be side-stepped to some extent by the use of expert or hypertext system shells but at the expense of flexibility in product design.

Developments in artificial intelligence, notably neural networks and genetic algorithms (Forsyth and Rada, 1986) have impinged on empirical modelling. They are suited to cases where large amounts of data relating to a problem can be collected, but there is little understanding of the relationships involved. For examples in entomology, see Moore (1991), where a neural network was used for the automated identification of insects from their wing-beat frequency spectra; Holt and Cheke (1996), where a genetic algorithm was used to relate time series data on desert locust plagues to rainfall patterns; and Holt and Chancellor (1996), where a genetic algorithm was used to search the parameter space of a simulation model for cases where particular simulated pest control strategies were effective.

13.4. Acknowledgements

The UK Overseas Development Administration provided financial support for the writing of this chapter. We thank Drs M.J. Jeger and M.C. Smith for useful comments on earlier drafts of the manuscript.

13.5. References

Begon, M., Bowers, R.G., Kadianakis, N. and Hodgkinson, D.E. (1992) Disease and community structure: the importance of host self-regulation in a host–host–pathogen system. *American Naturalist* 139, 1131–1150.

Bernstein, C. (1985) A simulation model for an acarine predator–prey system (*Phytoseilus persimilis – Tetranychus urticae*). *Journal of Animal Ecology* 54, 375–389.

Bosserman, R.W. and Ragade, R.K. (1982) Ecosystem analysis using fuzzy set theory. *Ecological Modelling* 16, 191–208.

Broadhead, E. and Cheke, R.A. (1975) Host spatial pattern, parasitoid interference and the modelling of the dynamics of *Alaptus fusculus* (Hym.: Mymaridae), a parasitoid of two *Mesopsocus* species (Psocoptera). *Journal of Animal Ecology* 44, 767–793.

Cartwright, T.J. (1994) *Modelling the World in a Spreadsheet*. Environmental simulation on a microcomputer. Johns Hopkins University Press, Baltimore and London.

Caswell, H. (1989) *Matrix Population Models*. Sinauer Associates, Massachusetts, 328pp.

Chan, M.S. and Jeger, M.J. (1994) An analytical model of plant virus disease dynamics with roguing and replanting. *Journal of Applied Ecology* 31, 413–427.

Cheke, R.A. (1977) A quantitative study of the grooming behaviour of the mymarid wasp *Alaptus fusculus*. *Physiological Entomology* 2, 17–26.

Cheke, R.A. (1978) Theoretical rates of increase of gregarious and solitarious populations of the desert locust. *Oecologia* 35, 161–171.

Cheke, R.A. and Tucker, M.R. (1995) An evaluation of potential economic returns from the strategic control approach to the management of African armyworm *Spodoptera exempta* (Lepidoptera: Noctuidae) populations in eastern Africa. *Crop Protection* 14, 91–102.

Comins, H.N. (1984) The mathematical evaluation of options for managing pesticide resistance. In: Conway, G.R. (ed.) *Pest and Pathogen Control: Strategic, Tactical and Policy Models*. John Wiley & Sons, New York, pp. 454–469.

Comins, H.N., Hassell, M.P. and May, R.M. (1992) The spatial dynamics of host–parasitoid systems. *Journal of Animal Ecology* 61, 735–748.

Crawley, M.J. (1983). *Herbivory*. Blackwell, Oxford, 437pp.

De Souza, K., Holt, J. and Colvin, J. (1995) Diapause, migration and pyrethroid resistance dynamics in the cotton bollworm, *Helicoverpa armigera* (Lepidoptera: Noctuidae). *Ecological Entomology*, 20, 333–342.

Dye, C. and Hasibeder, H. (1986) Population dynamics of mosquito-borne disease: effects of flies which bite some people more frequently than others. *Transactions of the Royal Society of Tropical Medicine and Hygiene* 80, 69–77.

Edwards-Jones, G. (1992) Knowledge-based systems for pest management: an applications-based review. *Pesticide Science* 36, 143–153.

Forsyth, R. and Rada, R. (1986) *Machine Learning: Applications in Expert Systems and Information Retrieval*. Ellis Horwood, Chichester.

Foster, M.A., Schultz, J.C. and Hunter, M.D. (1992) Modelling gypsy moth-virus-leaf chemistry interactions: implications of plant quality and pathogen dynamics. *Journal of Animal Ecology* 61, 509–520.

Garms, R., Walsh, J.F. and Davies, J.B. (1979) Studies on the reinvasion of the Onchocerciasis Control Programme in the Volta River Basin by *Simulium damnosum* s.l. with emphasis on the south-western areas. *Tropenmedizin und Parasitologie* 30, 345–362.

Geurrin, F. (1991) Qualitative reasoning about an ecological process: interpretation in hydroecology. *Ecological Modelling* 59, 165–201.

Gilpin, M. and Hanski, I. (eds) (1991) *Metapopulation Dynamics: Empirical and Theoretical Investigations*. Academic Press, London.

Gutierrez, A.P., Denton, W.H., Shade, R., Maltby, H., Burger, T. and Moorehead, G. (1974) The within-field dynamics of the cereal leaf beetle (*Oulema melanopus* (L.)) in wheat and oats. *Journal of Animal Ecology* 43, 627–640.

Hassell, M.P. (1978) *The Dynamics of Arthropod Predator–Prey Systems*. Princeton University Press, Princeton, New Jersey.

Hearne, J.W., van Coller, L.M. and Conlong, D.E. (1994) Determining strategies for the biological control of a sugarcane stalk borer. *Ecological Modelling* 73, 117–133.

Holling, C.S. (1959) Some characteristics of simple type of predation and parasitism. *Canadian Entomologist* 91, 385–398.

Holling, C.S. (1963) An experimental component analysis of population processes. *Memoirs of the Entomological Society of Canada* 32, 22–32.

Holling, C.S. (1965) The functional response of predators to prey density and its role in mimicry and population regulation. *Memoirs of the Entomological Society of Canada* 45, 5–60.

Holling, C.S. (1966) The functional response of invertebrate predators to prey density. *Memoirs of the Entomological Society of Canada* 48, 1–86.

Holt, J. and Chancellor, T.C.B. (1996) Simulation modelling of the spread of rice tungro virus: disease management by roguing? *Journal of Applied Ecology* 33(5), 927–936.

Holt, J. and Cheke, R.A. (1996) Successes and failures of a simple predictive rule for desert locust infestations in the Sahel. *Secheresse* 7, 151–154.

Holt, J. and Day, R.K. (1993) Rule-based models. In: Norton, G.A. and Mumford, J.D. (eds) *Decision Tools for Pest Management*. CAB International, Wallingford, pp. 147–158.

Holt, J., Cook, A. G., Perfect, T.J. and Norton, G.A. (1987) Simulation analysis of brown planthopper population dynamics on rice in the Philippines. *Journal of Applied Ecology* 24, 87–102.

Holt, J., King, A.B.S. and Armes, N.J. (1990) Use of simulation analysis to assess

Helicoverpa armigera control on pigeonpea in southern India. *Crop Protection* 9, 197–206.

Holt, J., Wareing, D.R. and Norton, G.A. (1992) Strategies of insecticide use to avoid resurgence of *Nilaparvata lugens* (Homoptera: Delphacidae) in tropical rice: a simulation analysis. *Journal of Economic Entomology* 85, 1979–1989.

Holt, J., Jeger, M.J. and Thresh, J.M. (1995) Theoretical models of African cassava mosaic virus disease dynamics to investigate possible virulence shifts in host–vector–virus interactions. *Proceedings of the 6th International Plant Virus Epidemiology Symposium*, Jerusalem, 23–27 April 1995. International Society of Plant Pathology.

Hughes, R.D. and Gilbert, N. (1968) A model of an aphid population – a general statement. *Journal of Animal Ecology* 37, 553–564.

Jeger, M.J. (1986) The potential of analytic compared with simulation approaches to modeling in plant disease epidemiology. In: Leonard, K.J. and Fry, W.E. (eds) *Plant Disease Epidemiology: Population Dynamics and Management*, vol.1. Macmillan, New York, pp. 255–281.

Johnson, C.G. (1969) *Migration and Dispersal of Insects by Flight*. Methuen, London.

Johnson, C.G. and Johnson, R.P.C. (1994) Computer modelling of the migration of *Simulium damnosum sensu lato* (Diptera: Simuliidae) across the Onchocerciasis Control Programme (OCP) area of West Africa. *Bulletin of Entomological Research* 84, 343–353.

Knight, J.D. and Mumford, J.D. (1992) Rule based models in crop protection. In: *Proceedings of the Brighton Crop Protection Conference – Pests and Diseases* 1992. British Crop Protection Council, Farnham, pp. 981–988.

Knight, J.D., Tatchell, G.M., Norton, G.A. and Harrington, R. (1992) FLYPAST. An information management system for the Rothamsted aphid database to aid pest control research and advice. *Crop Protection* 11, 419–426.

Labatte, J.M. (1994) Modelling the larval development of *Spodoptera frugiperda* (J.E. Smith), (Lep.: Noctuidae) on corn. *Journal of Applied Entomology* 118, 172–178.

Lefkovitch, L.P. (1966) The effects of adult emigration on populations of *Lasioderma serricorne* (F.) (Coleoptera: Anobiidae) *Oikos* 15, 200–210.

Legg, J.P. (1995) The ecology of *Bemisia tabaci* (Gennadius) (Homoptera: Aleyrodidae), vector of African cassava mosaic geminivirus in Uganda. PhD thesis, University of Reading.

Logan, J.A. and Hain, F.P. (eds) (1991) *Chaos and Insect Ecology.* Information series of the Virginia Polytechnic Institute and State University, College of Agriculture and Life Sciences, Blacksburg, Virginia. 108pp.

Madden, A.D., Holt, J. and Armes, N.J. (1995) The role of uncultivated hosts in the spread of pyrethroid resistance in *Helicoverpa armigera* populations in Andhra Pradesh, India: a simulation approach. *Ecological Modelling*, 82, 61–74.

Manly, B.F.J. (1974) Estimation of stage-specific survival rates and other parameters for insect populations developing through several stages. *Oecologia* (Berl.) 15, 277–285.

Manly, B.F.J. (1990) *Stage Structured Populations: Sampling, Analysis and Simulation*. Chapman & Hall, London.

Matthee, J.J. (1951) The structure and physiology of the egg of *Locustana pardalina* Walk. *Scientific Bulletin of the Department of Agriculture of South Africa* no. 316.

May, R.M. (1972) Limit cycles in predator–prey communities. *Science* 177, 900–902.

May, R.M. (1974a) Biological populations with nonoverlapping generations: stable points, stable cycles, and chaos. *Science* 186, 645–647.

May, R.M. (1974b) *Stability and Complexity in Model Ecosystems*, 2nd. edn. Princeton University Press, Princeton, New Jersey.

Mayer, D.G., Atzeni, M.G. and Butler, D.G. (1993) Spatial dispersal of exotic pests – the importance of extreme values. *Agricultural Systems* 43, 133–144.

Moore, A. (1991) Automated identification of insects in flight. In: Muniappin, R., Martani, M. and Denton, G.W.R. (eds) *Proceedings of the Workshop, Exotic Pests in the Pacific – Problems and Solutions*. Micronesia. Supplement No. 3, pp. 1229–1233.

Mumford and Holt, J. (1993) Modelling with spreadsheets. In: Norton, G.A. and Mumford, J.D. (eds) *Decision Tools for Pest Management*. CAB International, Wallingford, pp. 159–166.

Mumford, J.D. and Norton, G.A. (1989) Expert systems in pest management: implementation on an international basis. *AI Applications in Natural Resource Management* 3, 67–69.

Mumford and Norton (1993) Expert systems. In: Norton, G.A. and Mumford, J.D. (eds) *Decision Tools for Pest Management*. CAB International, Wallingford, pp. 167–179.

Nee, S. and May, R.M. (1992) Dynamics of metapopulations: habitat destruction and competitive coexistence. *Journal of Animal Ecology* 61, 37–40.

Norton, G.A. and Mumford, J.D. (1993) Descriptive techniques; Decision analysis techniques; Workshop techniques. In: Norton, G.A. and Mumford, J.D. (eds) *Decision Tools for Pest Management*. CAB International, Wallingford, pp. 23–44, 43–68, 69–78, respectively.

Rainey, R.C. (1951) Weather and the movement of locust swarms: a new hypothesis. *Nature, Lond.* 168, 1057–1060.

Rainey, R.C. (1989) *Migration and Meteorology. Flight Behaviour and the Atmospheric Environment of Locusts and other Migrant Pests*. Clarendon Press, Oxford.

Riley, J.R. and Reynolds, D.R. (1990) Nocturnal grasshopper migration in West Africa: transport and concentratim on by the wind, and the implications for air-to-air control. *Philosophical Transactions of the Royal Society of London* B 328, 655–672.

Rosen, R. (1970) *Dynamical System Theory in Biology*, vol.1. John Wiley & Sons, New York.

Roughgarden, J. (1979) *Theory of Population Genetics and Evolutionary Ecology: an Introduction*. Macmillan Publishing Co. Inc., New York.

Sands, W.A. (1992) The termite genus *Amitermes* in Africa and the Middle East. Natural Resources Institute Bulletin no. 51, Chatham, UK.

Steedman, A. (ed.) (1990) *Locust Handbook*, 3rd edn. Natural Resources Institute, Chatham, UK.

Sutherst, R.W. (1993) Role of modelling in sustainable pest management. In: Cory, S.A., Dall, D.J. and Milne, W.M. (eds) *Pest Control for Sustainable Agriculture*. CSIRO, Australia, pp. 66–71.

Takada, T. and Nakajima, H. (1992) An analysis of life history evolution in terms of the density-dependent Lefkovitch matrix model. *Mathematical Biosciences* 112, 155–176.

Wilder, J.W., Voorhis, N., Colbert, J.J. and Sharov, A. (1994) A three variable differential equation model for gypsy moth population dynamics. *Ecological Modelling* 72, 229–250.

Index